45⁰⁰
Const-HVAC
(3 Vols)

HEATING, VENTILATING AND AIR CONDITIONING LIBRARY VOLUME II

Oil, Gas and Coal Burners

Controls • Ducts • Piping

Valves

by James E. Brumbaugh

THEODORE AUDEL & CO.
a division of
THE BOBBS-MERRILL CO., INC.
Indianapolis/New York

SECOND EDITION
FIRST PRINTING

Published by The Bobbs-Merrill Company, Inc.
Indianapolis/New York

Manufactured in the United States of America

Library of Congress Cataloging in Publication Data

Brumbaugh, James E.
    Heating, ventilating, and air conditioning library.

    Includes indexes.
    1. Heating. 2. Ventilation. 3. Air conditioning.
I. Title.
TH7222.B78   1984   697   83-7064
ISBN 0-672-23389-4   (v. 1)
ISBN 0-672-23382-7   (v. 2)
ISBN 0-672-23383-5   (v. 3)
ISBN 0-672-23380-0   (set)

Foreword

The purpose of this series is to provide the layman with an introduction to the fundamentals of installing, servicing, and repairing the various types of equipment used in residential heating, ventilating, and air conditioning systems. Consequently, it was written not only for the engineer or skilled serviceman but also for the average homeowner. A special effort was made to remain consistent with the terminology, definitions, and practices of the various professional and trade associations involved in the heating, ventilating, and air conditioning fields.

Volume 1 begins with a description of the principles of thermal dynamics and ventilation, and proceeds from there to a general description of the various heating systems used in residences and small commercial buildings. Volume 2 contains descriptions of the working principles of various types of equipment and other components used in these systems. Volume 3 includes detailed instructions for installing, servicing, and repairing the different equipment and components. Those sections in Volume 3 that deal with air conditioning follow a similar format.

The author wishes to acknowledge the cooperation of the many organizations and manufacturers for their assistance in supplying valuable data in the preparation of this book. Every effort was made to give credit and courtesy lines for materials and illustrations used in this edition.

JAMES E. BRUMBAUGH

Contents

CHAPTER 1

CHAPTER 2

CHAPTER 3

CHAPTER 4

CHAPTER 5

CHAPTER 6

CHAPTER 7

diffusers—return air and exhaust air inlets—duct run fittings—air supply and venting—duct dampers—damper motors and actuators—installing damper motors—troubleshooting damper motors—blowers (or fans) for duct systems—designing a duct system—duct system calculations—duct heat loss and gain—air leakage—duct insulation—equal friction method—balancing an air distribution system—duct maintenance—roof plenum units—mobile home duct systems—proprietary air distribution systems—duct furnaces—electric duct heaters

CHAPTER 8

CHAPTER 9

CHAPTER 10

sion tanks—troubleshooting expansion tanks—air eliminators—pipe-line valves—steam-pressure-regulating valves—temperature regula-tors—electric control valves (regulators)—water tempering valves—hot-water heating control—balancing valves, valve adapters, and filters—pipeline strainers

CHAPTER 1

Oil Burners

An oil burner (Figs. 1-1 and 1-2) is a mechanical device used to prepare the oil for burning in heating appliances such as boilers, furnaces, and water heaters. The term "oil burner" is somewhat of a misnomer because this device does not actually burn the oil. It combines the fuel oil with the proper amount of air for combustion and delivers it to the point of ignition, usually in the form of a spray. Oil burners may be equipped with or without a primary safety control.

The fuel oil is prepared for combustion either by vaporization or by atomization. These two methods of fuel oil preparation are used in the three basic types of oil burners employed in commercial, industrial, and residential heating. These three oil burners are:

1. The atomizing, or gun-type, oil burner.
2. The vaporizing. or pot-type, oil burner.
3. The rotary oil burner.

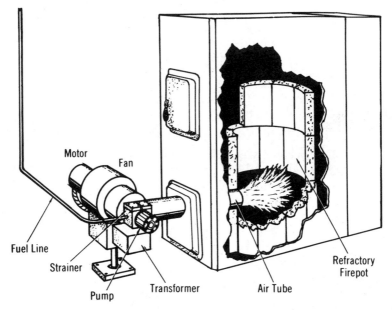

Courtesy U.S. Department of Agriculture

Fig. 1-1. Pressure-type gun oil burner.

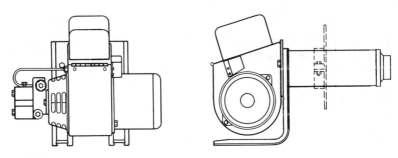

Courtesy Stewart-Warner Corp.

Fig. 1-2. Basic shape of a gun-type oil burner.

The atomizing oil burner can be of the low-pressure or high-pressure type. Both types are used in residential heating applications with the latter being by far the most popular of the two.

The advantage of the vaporizing, or pot-type, oil burner is that

10

it is the least expensive to use; however, it has limited heating applications and is currently effective only in small houses and buildings located in milder climates.

Vaporizing, or pot-type, burners can be divided into the three following kinds:

1. Natural-draft pot burners.
2. Forced-draft pot burners.
3. Sleeve burners.

Rotary oil burners are usually found in commercial or industrial buildings, although they can and have been used for residential heating applications. The types of rotary oil burners available for heating purposes are:

1. Vertical rotary burners.
2. Horizontal rotary burners.
3. Wall-flame rotary burners.

FUELS USED IN BURNERS

No. 1 and No. 2 fuel oil are both commonly used for residential heating purposes. The No. 2 is slightly more expensive, but the fuel oil gives more heat per gallon used. The lighter No. 1 fuel oil is used in vaporizing, or pot-type, oil burners. The No. 2 fuel oil is used in both atomizing and rotary oil burners.

The manufacturer of the oil burner will generally stipulate the grade of fuel oil to be used. If this information is unavailable, the label of Underwriters' Laboratories, Inc. and the Underwriters' Laboratories of Canada will stipulate the correct grade of fuel oil to be used.

The heavier the grade of fuel oil used in an oil burner, the greater the care that must be taken to ensure that the oil is delivered for combustion at the proper atomizing temperature. If the oil is not maintained at this temperature prior to delivery for combustion, the oil burner will fail to operate efficiently. An efficient oil burner is one that burns the fuel oil completely using the smallest amount of air necessary for combustion.

HIGH-PRESSURE ATOMIZING OIL BURNERS

High-pressure oil burners (Fig. 1-3) are sometimes called *sprayers* because they spray the fuel oil instead of vaporizing it. They are also referred to as *gun*, or *pressure*, oil burners because the oil is forced under pressure through a special gunlike atomizing nozzle. The liquid fuel is broken up into minute liquid particles or globules to form the spray. The essential parts of a high-pressure atomizing oil burner proper (or those parts within the casing) are shown in Fig. 1-4 and consist of the following:

1. Nozzle.
2. Nozzle tube.
3. Nozzle strainer.
4. Ignition electrodes.
5. Electrodes bracket.
6. Air entrance.

Courtesy Wayne Home Equipment Co., Inc.

Fig. 1-3. High-pressure atomizing oil burner.

7. Air adjustment collar.
8. Fan.
9. Rotary turbulator vanes

The essential parts outside the casing (i.e., the external parts) are:

1. Electric motor.
2. Oil pump.
3. Oil supply line.
4. Strainer.
5. Pressure-relief valve.
6. Bypass line.
7. Cutoff valve.
8. Oil line to the nozzle.

These internal and external parts of a high-pressure atomizing oil burner comprise the so-called *one-pipe system*. A *two-pipe system* is shown in Fig. 1-5. The principal difference between the

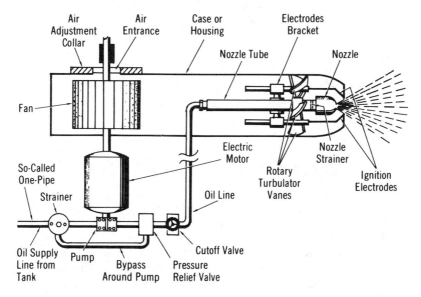

Fig. 1-4. Elementary high-pressure domestic oil burner showing essential parts.

13

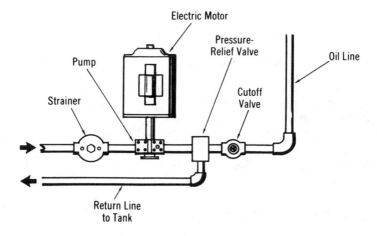

Fig. 1-5. Details of the two-pipe system.

two systems is the arrangements of the bypass line leading from the pressure-relief valve around the pump. In the one-pipe system, the bypass line connects the pressure-relief valve to the strainer (bypassing the pump) where it picks up the oil supply line to the tank. The bypass line in the two-pipe system functions as a return line to the tank. It extends from the pressure-relief valve around the pump, but does not connect to the strainer. It runs parallel to the oil supply line back to the tank. Hence the origin of the two-pipe system.

OPERATING FUNDAMENTALS

The operation of a high-pressure atomizing oil burner can be traced in Fig. 1-4. The fuel oil is drawn through a strainer from the supply tank by the pump, and is forced under pressure past the pressure-relief cutoff valve via the oil line where it eventually passes through the fine mesh strainer and into the nozzle. The amount of pressure required to pump the fuel oil through the line depends on the size and capacity of the oil burner and the purpose for which it is used. For example, residential oil burners

14

require 80 to 125 psi, whereas commercial and industrial oil burners operate on 100 to 300 psi.

As the fuel oil passes through the nozzle, it is broken up and sprayed in a very fine mist. The air supply is drawn in through the case opening (Fig. 1-4) and forced through the draft tube portion of the casing by the fan. This air mixes with the oil spray after passing through a set of vanes, called a *turbulator*. The turbulator gives a twisting motion to the air stream just before it strikes the oil spray, producing a more thorough mixture of the oil and air.

Ignition of the oil spray is supplied by a transformer that changes the house lighting current and feeds it to the electrodes to provide a spark at the beginning of each operating period (Figs. 1-6 and 1-7).

The starting cycle of the oil burner is initiated by the closing of the motor circuit. When the motor circuit is closed (automatically by room temperature control), the motor starts turning the fan and the pump. At the same time, the ignition transformer produces a spark at the electrodes ready to light the oil and air mixture.

The action of the pump draws the fuel oil from the tank through the strainer on the fuel line. Its flow is controlled by an oil cutoff valve, which prevents oil passing to the nozzle unless the pressure is high enough to spray the oil (approximately 60 lb. pressure). Because the pump in the oil burner pumps oil much faster than it can be discharged through the nozzle at that pressure (i.e., 60 lb. pressure), the oil pressure continues to rise very fast between the pump and the nozzle. When the pressure begins to rise above the normal operating pressure (100 lb.) a pressure-relief valve opens and allows the excess oil to flow through the bypass line to the inlet, as in the so-called one-pipe system, or to flow through a second or return line to the supply tank. The pressure-relief valve in either system maintains the oil at the correct operating pressure.

When the oil burner is turned off (i.e., when the motor stops), the oil pressure quickly drops below the operating pressure, and the relief or regulating valve closes. The flame continues until the pressure drops below the setting of the cutoff valve.

The cutoff and pressure-relief (regulating) valves may be either two separate units or combined into one unit. Fig. 1-8

15

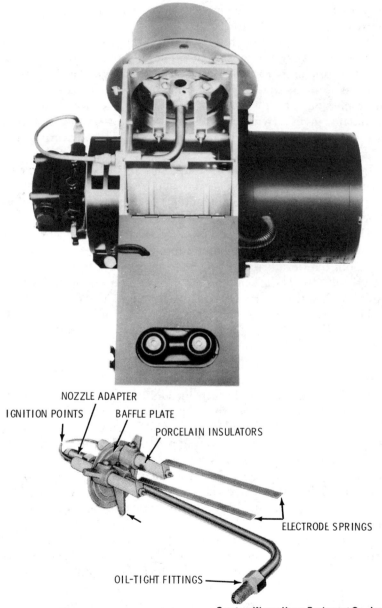

NOZZLE ADAPTER
IGNITION POINTS
BAFFLE PLATE
PORCELAIN INSULATORS
ELECTRODE SPRINGS
OIL-TIGHT FITTINGS

Fig. 1-6. Oil burner with transformer removed, revealing the gun assembly.

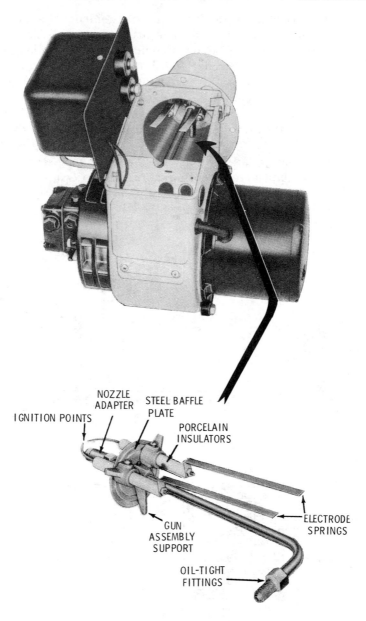

NOZZLE
ADAPTER   STEEL BAFFLE
PLATE
IGNITION POINTS
PORCELAIN
INSULATORS

GUN
ASSEMBLY
SUPPORT

ELECTRODE
SPRINGS

OIL-TIGHT
FITTINGS

Courtesy Wayne Home Equipment Co., Inc

Fig. 1-7. Gun assembly details in a high-pressure atomizing oil burner.

17

shows the essentials of the two-unit arrangement. These are, as shown, simply elementary mechanisms to illustrate basic principles. The cutoff needle valve is shown with spring inside of bellows, and the pressure-relief (mushroom) valve with exposed spring. In the cutoff valve arrangement, the spring acts against oil pressure on the head of the bellows (tending to collapse it); in the pressure-relief valve, the spring acts against the oil pressure, which acts on the lower face of the mushroom valve (tending to open it).

When the pump starts and the pressure in the line rises to about 60 lb. (depending upon the spring setting), this pressure acting on the head of the bellows overcomes the resistance of the spring, causing the cutoff valve to open. Since the pump supplies more oil than the nozzle can discharge, the pressure quickly rises to 100

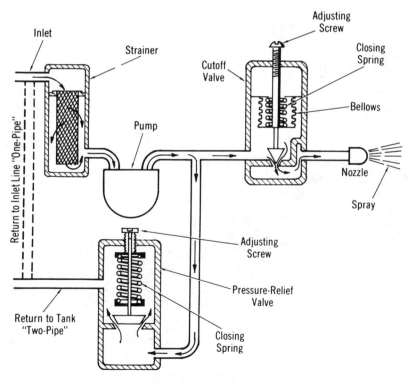

Fig. 1-8. Sectional view of separate unit cutoff valve and pressure-relief valve showing strainer, pump, and piping.

lb., overcoming the resistance of the relief valve spring and causing the valve to open. This allows excess oil to bypass or return to the tank.

The relief valve will open high enough to maintain the working pressure constant at 100 lb. When the oil burner is turned off, the oil pressure quickly drops, and the pressure-relief valve closes. However, oil will continue to discharge from the nozzle until the pressure drops below the cutoff valve setting when the cutoff valve closes and stops the nozzle discharge.

A passage to the return line is provided by a small slot cut in the seat of the mushroom valve. This causes any remaining pressure trapped in the line by the closing of the cutoff valve to be equalized.

Frequently the cutoff valve and pressure-relief valve are combined in a compact cylindrical casing (Fig. 1-9). Here the two valves are attached to a common stem with a flange, which comes in contact with a stop when moved upward by the pressure of the valve actuating the spring.

The position of the stop limits the valve movements to proper maximum lift. A piston, free to move in the cylindrical casing, has an opening in its head that forms the valve seat for the pressure-relief valve. The strong piston spring tends to move the piston downward and close the pressure-relief valve and then the cutoff valve.

When the pump starts and the pressure in the cylinder below the piston rises to about 60 lb. (depending on the piston spring setting), the piston and the two valves (i.e., the cutoff and pressure-relief valve) rise until the valve flange contracts with the stop. At this instant the cutoff valve is fully opened, allowing oil to flow to the nozzle, the pressure-relief valve still being closed. Since the nozzle does not have sufficient capacity to discharge all the oil that is pumped by the pump, the pressure below the piston will continue to rise.

HIGH-PRESSURE OIL BURNER CONSTRUCTION

The general external appearance of a typical high-pressure atomizing, or gun, domestic oil burner is shown in Fig. 1-10. The

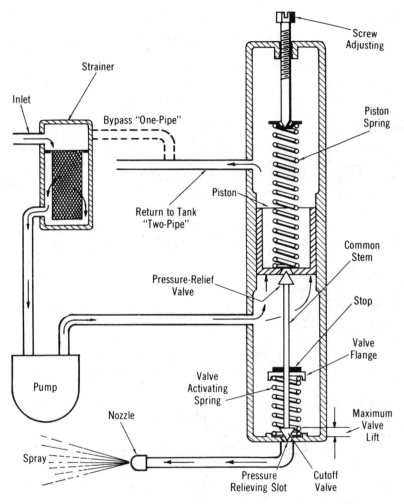

Fig. 1-9. Sectional view of combined cutoff valve and pressure-relief valve showing the strainer, pump, and piping.

location of the parts varies somewhat in different makes, but construction is now nearly standardized. Basically the typical oil burner contains the following parts:

1. Draft tube.
2. Counterbalanced draft shutter.

3. Transformer.
4. Ignition and firing assembly plate.
5. Pump strainer and valve unit.
6. Pyrex fire inspection holes.
7. Oil feed tube from pump to nozzle.
8. Motor.
9. Built-in thermal motor protector.
10. Legs.
11. Fan housing.

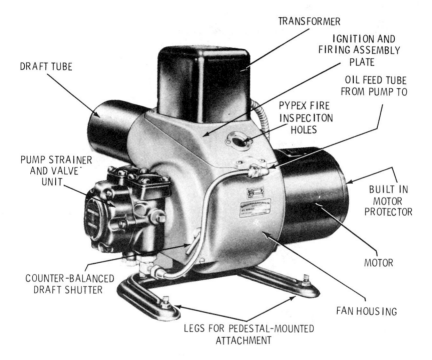

Courtesy S. T. Johnson Company

Fig. 1-10. High pressure oil burner.

The mechanism assembled inside the draft tube, comprising gun with nozzle, electrode placement, and tubular vanes, is plainly shown in Fig. 1-11. External draft tube assemblies are further shown in Fig. 1-12.

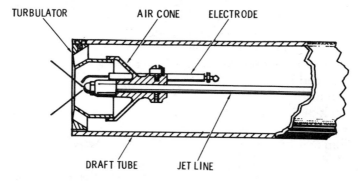

Fig. 1-11. Details of draft tube showing turbulator air cone, electrode, and jet line.

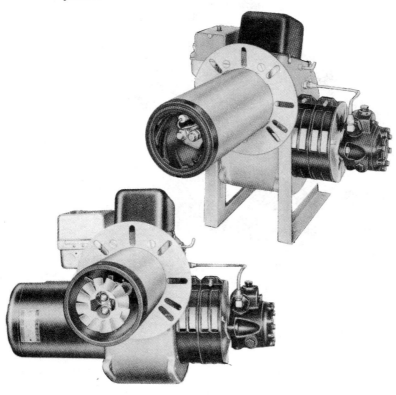

Courtesy Wayne Home Equipment Co., Inc.

Fig. 1-12. Oil burner turbulator designs.

ROTARY OIL BURNERS

Rotary burners operate with low-pressure gravity and are available in a number of designs depending on the different conditions of use. In each case the operating principle involves throwing the oil by centrifugal force.

Rotary oil burners can be classified either as rotary nozzle or rotary cup burners. The essential components of the rotary nozzle burner are shown in Fig. 1-13. Air pressure acting on the propeller causes the nozzle assembly to rotate at a very high speed. Oil is supplied through the hollow shaft to the nozzles, and the rotary motion causes the oil to be thrown off in a fine spray

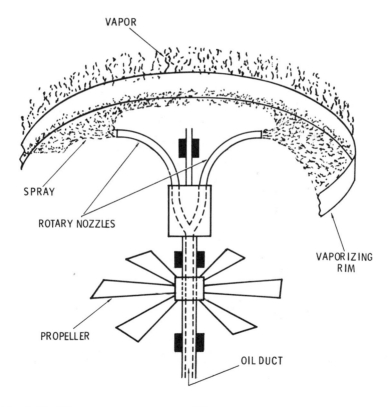

Fig. 1-13. Elementary rotary nozzle oil burner.

by centrifugal force. The flame from this spray heats up the metal vaporizing rim hot enough to vaporize the oil spray as it comes in contact with it. Being thoroughly mixed with air, a blue flame is produced. On some designs, the spray vaporized by the vaporizing rim is superheated by passing through grilles.

The rotary cup oil burner (Fig. 1-14) contains a cone-shaped cup that rotates on ball bearings carried by a central tube. The fuel is supplied to the cup through this tube. In operation, drops of oil, issuing from the oil feed tip, contact with the cup as shown and by centrifugal force; the drops are both flattened into a film and projected toward and off the rim of the cup, as shown in Fig. 1-15. The rim being surrounded by a concentric opening of the casing, the oil is met by the surrounding blast of primary air with which it mixes, giving the proper mixture for combustion.

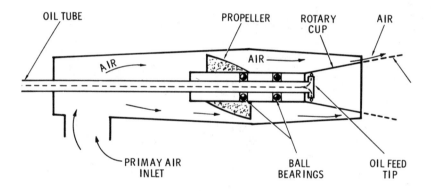

Fig. 1-14. Elementary rotary cup oil burner.

VAPORIZING, OR POT, OIL BURNERS

Figs. 1-16 and 1-17 shows a typical vaporizing or pot-type oil burner. The fuel oil is vaporized for combustion by heating it from below. The vaporized fuel oil rises vertically where it is burned at the top.

The two basic types of vaporizing, or pot, oil burners are:

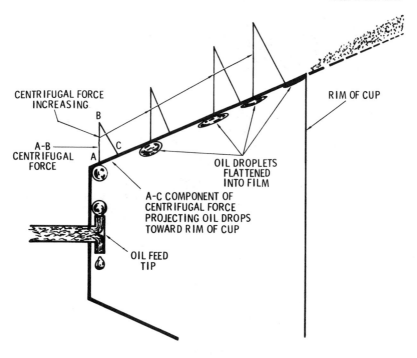

CENTRIFUGAL FORCE
INCREASING
B

RIM OF CUP

A-B
CENTRIFUGAL   C
FORCE   A

OIL DROPLETS
FLATTENED
INTO FILM

A-C COMPONENT OF
CENTRIFUGAL FORCE
PROJECTING OIL DROPS
TOWARD RIM OF CUP

OIL FEED
TIP

Fig. 1-15. Detail of cup showing centrifugal forces acting on the droplets of oil, which flatten them into a film and project them toward and off of rim.

1. The natural-draft pot burner.
2. The forced-draft pot burner.

In the former, the air necessary for combustion is provided by the chimney. The forced-draft pot burner relies on both the chimney and a mechanical device (e.g., a fan) for the air supply.

Sleeve burners (also referred to as *perforated sleeve burners*) represent a third type of vaporizing, or pot, burner. Although these burners are used mostly in conjunction with small oil-fired equipment (e.g., kitchen ranges and space heaters), they can also be employed to heat a small house, if outside temperatures do not become too low.

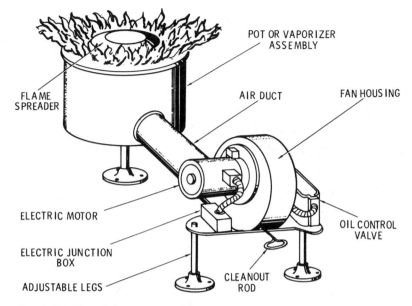

POT OR VAPORIZER
ASSEMBLY

FLAME
SPREADER

AIR DUCT

FAN HOUSING

ELECTRIC MOTOR

ELECTRIC JUNCTION
BOX

ADJUSTABLE LEGS

CLEANOUT
ROD

OIL CONTROL
VALVE

Fig. 1-16. Vaporizing, or pot, oil burner.

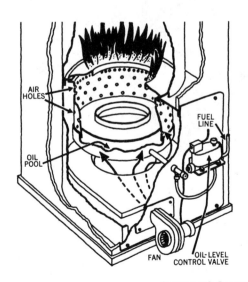

AIR
HOLES

FUEL
LINE

OIL
POOL

FAN

OIL-LEVEL
CONTROL VALVE

Courtesy U.S. Department of Agriculture

Fig. 1-17. Vaporizing, or pot, oil burner.

COMBINATION OIL AND GAS BURNERS

Some oil burners are available with combination oil and gas firing accessories that make it possible to use either of these fuels in the same burner. This is particularly advantageous in areas where low-cost gas is sometimes available.

The combination gas and oil burner illustrated in Fig. 1-18 contains independent ignition and control systems for gas or oil. One convenience built into these combination burners is that the oil burner components and parts are standard and require only conventional service procedures. The safety features include a standard cadmium sulfide detection cell and primary relay control.

Courtesy Wayne Home Equipment Co., Inc.

Fig. 1-18. Combination oil and gas burner.

FUEL UNITS (PUMPS)

Both single-stage and two-stage fuel units (pumps) are available in a number of sizes and designs (Figs. 1-19 and 1-20) that

27

meet the requirements of most *high-pressure atomizing oil burners.*

Sometimes the terms "fuel unit" and "pump" are used synonymously, but this is incorrect usage. For example, a *fuel unit* is an oil pumping device that consists of a pump, a pressure regulating valve, and a filter. Almost all modern oil burners used in residential heating installations are equipped with these fuel-units. A *pump*, on the other hand, is *only* an oil pumping device. A pressure-regulating valve and a filter can be attached to the mountings on the pump if so desired.

The single-stage unit illustrated in Fig. 1-21 is generally used for single-pipe, gravity feed installation or for two-pipe installations under low-lift conditions up to 10 in. of vacuum. The components of this fuel unit (see Fig. 1-22) are as follows:

1. Pumping gears.
2. Cutoff valve.

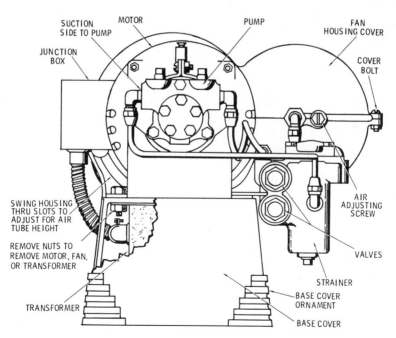

Fig. 1-19. Side view of an oil burner pump showing parts and connections.

3. Strainer.
4. Shaft seal.
5. Noise dampening device.
6. Shaft bearing.
7. Body.
8. Bleed valve.

The fuel oil first enters the unit by passing through the strainer where foreign particles such as dirt and line filter fibers are removed. The fuel oil then moves through the hydraulically balanced pumping gears and is pumped under pressure to the valve (see circuit diagram in Fig. 1-22). The pressure forces the piston away from the nozzle cutoff seat, and the fuel oil then flows out the nozzle port. Oil in excess of nozzle capacity is bypassed through the valve back to the strainer chamber in a single-pipe system, or in a two-pipe system is returned to the tank. Pressure is

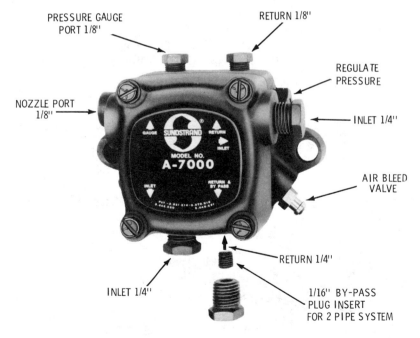

PRESSURE GAUGE
PORT 1/8"

RETURN 1/8"

REGULATE
PRESSURE

NOZZLE PORT
1/8"

INLET 1/4"

AIR BLEED
VALVE

RETURN 1/4"

INLET 1/4"

1/16" BY-PASS
PLUG INSERT
FOR 2 PIPE SYSTEM

Courtesy Sundstrand Hydraulics

Fig. 1-20. The fuel unit assembly.

(A.) Single-stage unit.

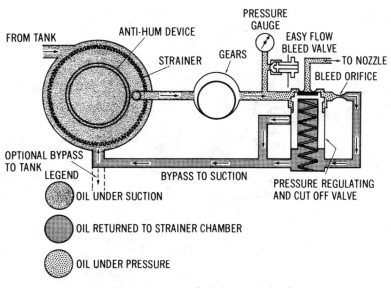

(B.) Circuit diagram of single-stage circuit.

Fig. 1-21. A single-stage fuel unit.

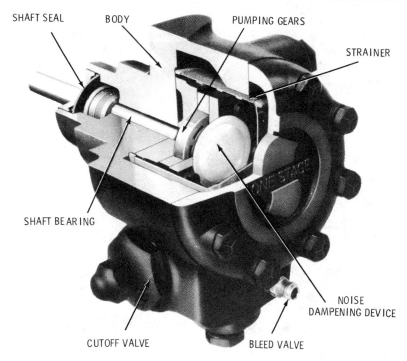

SHAFT SEAL   BODY   PUMPING GEARS

STRAINER

SHAFT BEARING

NOISE
DAMPENING DEVICE

CUTOFF VALVE   BLEED VALVE

Courtesy Sundstrand Hydraulics

Fig. 1-22. A single-stage full unit pump view.

reduced on the head of the piston when the pump motor is shut off. At this point, the piston snaps back, causing the nozzle port opening to close. A bleeder valve opening in the piston provides for automatic air purging on a two-pipe system, providing for fast cutoff.

The two-stage fuel unit differs from the single-stage type in that it contains two pumping gear sets rather than one.

A major advantage of a two-stage fuel unit is that all air is eliminated from the oil being delivered to the nozzle. The inlet of the first stage is located *above* the inlet for the second stage. As a result, any air drawn into the fuel unit after priming is picked up by the first stage and discharged to the tank before it reaches the second stage. Consequently, the second stage draws completely air-free oil.

31

As the air is being discharged into the tank by the first stage, pressure begins to build up in the second stage, causing the regulating valve to bypass excess oil back into the strainer. These operating principles are illustrated by the circuit diagram in Fig. 1-23.

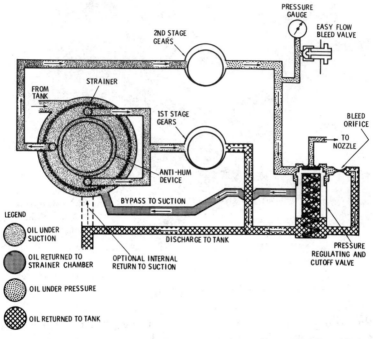

Fig. 1-23. Circuit diagram of a system using a two-stage fuel unit.

SERVICING A FUEL UNIT

A vacuum gauge and a pressure gauge are both used to service a fuel unit. With these two gauges, the individual can check the following:

1. Vacuum.
2. Lift.
3. Air leaks.

4. Pressure.
5. Cutoff.
6. Delivery.

Figs 1-24 and 1-25 illustrate the attachment of the vacuum and pressure gauges to a fuel unit. The pressure gauge (shown as the upper gauge in Fig. 1-24 and the gauge attached to the nozzle line opening in Fig. 1-25) will indicate whether a positive cutoff is operable or if an adequate and uniform buildup of pressure is present. When the pressure gauge is attached to the nozzle line opening, it should indicate a reading of 75 to 90 psi (see Fig. 1-25). Any drop of the pressure gauge reading to zero indicates leaky cutoff and probable difficulty with the shutoff valve in the nozzle line.

The existence of air leaks in the supply line can be determined by vacuum gauge readings once the gauge is attached to the optional inlet connection (Fig. 1-24). An evaluation of the gauge

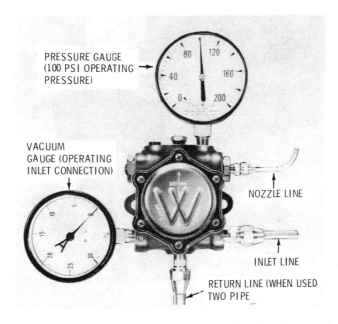

Courtesy Wayne Home Equipment Co., Inc.

Fig. 1-24. Attachment of vacuum and pressure gauge.

33

reading is itself determined by the location of the oil storage tank. If the tank is located above the burner, and the oil is supplied by gravity flow, the vacuum gauge must show a reading of zero, unless there is a problem in the system. These problems can take the following form:

1. A partially closed cutoff valve of the oil supply tank.
2. A kinked or partially blocked oil supply line.
3. A blocked line filter.

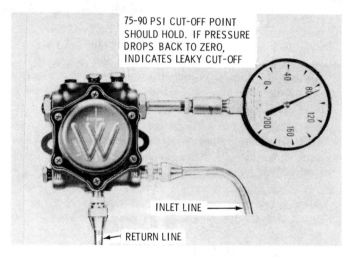

75-90 PSI CUT-OFF POINT SHOULD HOLD. IF PRESSURE DROPS BACK TO ZERO, INDICATES LEAKY CUT-OFF

INLET LINE ⟶

◀— RETURN LINE

Courtesy Wayne Home Equipment Co., Inc.

Fig. 1-25. Pressure gauge attached to the nozzle line.

A system with an oil supply tank located *below* the level of the oil burner that supplies the fuel oil through a line filter *must* produce a reading on the vacuum gauge if the system is operating properly. A zero reading will indicate the presence of an air leak.

The above are a few of the problems that may be encountered with fuel units and some suggested remedies for dealing with them. Most manufacturers of fuel units or gauges generally supply "troubleshooting" recommendations along with the installation and maintenance instructions.

OIL BURNER NOZZLES

An oil burner nozzle is a device designed to deliver a fixed amount of fuel to the combustion chamber in a uniform spray pattern and spray angle best suited to the requirements of a specific burner. The oil burner nozzle atomizes the fuel oil (i.e., breaks it down into extremely small droplets) so that the vaporization necessary for combustion can be accomplished more quickly.

The components in a typical nozzle (Fig. 1-26) consist of the following:

1. Orifice.
2. Swirl chamber.
3. Orifice disc.
4. Body.
5. Tangential slots.
6. Distributor.
7. Retainer.
8. Filter.

Fuel oil is supplied under pressure (100 psi) to the nozzle where it is converted to velocity energy in the swirl chamber by directing it through a set of tangential slots. The centrifugal force caused within the swirl chamber drives the fuel oil against the chamber walls, producing a core of air in the center. The latter effect moves the oil out through the orifice at the tip of the nozzle in a cone-shaped pattern.

The two basic spray cone patterns are:

1. The hollow cone.
2. The solid cone.

Each has certain advantages depending on its use.

The *hollow cone pattern* (Fig. 1-27) is recommended for use in smaller burners (those firing 1.00 GPH and under). As shown in Fig. 1-27, they are characterized by a concentration of fuel oil droplets all around the outer edge of the spray. There is little or no distribution of droplets in the center of the cone. The principal advantage of the hollow cone patterns is a more stable spray pattern and angle under adverse conditions than solid cone pat-

35

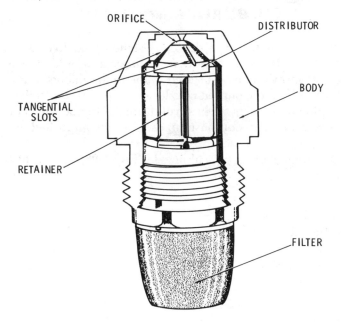

Courtesy Wayne Home Equipment Co., Inc.

Fig. 1-26. Details of an oil burner nozzle.

Courtesy Wayne Home Equipment Co., Inc.

Fig. 1-27. Hollow spray cone pattern

terns operating under the same conditions and at the same flow rate.

The *solid cone pattern*, illustrated in Fig. 1-28, is characterized by a uniform or near-uniform distribution of fuel oil droplets throughout the cone pattern. Nozzles producing this cone pattern are particularly recommended for smoother ignition in oil burners firing above 2.00 or 3.00 GPH. They are also recommended where long fires are required or where the air pattern or the oil burner is heavy in the center.

Courtesy Wayne Home Equipment Co., Inc.

Fig. 1-28. Solid spray cone pattern.

A combination cone pattern that is neither a true cone nor a true hollow cone can be used in oil burners firing between .40 GPH and 8.00 GPH.

Oil burner nozzles are also selected on the basis of the spray angle they produce (Fig. 1-29). The *spray angle* refers to the angle of the spray cone, and this angle will generally range from 30° to 90°. The angle selected will depend upon the requirement of the burner air pattern and combustion chamber. For example, 70° to 90° spray angles are recommended for round or square combustion chambers (Fig. 1-30); 30° to 60° spray angles for long, narrow chambers (Fig. 1-31). Recommended combustion chamber dimensions and spray angles for nozzles are given in Table 1-1.

37

Table 1-1. Recommended Combustion Chamber Dimensions

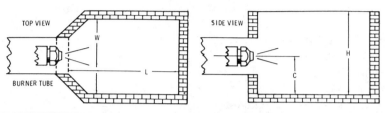

Nozzle Size of Rating (GPH)	Spray Angle	Sqaure or Rectangular Combustion Chamber				Round Chamber (Diameter in inches)
		L Length (in.)	W Width (in.)	H Height (in.)	C Nozzle Height (in.)	
0.50 – 0.65	80°	8	8	11	4	9
0.75 – 0.85	60°	10	8	12	4	*
	80°	9	9	13	5	10
1.00 – 1.10	45°	14	7	12	4	*
	60°	11	9	13	5	*
	80°	10	10	14	6	11
1.25 – 1.35	45°	15	8	11	5	*
	60°	12	10	14	6	*
	80°	11	11	15	7	12
1.50 – 1.65	45°	16	10	12	6	*
	60°	13	11	14	7	*
	80°	12	12	15	7	13
1.75 – 2.00	45°	18	11	14	6	*
	60°	15	12	15	7	*
	80°	14	13	16	8	15
2.25 – 2.50	45°	18	12	14	7	*
	60°	17	13	15	8	*
	80°	15	14	16	8	16
3.00	45°	20	13	15	7	*
	60°	19	14	17	8	*
	80°	18	16	18	9	17

*Recommend oblong chamber for narrow sprays.

Courtesy of Wayne Home Equipment Co., Inc.

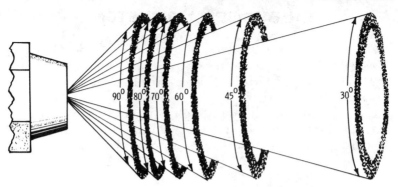

Fig. 1-29. Varieties of spray angles.

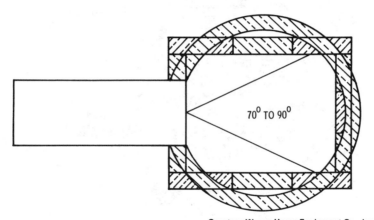

Fig. 1-30. Spray angles (70° to 90°) suitable for round or square chambers.

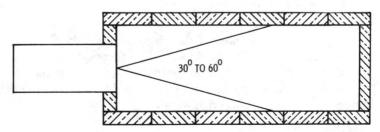

Fig. 1-31. Spray angles (30° to 60°) suitable for long, narrow chambers.

39

OIL BURNER AIR SYSTEM

The air system for the average oil burner is generally composed of the air shutter draft tube, the turbulator, and the fan. The draft tube and turbulator have already been shown (Figs. 1-11 and 1-12).

The fan construction consists of a (squirrel cage) series of vanes or blades mounted on the rim of a wheel. These vanes are slanted forward in such a manner as to provide the maximum discharge of air. Fig. 1-32 shows the construction of fan and flexible coupling.

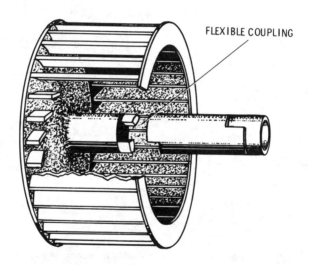

FLEXIBLE COUPLING

Fig. 1-32. Fan and flexible coupling.

The operating principles of the air system are fairly simple. The fan draws air into the fan housing and forces this air through the draft tube and turbulator and into the combustion chamber. The amount of incoming air can be regulated by adjusting the air shutter. As the air is forced through these vanes, it is given a swirling motion just before it strikes the oil spray. This motion provides a more thorough mixture of the oil and air, resulting in better combustion.

The shape of the turbulator varies in different models, but the purpose is the same: to thoroughly mix the air and oil spray. Fig. 1-33 shows a double turbulator consisting of an air impeller and nose piece.

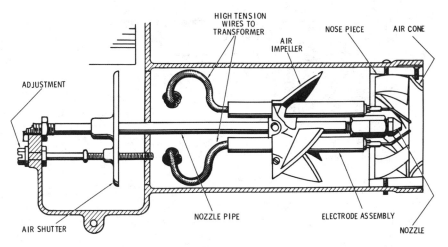

Fig. 1-33. Detail of a draft tube showing double turbulator consisting of air impeller and nose piece.

PRIMARY SAFETY CONTROL

The *primary safety control* is an automatic safety device designed to stop the flow of fuel oil at the burner should ignition or flame failure occur. It should be included on all automatic oil burners.

The cadmium detection cell is the most effective type of primary safety control used on oil burners. Malfunctions cause primary safety control to build up electrical resistance across the cell until the burner is automatically shut off.

The primary safety control can be tested by removing the motor lead from the burner and allowing the ignition circuit to be energized. Figs. 1-34 and 1-35 illustrate two typical wiring diagrams for primary safety controls.

41

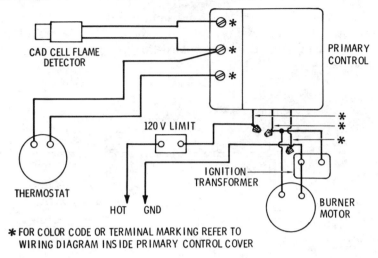

Courtesy Stewart-Warner Corp.

Fig. 1-34. Constant ignition wiring diagram.

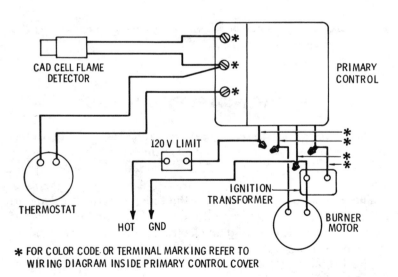

Courtesy Stewart-Warner Corp.

Fig. 1-35. Intermittent ignition wiring diagram.

INSTALLING AN OIL BURNER
(SAFETY REQUIREMENTS)

There are a number of requirements for the installation of oil burners that have nothing to do *per se* with the mechanical attachment of the unit. These requirements involve the proper installation of oil burners insofar as safety precautions are concerned and are detailed in booklets published by the National Fire Protection Association. These include the following:

1. "Installation of oil burners and oil-fired units" in *Installation of Oil Burning Equipment* 1972 (NFPA No. 31).
2. "Standard for Prevention of Furnace Explosions" in *Fuel Oil and Natural Gas-Watertube Boiler-Furnace with One Burner* (NFPA No. 85).
3. "Standard for Prevention of Furnace Explosions" in *Fuel Oil-Fired Multiple Burner Boiler-Furnaces* (NFPA No. 85D).

The safety requirements included in these booklets from the National Fire Protection Association must be carefully reviewed when installing oil burners. Furthermore, the instructions and recommendations provided by the manufacturers of the equipment, and the regulation and standards established by local authorities having jurisdiction for their use, must receive equal consideration.

Under most circumstances, oil burners and oil-fired units should be installed in rooms that provide adequate clearance from the combustible material. The only exception to this rule is when specific instructions are given otherwise. In this case, the manufacturer provides or specifies a suitable combustion chamber (stainless steel, firebrick, etc.).

Some sort of manual shutoff control should be provided for the oil burner in order to stop the flow of oil to the burner when the air supply is interrupted. This must be placed at a safe distance from the unit and in a convenient location. These manual shutoff valves generally consist of either a switch in the burner supply circuit (for electrically driven units) or a shutoff valve on the oil supply line.

Primary safety controls (automatic shutoff devices) must be provided for all oil burners and oil-fired units that operate auto-

43

matically without the need of an attendant on duty. In other words, those types of equipment found where a stationary engineer would not be employed (i.e., noncommercial and nonindustrial locations).

One problem encountered when converting solid-fuel heating equipment to oil use is the accumulation of potentially dangerous vapors in the ashpit of the unit. This can be avoided by removing the ash door or by providing bottom ventilation to the unit. This precaution is unnecessary if the ashpit also serves as a part of the combustion chamber.

Never install or permit the installation of an oil burner until the boiler or furnace has first been inspected and found to be in good condition. The flue gas passages must be tight and free of any leaks.

All oil burners listed by Underwriters' Laboratories, Inc. and Underwriters' Laboratories of Canada meet the safety requirements detailed in the various booklets of the National Fire Protection Association. Copies of these booklets can be purchased by writing to:

> National Fire Protection Association
> 470 Atlantic Avenue
> Boston, MA 02210

OIL SUPPLY TANKS

The fuel oil for an oil-fired heating unit is stored in an *oil supply tank*. The latter is connected directly or by pump to the oil burner.

The location of the oil supply tank is subject to local regulations. These must be consulted before installing a *new* tank. An existing tank has presumably already met the requirements of the local regulations.

The supply tank can be located inside or outside the building, above the level of the oil burner or below it. Furthermore, outside tanks can be located underground or above it. It is recommended that larger supply tanks be located outside and underground.

Figs. 1-36 through 1-39 illustrate four ways in which oil supply tanks can be located. When installing a fuel tank, the following suggestions should prove helpful:

1. The filler pipe should be a *minimum* of 2 in. in diameter; the vent pipe 1¼ in.
2. Use wrought-iron pipe with malleable-iron fittings for both the filler and vent pipes.
3. Coat only the *male* thread of the pipe with a pipe compound suitable for use with oil burning equipment.
4. Oil supply lines between the oil supply tank and the oil burner should be made of copper tubing (diameters will vary depending on local regulations, pipe length, and the specifications of the oil burner being used).
5. Use a floor-level tee if the oil supply lines run overhead.
6. The oil burners for the water heater and furnace (or boiler) may be connected to a common feed line in a conventional gravity feed installations (Fig. 1-39).
7. No return line is necessary when the supply tank is installed above the level of the oil burner and the oil is fed by gravity to the burner.

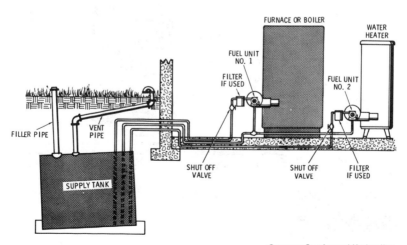

Courtesy Sundstrand Hydraulics

Fig. 1-36. Outside tank installation with the tank located below the level of the oil burner.

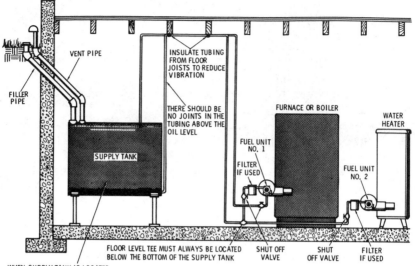

WHEN SUPPLY TANK IS LOCATED
BELOW THE LEVEL OF THE BURNERS,
INDIVIDUAL SUCTION LINES MUST BE USED

Courtesy Sundstrand Hydraulics

Fig. 1-37. Inside tank installation.

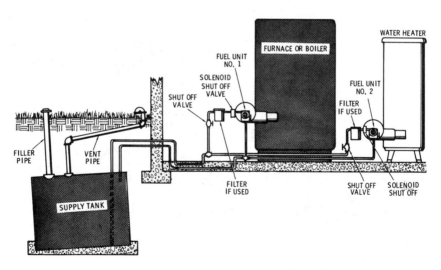

Courtesy Sundstrand Hydraulics

Fig. 1-38. Outside tank installation involving lift.

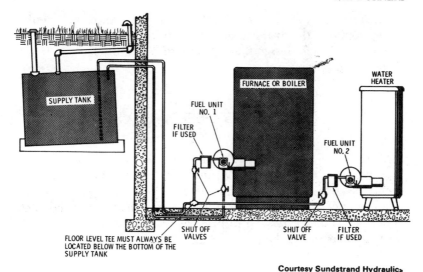

Fig. 1-39. Outside tank installation with burner located below the level of the tank.

8. Use a single-line system if the oil is gravity fed from the supply tank to the burner.
9. Use a two-line system if the oil tank is buried and below the level of the burner.

FILLER PIPE

The *filler pipe* is the filling connection attached to the top of the oil supply tank and terminating above the ground level at least 2 ft. from the outside building wall. This is commonly a 2-in. pipe of a design and material specified by the local authorities. Generally the specifications will mandate the use of a corrosion resistant material.

It is important that the filler pipe be connected to one opening and the vent pipe to a separate opening on the tank. No cross connection of vent pipe is permitted with the filler pipe or the return line from the oil burner. Some authorities demand double swing connections at the oil supply tank.

47

The termination point of the filler point at ground level should be equipped with a watertight metal cap. The termination point should be of such design that oil spillage is minimized when the oil hose is disconnected.

TANK VENT PIPE

The vent pipe is attached to the oil supply tank at a point separate from that of the filler pipe, and it should terminate at a point above the ground at least 2 ft. from the outside wall of the building. The vent pipe must be equipped with an approved vent hood or weatherproof cap to prevent water and other contaminants from entering the pipe and gaining access to the oil supply tank.

TROUBLESHOOTING OIL BURNERS

Individuals involved in the installing and repairing of oil burners should be aware of a number of different indicators of malfunctions in the equipment, their probable causes, and some suggested remedies.

The average individual is most aware of malfunctions that warn the senses through excessive noise, smoke, or odor. These are *external* warning signals that require immediate investigation. Their nature is such that tracing the probable cause of the malfunction is made easier.

Excessive noise (pulsation, thumping, rumbling, etc.) in the heating unit is generally caused by a problem with the oil burner nozzle. It can usually be corrected by any one of the following methods:

1. Replace the nozzle with one having a wider spray angle.
2. Replace the nozzle with one having the next size smaller opening.
3. Install a delayed-opening solenoid on the nozzle line (this reduces pulsation).

Sometimes a noisy fire is caused by cold oil originating from outside storage tanks. This noise may be greatly reduced or eliminated by pumping the fuel oil under 120 to 125 psi through the next size smaller nozzle.

Excessive smoke has a number of possible causes, including:

1. The air handling parts of the oil burner may be too dirty to operate efficiently.
2. The combustion chamber or burner tube may be damaged by burn-through or loose materials.
3. The oil burner nozzle may be the wrong size.

The dirty air handling parts (i.e., the fan blades, air intake, and air vanes in the combustion head) can be made to operate more efficiently by a thorough cleaning. If the excessive smoke is caused by the oil burner nozzle, this can be corrected by replacing the nozzle with one a size smaller or one having the next narrower spray angle. A damaged combustion chamber is a more difficult problem to correct than the other two. In any event, all leakage through the walls must be eliminated before the oil burner can be expected to operate efficiently.

Excessive odors can be caused by flue obstructions or poor chimney draft. If the draft over the fire is lower than .02 to .04, it is usually an indication that the problem lies with the flue or chimney draft. Obstructions in the flue or poor chimney draft are usually the causes. Other causes of excessive odor include:

1. Delayed ignition.
2. Too much air through the burner.

Delayed ignition is commonly traced to a problem with the electrodes. This condition can result from a variety of causes, including:

1. Improper electrode setting.
2. Insulator cracks.
3. A coating of soot or oil on the electrode.
4. Incorrect pump pressure setting.
5. Incorrect spray pattern in the nozzle.
6. Clogged nozzle.
7. Air shutter open too far.

Table 1-2 lists a number of recommended electrode settings that should eliminate delayed ignition if the electrode setting is the cause of the problem. The type of nozzle spray pattern can also result in delayed ignition. This is particularly true when using a hollow spray pattern in oil burners firing 2.00 GPH and above. It can be corrected by replacing the nozzle with one having a solid spray pattern.

Table 1-2. Recommended Electrode Settings

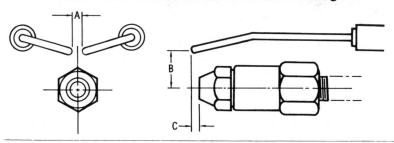

Nozzle	GPH	A	B	C
45°	(0.75 to 4.00)	1/8" to 3/16"	1/2" to 9/16"	1/4"
60°	(0.75 to 4.00)	1/8" to 3/16"	9/16" to 5/8"	1/4"
70°	(0.75 to 4.00)	1/8" to 3/16"	9/16" to 5/8"	1/8"
80°	(0.75 to 4.00)	1/8" to 3/16"	9/16" to 5/8"	1/8"
90°	(0.75 to 4.00)	1/8" to 3/16"	9/16" to 5/8"	0

Table 1-3 lists a variety of problems encountered with oil burners, many of which are of an *internal* nature and require a great degree of experience and training to correct.

Table 1-3. Oil Burner Troubleshooting

Cause	Remedy
NO Oil level below intake line	
OIL in supply tank	Fill tank with oil.
FLOW Clogged strainer or filter	Remove and clean strainer. Replace
AT	filter element.
NOZZLE Clogged nozzle	Replace nozzle.

Table 1-3. Oil Burner Troubleshooting (Cont'd.)

Cause	Remedy
Air leak in intake line	Tighten all fittings in intake line. Tighten unused intake port plug. Check filter cover and gasket.
Restricted intake line . . . . (High vacuum reading)	Replace any kinked tubing and check any valves in intake line.
A two-pipe system that becomes airbound	Check for and insert bypass plug. Make sure return line is below oil level in tank.
A single-pipe system that becomes airbound	Loosen gauge port plug or easy flow valve and bleed oil for 15 seconds after foam is gone in bleed hose. Check intake line fittings for tightness. Check all pump plugs for tightness.
Slipping or broken coupling	Tighten or replace coupling.
Rotation of motor and fuel unit is not the same as indicated by arrow on pad at top of unit	Install fuel unit with correct rotation.
Frozen pump shaft	Return unit to approved service station or factory for repair. Check for water and dirt in tank.

	Cause	Remedy
OIL LEAK	Loosen plugs or fittings	Dope with good-quality thread sealer. Retighten.
	Leak at pressure adj. screw or nozzle plug	Washer may be damaged. Replace the washer or O-Ring.
	Blown seal (single-pipe system)	Check to see if bypass plug has been left in unit. Replace fuel unit.
	Blown seal (two-pipe system)	Check for kinked tubing or other obstructions in return line. Replace fuel unit.
	Seal leaking	Replace fuel unit.
	Cover	Tighten cover screws or replace damaged gasket.

	Cause	Remedy
NOISY OPERATION	Bad coupling alignment	Loosen fuel-unit mounting screws slightly and shift fuel unit in different positions until noise is eliminated. Retighten mounting screws.
	Air in inlet line	Check all connections. Use only good flare fittings.
	Tank hum on two-pipe system and inside tank	Install return-line hum eliminator in return line

51

Table 1-3. Oil Burner Troubleshooting (Cont'd.)

Cause	Remedy
PULSATING Partially clogged strainer or filter. . . .	Remove and clean strainer. Replace
PRESSURE	filter element.
Air leak in intake line	Tighten all fittings.
Air leaking around cover	Be sure strainer cover screws are tightened securely. Check for damaged cover gasket.
LOW Defective gauge	Check gauge against master gauge
OIL	or other gauge.
PRESSURE Nozzle capacity is greater	
than fuel unit capacity	Replace fuel unit with unit of correct capacity.

IMPROPER To determine the cause of improper cutoff, insert a pressure gauge in the
NOZZLE nozzle port of the fuel unit. After a minute of operation, shut burner down.
CUTOFF If the pressure drops from normal operating pressure and stabilizes, the
fuel unit is operating properly and air is the cause of improper cutoff. If,
however, the pressure drops to 0 psi, fuel unit should be replaced.

Filter leaks	Check face of cover and gasket for damage.
Strainer cover loose	Tighten 4 screws on cover.
Air pocket between cutoff	
valve and nozzle	Run burner stopping and starting unit until smoke and afterfire disappear.
Air leak in intake line	Tighten intake fittings. Tighten unused intake port and return plug.
Partially clogged nozzle strainer	Clean strainer or change nozzle.
Leak at nozzle adapter	Change nozzle and adapter.

Courtesy Sunstrand Hydraulics

CHAPTER 2

Gas Burners

A *gas burner* (Fig. 2-1) is a device for supplying gas, or a mixture of gas and air, to the combustion area. Burners used for domestic heating can be of the atmospheric, yellow (luminous) flame, or power burner types. Only those gas burners bearing the seal of approval of the American Gas Association should be used.

Gas is used as a heating fuel in both urban and rural areas. Manufactured, natural, and bottled gas are the three types used as heating fuels. Each of these gases has different combustion characteristics and will have different heat values when burned. Because of this, a gas burner must be adjusted for each gas fuel, particularly when changing from one type to another. The principal types of bottled gas used as heating fuels are propane and butane. Bottled gas is frequently called LPG (liquefied petroleum gas) and is widely used as a heating fuel in rural areas. A more detailed description of heating fuels is found in Chapter 5 of Volume 1 (Heating Fuels).

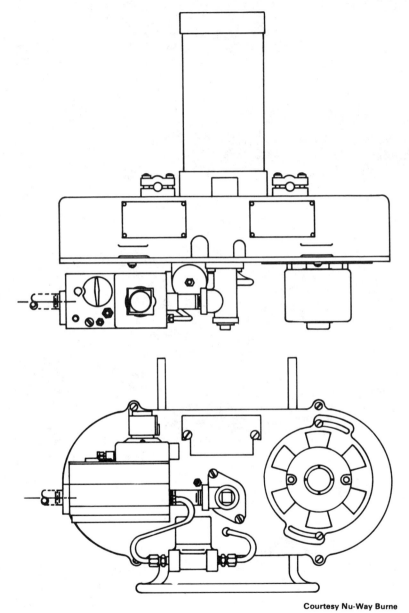

Fig. 2-1. Power gas burner.

54

OPERATING PRINCIPLES

Gas burners used for domestic heating purposes are most commonly the atmospheric injection type that operate on the same principle as the Bunsen burner.

The essential features of the Bunsen burner are shown in Fig. 2-2. The burner consists of a small tube or burner, which is placed inside a larger tube. The latter has holes positioned slightly below the top of the small tube. The gas escaping from the small tube draws the air in through the holes and produces what is called an *induced current* of air in the large tube. This air enters through the holes and is mixed with the gas in the tube. The mixture is

Fig. 2-2. Essential features of the Bunsen burner.

burned at the top of the larger tube. The flame from such a burner gives hardly any light, but the heat is intense. The intensity of the heat can be illustrated by holding a metal wire over the flame for a few seconds. It will glow with heat in a very short time.

The air supply in an atmospheric injection burner is classified as either *primary* or *secondary* air and is commonly introduced and mixed with the gas in the throat of the mixing tube. The cutaway of an upshot atmospheric gas burner in Fig. 2-3 illustrates this principle of operation. The gas passes through the small orifice in the mixer head, which is shaped to produce a straight-flowing jet moving at high velocity. As the gas stream enters the throat of the venturi, or mixing tube, it tends to spread and induce air in through the opening at the adjustable air shutter. The energy in the gas stream forces the mixture through the mixing tube into the burner manifold casting, from which it issues through ports where additional air must be added to the flame to complete combustion. The air coming in through the venturi is the *primary* air and that supplied around the flame is the *secondary* air.

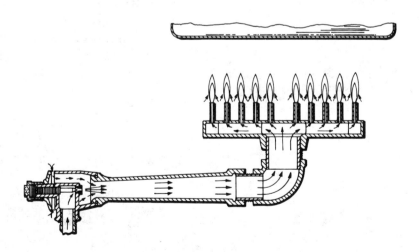

Fig. 2-3. Cutaway of the venturi, or mixing tube, of an upshot gas burner.

The primary air is admitted at a ratio of about 5 parts primary air to 1 part gas for manufactured gas, and a 10 to 1 ratio for natural gas. These ratios are generally used as theoretical values of air for purposes of complete combustion. Most atmospheric injection burners operate efficiently on 40 to 60 percent of the theoretical value.

The excess air required depends on several factors, notably:

1. Uniformity of air distribution and mixing.
2. Direction of gas travel from the gas burner.
3. The height and temperature of the combustion chamber.

The secondary air is drawn into the burner by natural draft. Excess secondary air constitutes a loss and should be reduced to a proper minimum (usually not less than 25 to 35 percent). All yellow flame gas burners depend exclusively on secondary air for combustion.

The Bunsen burner flame is bluish and practically nonluminous. A yellow flame indicates dependence entirely on secondary air for combustion. Primary air is regulated by means of an adjustable shutter. For manufactured gas, the air supply is regulated by closing the air shutter until yellow flame tips appear and then by opening the air shutter to a final position at which the yellow tips just disappear. This type of flame obtains ready ignition from port to port and favors quiet flame extinction. When burning natural gas, the air adjustment is generally made to secure as blue a flame as possible.

The division of air into primary and secondary types is a matter of burner design, the pressure of gas available, and the type of flame desired.

The gas should flow out of the burner ports fast enough so that the flame cannot travel or flash back into the burner head. The velocity must not be so high that it blows the flame away from the port. In an all-yellow flame, flame flashback cannot occur, and a much higher velocity is needed to blow off the flame.

A draft hood is used to ensure the maintenance of constant low-draft conditions in the combustion chamber with a resultant stability of air supply. A draft hood will also control back drafts that tend to extinguish the gas burner flame and the amount of excess air. These draft hoods must conform to American Standard Requirements.

TYPES OF GAS BURNERS

Gas burners for domestic use may be atmospheric injection, yellow (luminous) flame, or power burner types. Figs. 2-4, 2-5, and 2-6 show examples of some of these types. This classification is based on the firing method and the air supply used. Gas burners can also be divided into two broad classifications based on whether they are specifically designed as integral parts of gas-fired heating equipment or serve to convert coal or oil furnaces or boilers to gas use (i.e., conversion gas burners). Conversion burners are generally complete burners and controls so designed that they can be installed as a unit in the existing furnace or boiler (Figs. 2-7 and 2-8).

Courtesy Adams Manufacturing Co.

Fig. 2-4. Upshot conversion gas burner.

58

Gas burners are also classified as *inshot* and *upshot* types, depending on the design of the burner tube. The burner tube of an inshot gas burner is commonly a straight, adjustable venturi

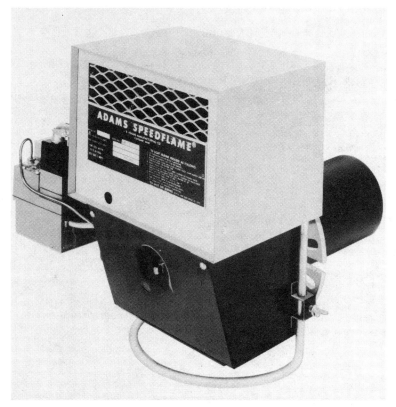

Fig. 2-5. Inshot power burner.

that extends horizontally from the unit (Fig. 2-9). An upshot gas burner is characterized by a burner tube that extends horizontally from the unit and then bends to assume a vertical position (Fig. 2-10).

Courtesy Adams Manufacturing Co.

Fig. 2-6. Forced-draft gas burner.

GAS BURNER COMPONENTS

A typical gas burner that functions as an integral part of a gas-fired heating system consists of the following principal components (Fig. 2-11):

1. Pilot light.
2. Pilot gas line.
3. Main gas line.
4. Burner head.
5. Automatic controls.
   a. Room thermostat.
   b. Safety thermostat element.
   c. Safety control valve.
   d. Gas pressure regulator.
   e. Main gas valve.
   f. Limit control
   g. Low-water cutoff control.

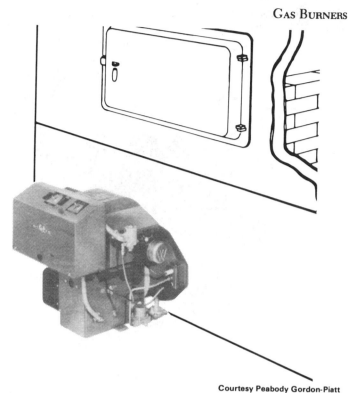

Courtesy Peabody Gordon-Piatt

Fig. 2-7. Gas burner attached to furnace.

Some gas burners are equipped with direct spark ignition rather than a pilot flame. As a result, the list of basic components will be slightly different (particularly in the case of items 1 and 2 above).

GAS BURNER AUTOMATIC CONTROLS

The automatic controls on domestic gas burners are designed to give either intermittent or constant operation. With the intermittent cycle, the burner is started and stopped by the room thermostat, which in turn opens and closes a control valve in the gas supply line that provides the fuel gas to the burner head. The constant operation control varies the fire by means of automatic devices from high to low depending on the heat load.

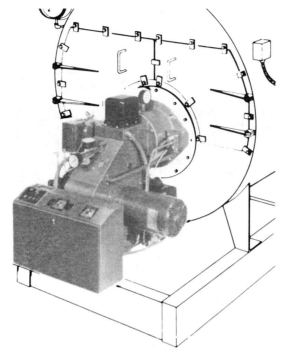

Courtesy Peabody Gordon-Piatt

Fig. 2-8. Gas burner attached to boiler.

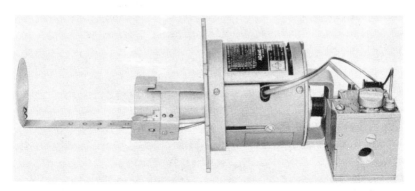

Courtesy Adams Manufacturing Co.

Fig. 2-9. Inshot conversion gas burner for furnaces or boilers.

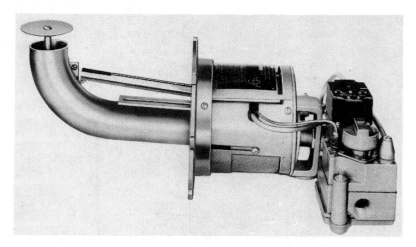

Courtesy Adams Manufacturing Co.

Fig. 2-10. Upshot conversion gas burner for furnaces or boilers.

The room thermostat should be installed on an inside wall at a location not subject to sudden changes in temperature and in the natural circulating path of room air. Locations subject to cold air drafts or heat radiation must be avoided. The proper installation and adjustment of the room thermostat is of primary importance to the efficient operation of a heating system.

Ignition is secured in the on and off control by the use of a pilot light. The *pilot light* (Figs. 2-12 and 2-13) is a small, continuously burning flame that provides ignition for the gas burner in much the same manner as does the electric spark for the high-pressure atomizing oil burner.

A *safety thermostat element* located near the burner head and connected by a line to the safety control valve (Fig. 2-11) prevents the gas valve from opening when the pilot light is extinguished for any reason. All gas burners should be equipped with this safety device. For satisfactory operation, automatically fired gas burners should be provided with pressure regulators on the gas supply line.

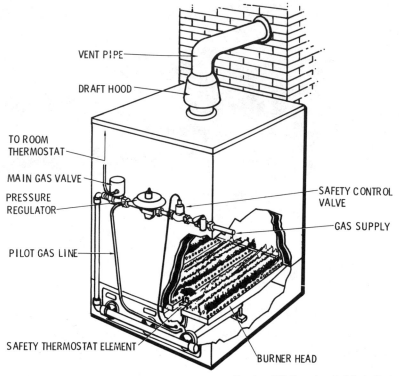

VENT PIPE

DRAFT HOOD

TO ROOM
THERMOSTAT

MAIN GAS VALVE

PRESSURE
REGULATOR

PILOT GAS LINE

SAFETY CONTROL
VALVE

GAS SUPPLY

SAFETY THERMOSTAT ELEMENT

BURNER HEAD

Courtesy U.S. Department of Agriculture

Fig. 2-11. Standard gas burner components.

Although a pilot light can be shut off during the nonheating season, continuous year-round operation greatly reduces or eliminates the possibility of corrosion occurring in the system. The amount of fuel gas consumed by the continuous operation of a pilot light is insignificant in terms of cost.

Limit controls are of two types:

1. Automatic temperature-limit controls.
2. Automatic pressure-limit controls.

Each limit control is designed to shut off the gas burner when a predetermined temperature or pressure is reached. Recommended limit control settings are given in Table 2-1.

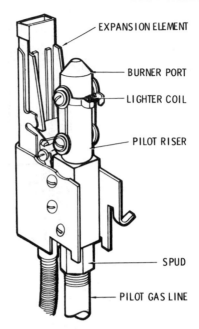

EXPANSION ELEMENT

BURNER PORT

LIGHTER COIL

PILOT RISER

Fig. 2-12. Automatic pilot for gas burner.

SPUD

PILOT GAS LINE

An automatic temperature-limit control is mounted on the warm-air plenum of gravity and forced warm-air furnaces. This is a low-voltage switch that opens the electrical circuit (and turns off the gas burner) on temperature rise. Forced and gravity hot-water boilers are also equipped with this type of limit control.

Steam and vapor boilers are equipped with automatic pressure-limit controls. These are also low-voltage switches that shut off the gas burner by opening the electrical circuit, but these are activated by pressure rise rather than temperature rise.

Table 2-1. Recommended Limit Control Settings

Hot-water system (gravity)	180°F
Hot-water system (forced)	160°F
Warm air .... (forced)	200°F
Warm air .... (gravity)	300°F
Steam system	"off" 3 lb.-"on" 1 lb.
Vapor system	"off" 4 oz.-"on" 2 oz.

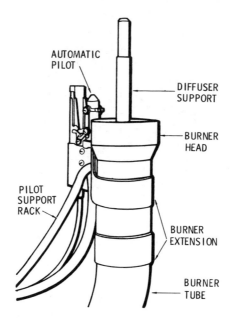

Fig. 2-13. Gas burner extension showing burner head, diffuser support, and automatic pilot.

All steam or vacuum vapor boilers should be equipped with low-water cutoff controls. This control prevents the gas burner from firing if the boiler is dry or the water level is too low. A thermostat inserted in the bonnet of a warm-air furnace functions as a *high* limit control, shutting off the gas burner when the furnace temperature exceeds a predetermined point.

WIRING AUTOMATIC CONTROLS

The limit control, thermostat, safety pilot light, and automatic main gas valve should be wired in series as shown by the two examples in Figs. 2-14 and 2-15. Note the position of the low-water cutoff unit in both wiring diagrams.

GAS BURNERS

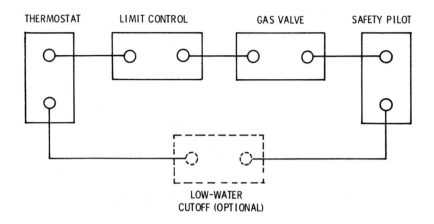

Fig. 2-14. Typical wiring diagram when using automatic gas valve operated by generator pilot.

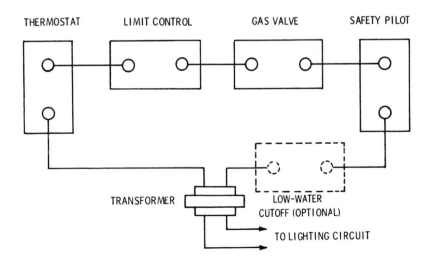

Fig. 2-15. Typical wiring diagram when using automatic gas valve operated by transformer.

67

All wiring must conform to local standards and regulations as well as the rules of the National Electrical Code. If the wires parallel the gas supply lines, they should be securely attached to it. Never allow the wires to sag. Furthermore, never permit the wires to be carried through cold air returns, clothes chutes, or leaders.

If the automatic main gas valve is operated by a transformer (as opposed to a generator pilot), it should be mounted with a knock-out box and an on-off switch (clearly marked) mounted on the hot wire of the supply circuit.

GAS CONVERSION BURNERS

A *gas conversion burner* is used to convert heating equipment designed for coal or oil to gas fuel use (Fig. 2-16). The boiler or

Courtesy MIDCO International, Inc.

Fig. 2-16. Residential spark-ignition gas conversion burner.

furnace must be properly gas tight and must have adequate heating surfaces.

A characteristic of gas conversion burners is that pressure will sometimes build up in a furnace due to puffs or backfire resulting from delayed ignition and other causes. Local heating codes and regulations usually stipulate that furnace doors be held tightly closed by spring tension only (in other words, not permanently closed) in order to provide a means for relieving pressure. Fig. 2-17 shows an example of a door spring that can be used for this purpose.

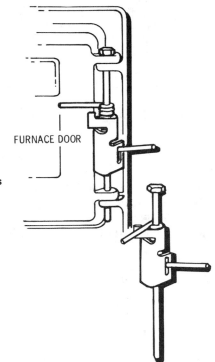

FURNACE DOOR

Fig. 2-17. Door-closing springs for furnace doors.

Courtesy Magic Servant Products Co.

Gas Conversion Burner Combustion Chambers

The combustion chamber for a gas conversion burner is commonly located in the ashpit of a boiler or furnace originally

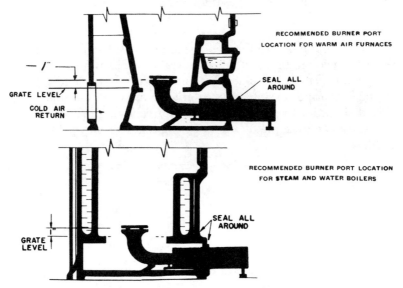

RECOMMENDED BURNER PORT
LOCATION FOR WARM AIR FURNACES

SEAL ALL
AROUND

GRATE LEVEL

COLD AIR
RETURN

RECOMMENDED BURNER PORT LOCATION
FOR STEAM AND WATER BOILERS

SEAL ALL
AROUND

GRATE
LEVEL

Courtesy Magic Servant Products Co.

Fig. 2-18. Positioning an upshot gas burner.

designed for coal or oil firing. Fig. 2-18 illustrates the positioning of an upshot gas conversion burner. Note that the burner head port is located 1 in. (plus or minus ¼ in.) *above* the grate level. This is a standard measurement when installing an upshot burner in the ashpit of a furnace.

The manufacturer's installation instructions provided with a conversion gas burner generally include specific instructions on the preparation of the combustion chamber. The main points to be remembered are as follows:

1. All openings in the boiler must be sealed.
2. The combustion chamber must be thoroughly cleaned.
3. Heat exchanger surfaces must be protected against concentrated heat.
4. All nonheat transfer surfaces must be protected.
5. The combustion chamber must be designed to contain combustion, to radiate heat, and to insulate the ashpit.

Additional information concerning gas conversion burners can be found in Chapter 16 of Volume 1 (Boilers and Furnace Conversions).

GAS PIPING FOR CONVERSION BURNERS

Gas piping is generally wrought iron or steel with malleable-iron pipe fittings. Joint compound (pipe dope) is applied sparingly to the male threads only. Make certain before applying the joint compound that it is approved for all types of gas. Never use aluminum tubing or cast-iron fittings on the main gas line. Soldered or sweated connections are also not recommended.

Tables 2-2 and 2-3 can be used to determine the size pipe to use from the meter to the gas burner. The correct number of threads for any particular length of pipe is given in Table 2-4.

Use pipe fittings at all turns in the gas line. Never bend or lap welded pipe because it will pinch or weaken it. Support the pipe with straps, bands, pipe hooks, or hangers (never allow one pipe to rest on another or to sag).

Pitch all horizontal pipe so that it grades toward the meter without the occurrence of sags. The piping should be protected against freezing and the accumulation of condensation.

Pipes and pipe fittings that are defective should be replaced, never repaired. Every effort must be made to eliminate any possibility of gas leakage. Gas leaks on pipes and pipe fittings should be located by spreading a soap solution over the surface. *Never try to locate a gas leak with a flame.* The results could be extremely hazardous.

Fig. 2-19 illustrates the general configuration of the main shutoff valve, pilot shutoff valve, and riser installation for a gas conversion burner. The following suggestions are worth noting:

1. Install the main manual gas shutoff valve on the riser *at least 4 ft. above the floor level.*
2. Install the pilot valve *on the inlet side* of the main manual gas shutoff valve.
3. Install a tee fitting at the bottom of the riser to catch any foreign matter in the pipe. The bottom of the tee fitting should be plugged or capped.

Table 2-2. Pipe Capacity Table

Length of Pipe in Feet	Nominal Diameter of Pipe in Inches				
	³⁄₄	1	1¼	1½	2
	Capacity—Cu. Ft. Per Hr. with a 0.6 Sp Gr Gas and Pressure Drop of 0.3″ Water Column				
15	172	345	750		
30	120	241	535	850	
45	99	199	435	700	
60	86	173	380	610	
75	77	155	345	545	
90	70	141	310	490	
105	65	131	285	450	920
120		120	270	420	860
150		109	242	380	780
180		100	225	350	720

Courtesy Magic Servant Products Co

Table 2-3. Multipliers for Various Specific Gravities

Specific Gravity	Multiplier	Specific Gravity	Multiplier
.35	1.31	1.00	.775
.40	1.23	1.10	.740
.45	1.16	1.20	.707
.50	1.10	1.30	.680
.55	1.04	1.40	.655
.60	1.00	1.50	.633
.65	.962	1.60	.612
.70	.926	1.70	.594
.75	.895	1.80	.577
.80	.867	1.90	.565
.85	.841	2.00	.547
.90	.817	2.10	.535

Courtesy Magic Servant Products Company

Table 2-4. Specifications for Threading Pipe

Nominal Size of Pipe (inches)	Approx. Length of Threaded Portion (inches)	Approx. Number of Threads to be Cut
³⁄₄	³⁄₄	10
1	⁷⁄₈	10
1¼	1	11
1½	1	11
2	1	11

*See AGA Requirements and Recommended Practice for House Piping and Appliance Installation.

Courtesy Magic Servant Products Company

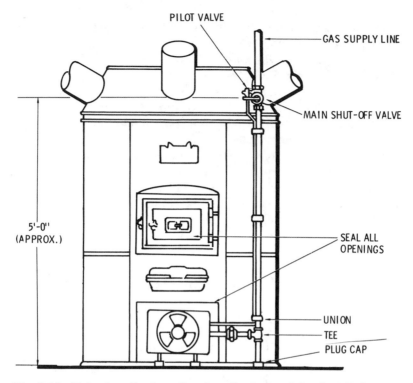

PILOT VALVE

GAS SUPPLY LINE

MAIN SHUT-OFF VALVE

5'-0"
(APPROX.)

SEAL ALL
OPENINGS

UNION

TEE

PLUG CAP

Fig. 2-19. Main shutoff valve, pilot shutoff valve, and riser installation.

4. Install a ground joint union in the gas line between the burner air duct box and the tee fitting in the riser.

5. Install a pilot supply line ($\frac{1}{4}$-in. OD tubing) between the pilot valve and a point located on the upper right side of the air duct box on the gas burner. This line will run parallel to the riser.

VENTING AND VENTILATION

The venting system for gas heating equipment consists of:

1. The chimney or smoke elimination pipe.
2. The draft diverter or draft diverter hood.

73

Gas-burning equipment must be vented to the outside. All pipes leading from the equipment must be fitted so that the joints are tight and free from leaks.

A *draft diverter* or *draft hood* is a wind deflector placed in the chimney to prevent downdrafts of air (i.e., air moving down the chimney from the outside) from blowing out the pilot light. Many draft diverters are designed and positioned so that the downdraft is deflected into the room containing the heating unit.

Any room containing gas heating equipment must have adequate ventilation. Provisions for incoming air (i.e., air necessary for combustion) is especially important in rooms or buildings of particularly tight constuction. The minimum area requirements (in relation to each 1000 Btu per hour input) for both ventilating air openings and air inlet openings can be obtained from most manufacturers of the gas equipment.

SAFETY PRECAUTIONS

The annual cleaning and inspection of gas heating equipment is important not only because it contributes to its efficient operation but also because it provides an additional safety factor.

Electrical controls should be connected on a separate switch. This enables the circuit to be broken should the equipment malfunction.

Because fuel gas is extremely volatile, it should be handled and stored with the utmost care. Bottled gas is particularly dangerous when it leaks. Because it is heavier than air, bottled gas will accumulate at low points in a room and present an explosion hazard.

Be sure to observe the following basic safety rules when working with heating gases and gas controls:

1. Always shut off the gas supply to the device when installing, modifying, or repairing it. Allow at least 5 minutes for any unburned gas to leave the area before beginning work. Remember that LP gas is heavier than air and does not vent upward naturally.

2. Always conduct a *gas leak test* after completing the installation, modification, or repair. To test for a gas leak, coat the pipe joint, pilot gas tubing connections, and valve gasket lines with a soap-and-water solution. Then, with the main burner in operation, watch for bubbles at those points. The bubbles will indicate a gas leak, which can normally be eliminated simply by tightening joints or screws or by replacing the gasket. Note: *Never use a flame to check for a gas leak.*

3. Always disconnect the power supply to prevent electrical shock or equipment damage before connecting or disconnecting any wiring.

4. Change the main burner and pilot orifice(s) to meet the appliance manufacturer's instructions when converting a gas system from one type of gas to another.

5. Always read and carefully follow the installation and operating instructions supplied with the appliance or component. Failure to follow them could result in damage or cause a hazardous condition.

6. Make certain the appliance or component is designed for your application. Check the ratings given in the instructions and on the appliance or component.

7. Check the operation of the appliance or component with the manufacturer's instructions after installation is completed.

8. Do not bend the pilot tubing at the control after the compression nut has been tightened. This could cause a gas leak at the connection.

9. Never jump (or short) the valve coil terminals on 24-volt controls. Doing so could short out the valve coil or burn out the heat anticipator in the thermostat.

10. Never connect millivoltage controls to line voltage or to a transformer because doing so will burn out the valve operator or the thermostat anticipator.

11. Do not remove the seals covering control inlets or outlets until you are ready to connect the piping. The seals are there to prevent dirt and other materials from getting into the gas control and interfering with its operation.

TROUBLESHOOTING GAS BURNERS

As is the case with all mechanical and electrical equipment, it is recognized that occasional repair and adjustment may be necessary on any burner. Table 2-5 represents a very general list of troubles and causes that can occur with gas burners and gives some suggested remedies.

Table 2-5. Troubleshooting a Gas Burner

SYMPTOMS	CAUSE
1. PILOT DOES NOT LIGHT.	a. Air in gas line. b. High or low gas pressure. c. Blocked pilot orifice. d. Flame runner improperly located.
2. PILOT GOES OUT FREQUENTLY DURING STANDBY, OR SAFETY SWITCH NEEDS FREQUENT RESETTING.	a. Restriction in pilot gas line. b. Low gas pressure. c. Blocked pilot orifice. d. Loose thermocouple connection on 100 percent shutoff. e. Defective thermocouple or pilot safety switch. f. Poor draft condition. g. Draft tube set into or flush with inner wall of combustion chamber.
3. PILOT GOES OUT WHEN MOTOR STARTS.	a. Restriction in pilot gas line. b. High or low gas pressure. c. Excessive pressure drop when main gas valve opens.
4. MOTOR DOES NOT RUN.	a. Burned-out fuse or current off. b. Thermostat or limit defective or improperly set. c. Relay or transformer defective. d. Motor burned out. e. Tight motor bearings from lack or oil. f. Improper wiring.
5. MOTOR RUNNING BUT NO FLAME.	a. Pilot out. b. Pilot safety switch needs to be reset. c. Thermocouple not generating sufficient voltage. d. Very low or no gas pressure. e. Motor running too slow.
6. SHORT, NOISY BURNER FLAME.	a. Pressure regulator set too low. b. Air shutter open too wide. c. Too much pressure drop in gas line. d. Vent in regulator plugged. e. Defective regulator.

Table 2-5. Troubleshooting a Gas Burner (Cont'd.)

SYMPTOMS	CAUSE
7. LONG, YELLOW FLAME.	a. Air shutter not open enough.
	b. Air openings or blower wheel clogged.
	c. Too much input.
8. MAIN GAS VALVE DOES NOT CLOSE WHEN BLOWER STOPS.	a. Defective valve.
	b. Obstruction on valve seat.
9. REGULATOR VENT LEAKING GAS.	a. Hole in diaphragm.

Courtesy Magic Servant Products Company.

CHAPTER 3

Coal Firing Methods

The fuel for a coal-fired furnace or boiler may be fed either automatically or by hand. Both methods have certain advantages and disadvantages. For example, automatic (stoker) firing is initially more expensive because it requires the purchase and installation of a suitable mechanical stoker to feed the coal to the furnace or boiler. As a result, stoker firing is a common practice in larger buildings (e.g., stores, hotels) where the initial high cost of the equipment can be more easily absorbed into the total cost of the structure. Despite the relative high cost of stokers, there are some designed for use in single-family residences. These will be considered in detail at a later point in this chapter.

Because there is no need to invest in special and expensive coal handling equipment, hand firing the coal has been the traditional method used for firing house heating furnaces and boilers. Although hand firing coal *is* less expensive than stoker firing for these smaller installations, the following objections to the hand firing method should be noted:

1. The frequent opening of the furnace or boiler doors allows a large excess of air to enter and chill the flame. The combustion efficiency of the flame therefore tends to fluctuate.
2. The dumping of a lot of fuel at each firing results in a smoke period until normal combustion conditions are restored.
3. Hand firing coal is by its nature an intermittent firing method. The flame often reaches a low and inefficient level or is extinguished before new fuel is added.

COAL FIRING DRAFT REQUIREMENTS

The amount of draft required for proper combustion is an important consideration, and it depends on a number of different considerations, including:

1. Grate area.
2. Fuel size and type.
3. Fuel-bed thickness.
4. Boiler pass resistance.

The degree of resistance offered by the boiler passes to the flow of the gases is an important consideration in determining the required amount of draft. These gases *must* exist at a speed sufficient to prevent them from backing up into the combustion chamber and robbing the fire of necessary oxygen.

The total area of the grate, the type of coal burned (e.g., bituminous, semibituminous), the size of the coal, and the thickness of the coal bed all affect the amount of draft required for proper combustion.

Insufficient draft usually results in the accumulation of excess ashes in the ashpit. Moreover, it necessitates additional attention to the fire including more frequent cleaning. These and other aspects of improper firing contribute to fuel waste and higher operating costs.

FIRING ANTHRACITE COAL

Anthracite coal is preferred over other coal for domestic heating purposes because it produces a steadier and cleaner flame.

Furthermore, it burns longer and with greater heat than the others. The principal objections to using anthracite coal are that it requires more heat than other coals to start combustion, and it is slightly more expensive.

Anthracite coal is available in a number of different standardized sizes, each suited to a different size of grate and firepot. Some of these coal sizes and their descriptive names are:

1. Buckwheat.
2. Egg.
3. Stove.
4. Chestnut.
5. Pea.
6. Broken.

Buckwheat-size coal is available in four grades or sizes: (1) Buckwheat No. 1; (2) Buckwheat No. 2, or Rice size; (3) Buckwheat No. 3, or Barley size; (4) Buckwheat No. 4; and (5) Buckwheat No. 5. Buckwheat Anthracite No. 2 (or Rice size) finds the widest use of automatic coal firing equipment, being used in domestic, commerical, and industrial stokers.

There are certain recommended practices to be followed when firing buckwheat coal. For best results, the following techniques should be employed:

1. Always maintain a uniform low fire. This reduces clinker formation to a minimum and enables those clinkers that do form to be broken up more easily.
2. Use a smaller mesh grate when possible or a domestic stoker. Because of its relatively small size, buckwheat anthracite coal often falls through the openings on ordinary grates when they are shaken.
3. Immediately after coaling, push a poker down through the fresh bed of buckwheat anthracite coal and expose a portion of the hot fire. This tends to prevent delayed ignition and such undesirable accompanying side effects as furnace or boiler doors being blown open.
4. Keep the heating system warm at all times. Allowing it to cool down and then having to warm it up results in burning fuel at a higher rate.

81

Egg-size coal should be used in large firepots (24-in. grates or larger). This is a deep firing coal, and the most suitable results are obtained with fuel beds that are 16 in. or more deep.

Stove-size coal was extensively used in heating buildings, although today it has been largely replaced by gas or oil. It is used on grates 16 in. and over. The fuel bed should be at least 12 in. deep.

Chestnut-size coal is used for firepots as large as 20 in. in diameter. Fuel beds for this anthracite coal range in depth from 10 to 15 in.

With careful firing, *pea-size coal* can be burned on standard grates. Care should be taken not to over shake the grates (shake only until the first bright coals begin to fall through the grates). After a pea-coal fire has been built, the thickness of the fuel bed should be increased by the addition of small charges until it is at least level with the sill of the fire door. A common method of firing pea coal consists of drawing the red coals toward the front of the firebox and piling fresh fuel toward the back of the firebox.

A strong draft is required when burning pea coal. Fig. 3-1 illustrates a satisfactory method of burning this coal size in a boiler. The choke damper is kept open and regulated by means of the cold air check and air inlet dampers. In stoker firing, the air setting is generally kept lower for pea coal because an excess of air under this kind of coal will burn up the retort. Forced draft and small mesh grates are frequently used for burning buckwheat anthracite coal.

FIRING BITUMINOUS COAL

Bituminous coal is a broad category encompassing many different burning characteristics and properties. Generally speaking, bituminous coals ignite and burn easily with a relatively long flame. They are also characterized by excess smoke and soot when improperly fired.

The *side bank* method is commonly recommended for firing bituminous coal. It consists of moving the live coals to one side or the other of the grate and placing a fresh fuel charge on the

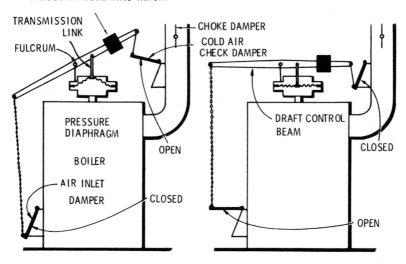

MINIMUM   MAXIMUM

PRESSURE REGULATING WEIGHT

TRANSMISSION LINK

FULCRUM

CHOKE DAMPER

COLD AIR CHECK DAMPER

PRESSURE DIAPHRAGM

BOILER

AIR INLET DAMPER

OPEN

CLOSED

DRAFT CONTROL BEAM

CLOSED

OPEN

Fig. 3-1. Dampers used on boilers and controls.

opposite side. Variations of this firing method call for placing the live coals at the back of the grate or covering the fresh fuel charge with a layer of fine coal. The side bank method results in a slower and more uniform release of volatile gases.

Other recommendations that should be followed when firing bituminous coal include:

1. Fire bituminous coal in small quantities at short intervals. This results in a better combustion because the fuel supply is maintained more nearly proportional to the air supply.

2. *Never* fire bituminous coal over the entire fuel bed at one time. A portion of the glowing fuel should always be left exposed to ignite the gases leaving the fresh fuel charge.

3. Use a stoking bar to break up a fresh charge of coking coal approximately 20 minutes to 1 hour after firing.

4. Do *not* bring the stoking bar up to the surface of the fuel. Doing so will bring ash into the high-temperature zone at the top of the fire, where it will melt and form clinkers.

A stoking bar should always be kept as near the grate as possible and should be raised only enough to break up the fuel. The ash will usually be dislodged when stoking, making it unnecessary to shake the grates.

Alternate or *checker firing* is a bituminous coal firing method in which the fuel is fired alternately on separate sides of the grate. This method tends to decrease the amount of smoke and to maintain a higher furnace or boiler temperature.

A similar effect is produced by the *coking method* of firing bituminous coal. The coal is first fired close to the firing door, and the coke is moved back into the furnace just before firing again.

FIRING SEMIBITUMINOUS COAL

Semibituminous coal burns with far less smoke than the bituminous type. It ignites with more difficulty than bituminous coal, but produces far less smoke.

The *central cone method* is recommended for firing semibituminous coal. In this method, the coal is heaped onto the center of the bed, forming a cone, the top of which should be level with the middle of the firing door. This allows the larger lumps to fall to the sides and the fine cones to remain in the center and be coked.

The poking should be limited to breaking down the coke without stirring, and to gently rocking the grates. It is recommended that the slides in the firing door be kept closed, as the thinner fuel bed around the sides allows enough air to get through.

STOKER FIRING

A *stoker* is a mechanical device designed and constructed to automatically feed fuel to a furnace. Stokers are used in commercial, industrial, and domestic heating systems. Their use results in more efficient combustion owing to constant instead of intermittent firing.

According to the *ASHRAE Guide* (1960), coal stokers can be

84

divided on the basis of their coal-burning capacity into the following four classes:

1. Class 1 stokers (10 to 100 lb. per hour).
2. Class 2 stokers (100 to 300 lb. per hour).
3. Class 3 stokers (300 to 1200 lb. per hour).
4. Class 4 stokers (over 1200 lb. per hour.)

Class 1 stokers are used most commonly in domestic heating installations. The other three classes of stokers are used in commercial and industrial heating systems.

Class 1 stokers are usually the underfeed type and are designed to burn anthracite, bituminous, semibituminous, and lignite coal, and coke. Ash can be removed automatically or manually, with the latter method being the most popular.

Stokers can also be classified on the basis of whether the coal is stored in a hopper or bin. The disadvantage of the hopper design (Fig. 3-2) is that it must be refilled at least once each day. The bin stoker design (Fig. 3-3) eliminates coal handling. The coal is delivered by the supplier and placed directly into the bin.

The underfeed stoker (Fig. 3-4) is generally used for house heating furnaces and boilers. This type of stoker is one in which the fuel is fed upward from underneath the furnace or boiler. The action of a screw or worm carries the fuel back through a retort from which it passes upward as the fuel above is being consumed. The ash is generally deposited on dead plates on either side of the retort, from which it can be removed.

Underfeed stokers can be designed for use with either anthracite or bituminous coal, but the individual pieces of coal should be uniform in size and no larger than 1 in. in diameter. As mentioned elsewhere in this chapter, it is desirable to treat the coal with oil in order to eliminate dust. The worm feed mechanism can be regulated to feed coal at variable rates.

STOKER CONSTRUCTION

Although there are variations in the type and design of domestic stokers, the general features are much the same. An elemen-

85

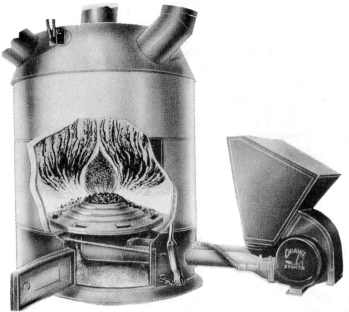

Courtesy Drawz Stoker Mfg. Co.

Fig. 3-2. Hopper-fed conical grate. Coal is underfed into the furnace and overfed to the fire in a slow movement.

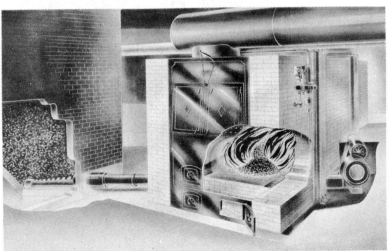

Courtesy Drawz Stoker Mfg. Co.

Fig. 3-3. Bin-fed stoker equipped with conical grate.

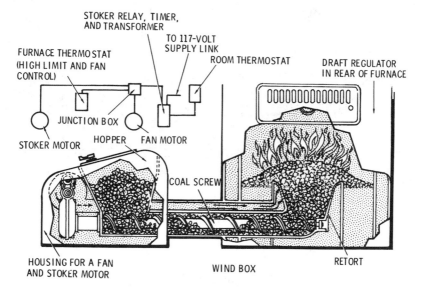

Fig. 3-4. Underfeed stoker showing principal parts and typical wiring diagram.

tary stoker is shown in Figs. 3-5 and 3-6, which gives the essentials and the names of parts. These parts may be listed as follows:

1. Retort.
2. Fan.
3. Motor.
4. Transmission.
5. Air duct.
6. Air control.
7. Hopper.
8. Feed worm.
9. Bin.

The *retort* is a firepot cast in a round or rectangular trough-like shape in which the coal is burned. It is made of cast iron and is surrounded by the windbox. The retort is provided with a number of air ports, or tuyeres, through which air for combustion is supplied.

The purpose of the *fan* is to supply forced draft, which is

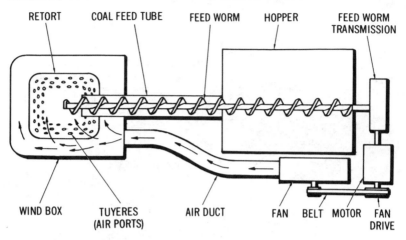

RETORT   COAL FEED TUBE   FEED WORM   HOPPER   FEED WORM TRANSMISSION

WIND BOX   TUYERES (AIR PORTS)   AIR DUCT   FAN   BELT   MOTOR   FAN DRIVE

Fig. 3-5. Underfeed stoker.

directed to the windbox that surrounds the air ports in the retort. This fan is commonly of the squirrel-cage type.

The air enters the retort through the ports via the *air duct* from the fan and the windbox that surrounds the retort. The fan is equipped with either manual or automatic control in the form of a damper at either the discharge or intake end. The air supply is controlled by means of these fan controls.

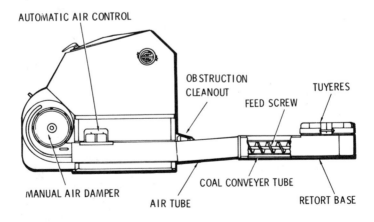

AUTOMATIC AIR CONTROL

OBSTRUCTION CLEANOUT   TUYERES

FEED SCREW

MANUAL AIR DAMPER   COAL CONVEYER TUBE

AIR TUBE   RETORT BASE

Fig. 3-6. Domestic underfeed stoker.

Coal stokers are designed to operate on either high or low air-pressure systems. In the high air-pressure system, the air is forced in small jets into the fire area. A major disadvantage of this type of system is that the coal sometimes tends to fuse, causing clinkers to form and wasting some of the combustible matter of the fuel. A low air-pressure system tends to produce a more complete circulation of burning gases to all heat inducting surfaces.

A stoker is usually powered with an *electric motor*, which operates both the coal feed worm and the fan. The stoker drive consists of a *transmission*, a shear pin (or clutch throw-out), pulleys, belts, and related components. The purpose of the shear pin is to protect the driving mechanism against damage in case large-size foreign objects get mixed up with the coal.

The transmission rotates the coal feed worm at the proper speed to feed the amount of coal required. The construction is such that the rate of feed can be changed as desired.

The two kinds of transmission usually employed in stokers are the *continuous drive*, which is operated by means of reduction gears, and the *intermittent drive*, which operates with a ratchet. Another drive used in stoker transmissions is the *hydraulic* (usually referred to as an *oil drive*), which operates by regulating the oil pressure on the driving mechanism to control the number of revolutions the feed screw makes per minute.

The *feed worm* (sometimes called the *feed screw*) carries the coal from the hopper to the retort (firepot). It is geared to the transmission, its rate of revolution depending on the desired feed rate. The feed worm extends from the coal supply in the hopper or bin, through the coal feed tube into the retort, where the coal it carries is discharged.

Ashpits can be constructed so that they are located directly below the furnace or boiler (Fig. 3-7). The ashes are automatically deposited into the pit as the coal is burned. If the pit is designed large enough, the ashes will need to be removed only once or twice a year. It is recommended that the ashpit be constructed so as to permit removal of ashes from outside the house. This will result in a much cleaner and more convenient operation in the long run. *Always* vent the ashpit to the chimney or outdoors.

Fig. 3-7. Bin-fed conical grate stoker.

Some stokers (e.g., the Drawz stoker illustrated in Figs. 3-2 and 3-3) are designed to permit hand firing in case of power failure. In these situations, a natural draft may be provided by opening the grate and ashpit door.

STOKER AUTOMATIC CONTROLS

Fig. 3-8 illustrates typical controls for a stoker in a forced warm-air heating system. There are approximately three *basic* automatic controls necessary for satisfactory operation of the stoker. These three controls are:

1. Thermostat.
2. Limit control.
3. Hold-fire control.

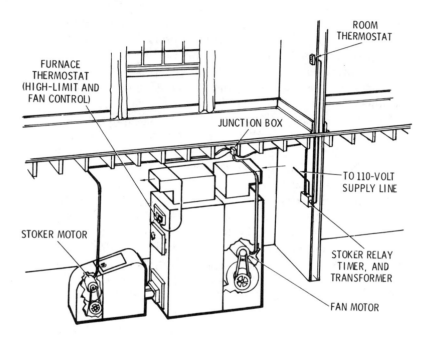

Fig. 3-8. Typical controls for a stoker-fired coal burner in a forced warm-air heating system.

The purpose of the *thermostat* is to start the fire when the room temperature falls below a predetermined point and to stop it when the temperature again rises to normal. The thermostat setting should be adjusted to give comfortable room temperature. Usually a setting between 72° and 75° is desirable.

The *limit control* stops the stoker should the furnace or boiler pressure become greater than the setting of the control. Furnace limit switches on warm-air gravity installations usually require setting above 300°. Hot-water limit switches on hot-water systems usually require settings above 160°. Steam pressure controls on steam pressure installations usually require settings of 2 to 5 lb.

The purpose of the *hold-fire control* is to produce a stoker operation at intervals during mild weather in order to maintain fire when the thermostat is not demanding heat. Sometimes the hold-fire control feeds either too much or too little coal to the retort. The former results in overheating, and the latter may cause the fire to go out. It is best to call a service representative of the stoker manufacturer to adjust the hold-fire control, because this is a complicated mechanism. The two types of hold-fire controls are: (1) interval timers and (2) stack temperature control switches.

Timers may be adjusted to give various-length firing periods so that the stoker is operated for a few minutes at preset intervals. This is done to keep the fire alive during cool weather when little heat is required. The cycle of operation may be set for either 30-minute or 1-hour intervals.

Figs. 3-9 and 3-10 show two examples of typical timers used on stokers. The combination switching relay and synchronous motor-driven timer shown in Fig. 3-9 provides periodic burner operation so that the fire can be maintained during times when the thermostat is not demanding heat. It may be used with any 2-wire, 24-volt thermostat or operating controller. This particular timer is adjustable from ½ to 7½ minutes every 30 or 60 minutes.

The timer shown in Fig. 3-10 is designed for line voltage switching. When used with a line voltage controller, it maintains the stoker fire by providing short *on* periods between the controller *off* periods. Timing may be adjusted from 1 to 7½ minutes at 30- or 60-minute intervals.

The *Stack switch* (or *stack thermostat*) starts the stoker when the stack temperature becomes lower than a predetermined point and operates it until the fire is again kindled to a degree that will guarantee that it will not go out.

Stack switches are not found on all stokers, but they should be required in areas where electric power failures are long enough to let the fire go out. The stack switch will keep the stoker from filling the cold firepot with coal as soon as the electricity goes on again. Sometimes a light-sensitive electronic device (such as an electric eye) is used instead of a stack switch.

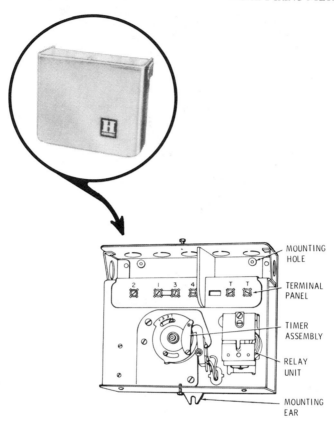

Fig. 3-9. Honeywell combination switching relay and synchronous motor-driven timer.

STOKER OPERATING INSTRUCTIONS

The stoker operating instructions found in the paragraphs that follow should be regarded as generalized suggestions or recommendations rather than specific instructions. They may prove useful in those situations in which no operating manual from the manufacturer can be found. When possible, *always* consult the manufacturer's operating manual.

93

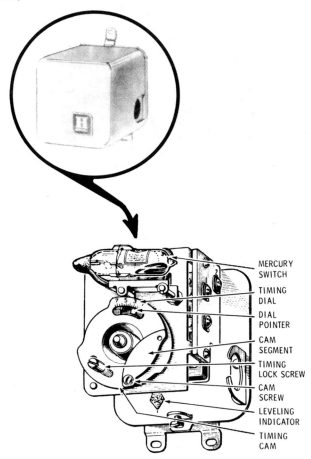

MERCURY
SWITCH

TIMING
DIAL

DIAL
POINTER

CAM
SEGMENT

TIMING
LOCK SCREW

CAM
SCREW

LEVELING
INDICATOR

TIMING
CAM

Courtesy Honeywell, Inc.

Fig. 3-10. Internal view of a Honeywell stoker timer.

COAL SELECTION

A bituminous coal, low in ash (6 percent or less), with an ash fusion temperature of 2200°F to 2600°F and 1¼-in. to ⅜-in. maximum is ideal for stoker operation. Oil treatment of the coal is recommended in order to eliminate dust and add lifetime to the stoker. Generally speaking, in most sections of the country, a high-quality coal is generally most satisfactory and most economical. The annual coal tonnage for domestic stokers is usually low. Convenience and satisfaction are usually the primary factors considered when making the decision to install a stoker; therefore, good coal is recommended. Consult your local coal dealer or the stoker manufacturer for recommendations.

STARTING THE FIRE

Generally, the procedure involved in starting a fire in a stoker-equipped heating installation includes the following steps:

1. Set the room thermostat above the room temperature.
2. Set the coal feed and air setting to the proper rate.
3. Throw the line switch to the *on* position so that the stoker starts.
4. Open the hopper lid, and watch the feed screw to make certain that it is turning. Sometimes in shipping or installing the stoker, the feed screw may slip off the shaft on the gear case. Be certain that the feed screw is engaged *before* putting any coal in the hopper.
5. Fill the hopper with coal.
6. Set the over fire air door on the furnace ¼ to ½ open and lock in this position.
7. Let the stoker run until the retort (inside the furnace or boiler) is filled with coal.
8. Place a quantity of paper, kindling wood, and a small amount a coal on top of the retort and ignite it.

NATURAL STACK DRAFT

Natural draft has a decided effect on the operating economy of the stoker installation. Check the draft and baffle damper, which should be adjusted to give the lowest possible draft without causing smoking from the fire door.

The check damper in the smoke outlet to the furnace can also be used advantageously when extreme natural draft conditions exist. The ideal arrangement is obtained by limiting draft just to the point that smoke or fumes are not emitted from the fire door when the stoker is in operation.

MANUAL AIR ADJUSTMENT

As the fuel bed builds up to the desired condition, the air adjustment should be made in the following manner: Open or close the manual air damper to give a yellow and practically smokeless flame (not white hot) and a fire fed with no intense blasts from air ports in the burner. Sufficient air must be delivered to maintain an even-burning fuel bed with a fairly consistent depth.

AUTOMATIC AIR CONTROL

Each stoker will usually have some means of automatically controlling the pressure and volume of air delivered by the fan so that the correct amount is supplied to the fire as burning conditions vary. Usually no adjustment is necessary as the setting made at the factory will enable this control to function properly on most installations.

CHANGING COAL FEEDS

On some stokers, the coal feed change is easily made by altering the position of the drive belt from the smaller to the larger or from the larger to the smaller pulleys of the motor and transmission.

Instructions:

1. Cut off the stoker line switch.
2. Move the belt change lever down to reduce the tension on the belt.
3. Move the belt to the pulley desired.
    a. Belt on the large pulley of the motor gives maximum feed.

b. Belt on the center pulley of the motor gives intermediate feed.

c. Belt on the small pulley of the motor gives minimum feed.

4. Move the belt tightening lever up to the original position.

5. Throw in the line switch.

MOTOR OVERLOAD PROTECTION

The stoker motor will have a built-in device for protection against excessive motor temperatures. Should the motor become overheated, the protection device on the motor will prevent damage by breaking the electrical circuit. Motor overloads are usually caused by lack of bearing lubrication, low voltage, or excessive belt tension. To reset, push in reset button on motor after the motor has cooled sufficiently.

TRANSMISSION OVERLOAD PROTECTION

A stoker transmission will also include an overload protection device that automatically breaks the electrical circuit to the motor in the event that an obstruction should become lodged in the conveying mechanism of the unit. To reset (after removal of the obstruction), push in reset button on the side of the transmission.

REMOVAL OF OBSTRUCTION

Read the manufacturer's instructions for removing obstructions from the conveying mechanism of the transmission. If these are not available, then you will have to determine the best way to gain access to the obstruction and remove it. There is usually an obstruction clean-out panel located in back of the hopper. Full access to the feed screw is obtained by removal of this panel. It may be necessary to reverse the rotation of the feed screw manually to relieve the obstruction. To do this, the transmission must be placed in neutral. This can be done by disengaging the transmission from the conveying mechanism.

LUBRICATION

The electric motor should be lubricated at the beginning of the heating season and twice during the season. Use a good grade of medium engine oil.

The transmission will require approximately one pint of a suitable grade of engine oil. This should be checked once each season. The oil should be removed and replaced at the end of two heating seasons provided these has been no flooding. Should the transmission become submerged in water, it is recommended that it be serviced by a representative of the stoker manufacturer.

SUMMER SERVICE

It is recommended that the stoker be prepared for the next heating season just after the spring heating has been completed. The stoker should be prepared in the following manner:

1. Remove the coal from the hopper.
2. Paint or grease the inside of the hopper.
3. Open the hopper lid for air circulation.
4. Remove any siftings from the retort base, and any ash or clinker formation from the burner.
5. Clean and oil the electric motor and adjust the belts.
6. Oil the stoker screw (or worm).
7. Replace oil in the transmission, if necessary.
8. Run a heavily oiled coal or sawdust through the stoker, leaving the feed screw and coal tube full, over the summer. This prevents corrosion and rusting.

HOW TO REMOVE CLINKERS

You will find it easier to remove the clinker if you will let the fire cool off for 5 to 10 minutes before removing it. Turn the stoker *off* and open the fire door to cool the fire. Fill the hopper while the clinker is cooling. The clinker normally forms *around* the retort. Use an iron bar or poker to raise the clinker. *Do not dig in the retort.* After you have raised the clinker, use the clinker tong to lift it from the furnace. It may be in one piece or several pieces, but remove all of it. Keep the fuel bed clean. Remove clinkers as often as necessary.

HOW TO ADJUST COAL FEED

Fig. 3-11 illustrates the steps involved in a typical coal feed adjustment. Their order (in sequence) is as follows:

1. Select the proper amount of coal feed for the furnace (refer to the coal feed chart provided by the stoker manufacturer).
2. When the proper coal feed is selected, the opposite side of the pointer indicates the proper air setting (Fig. 3-11A).
3. When the coal meter is set on proper coal feed, lock the meter with a wrench at the locknut shown (Fig. 3-11B).

HOW TO ADJUST AIR SUPPLY

From information on the coal meter, set the air selector knob to the proper point and the automatic damper will furnish the proper amount of air for the amount of coal fed to the furnace. The automatic air damper opens slowly after the stoker starts feeding coal, thus preventing "puff backs" out of the fire door and closes when the stoker stops and automatically banks the fire. When this occurs, the motor stops running.

TROUBLESHOOTING COAL STOKERS

All mechanical devices occasionally malfunction or operate below a commonly accepted level of efficiency. Coal stokers are no exception to this rule. The following is the list of conditions that indicate faulty operation:

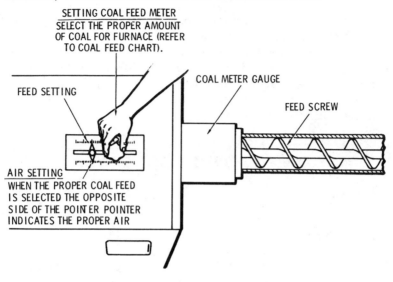

SETTING COAL FEED METER
SELECT THE PROPER AMOUNT
OF COAL FOR FURNACE (REFER
TO COAL FEED CHART).

COAL METER GAUGE

FEED SETTING

FEED SCREW

AIR SETTING
WHEN THE PROPER COAL FEED
IS SELECTED THE OPPOSITE
SIDE OF THE POINTER POINTER
INDICATES THE PROPER AIR

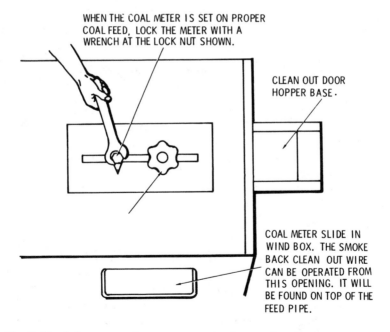

WHEN THE COAL METER IS SET ON PROPER
COAL FEED, LOCK THE METER WITH A
WRENCH AT THE LOCK NUT SHOWN.

CLEAN OUT DOOR
HOPPER BASE.

COAL METER SLIDE IN
WIND BOX. THE SMOKE
BACK CLEAN OUT WIRE
CAN BE OPERATED FROM
THIS OPENING. IT WILL
BE FOUND ON TOP OF THE
FEED PIPE.

Fig. 3-11. Adjustment of a coal feed.

Symptom and Possible Cause   *Possible Remedy*

Abnormal Noises

1. Loose pulleys or belt.	1. Tighten or replace.
2. Dry motor bearings.	2. Oil the bearings.
3. Worn gears.	3. Oil or replace.
4. Gears lack oil.	4. Oil the gears.

Motor Will Not Start

1. Hard clinkers over or on retort.
2. Foreign matter caught in the feed screw.
3. Packing of coal in the retort caused by the end of the feed screw being worn.

1. Remove the clinkers.

2. Remove the foreign matter.

3. Remove packed coal from the retort and replace feed screw.

Stoker Operates Continuously

1. Controls out of adjustment.
2. Dirty fire.
3. Fire out.
4. Dirty furnace or boiler.

1. Contact manufacturer for a service call.
2. Rebuild or clean fire.
3. Rebuild fire.
4. Clean furnace or boiler.

Furnace Filled with Unburned Coal

1. Clinkers clogging the retort.
2. Coal feed set too high.
3. Insufficient air getting to the fire.

1. Remove the clinkers.

2. Reduce coal feed setting.
3. Open manual damper. If this does not help, check air ports for clogging. Also check the wind box. If it is full of siftings, they should be removed.

101

Symptom and Possible Cause   *Possible Remedy*

Stoker Will Not Run

1. Limit control has shut off furnace or boiler due to overheating.
2. Low-water cutoff has shut down the boiler.
3. Gear case has been exposed to water.

1. Allow limit control time to cool off.
2. Check water level in boiler and correct.
3. *Do not* try to operate the stoker. Drain and flush out the gear case immediately and refill with oil.

Smoke Backed Into Hopper

1. Hopper empty or low in coal.
2. Clinker obstructing the retort.
3. Clogged smoke back connection.
4. Fire burning down in the retort.

1. Fill the hopper to the proper level.
2. Remove clinker.
3. Remove obstruction.
4. Check air supply (fire may be getting too much) or rate of coal feed (may be too low).

Fire Is Out

1. Empty hopper.
2. Clinkers obstructing the retort.
3. Switch may be off.
4. Blown fuse.
5. Failure in electric controls.

1. Refill to proper level.
2. Remove clinkers.
3. Place in *on* position.
4. Replace fuse.
5. Contact manufacturer for a service call.

CHAPTER 4

Thermostats
and Humidistats

This chapter, and the two that immediately follow it, describe the principal components of the automatic control systems used in heating, ventilating, and air conditioning. The index should also be checked because additional information about automatic controls has been included in other chapters.

AUTOMATIC CONTROL SYSTEMS

An automatic control system consists primarily of the following two basic components:

1. Controller.
2. Controlled device.

A *controller* is any device that can measure changes in temperature, humidity, or pressure and respond to these changes by activating a controlled device.

A *controlled device* may be a valve or a damper, or a motor that drives either of the two. It may also be a pump, fan, electric relay, or any other device used to regulate the flow of air, steam, water, gas, or oil.

Automatic control systems can be classified as either closed loop or open loop. A closed-loop system (Fig. 4-1) is the more common type and involves the following stages in the control sequence:

1. The controller measures a change in a variable condition (e.g., temperature) and actuates the controlled device.
2. The controlled device compensates for the change in the variable condition by regulating the flow rate of the medium (e.g., water, air, steam) carried in the system.
3. The result of the action of the controlled device is measured, and this information is fed back to the controller completing (i.e., closing) the loop.

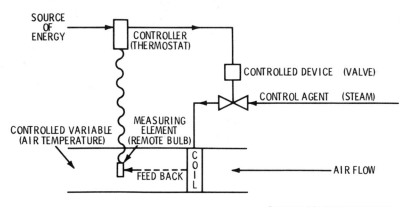

Courtesy *ASHRAE 1960 Guide*

Fig. 4-1. Closed-loop automatic control system.

An example of an open-loop system is one using an outdoor thermostat. It is characterized by having no means of feedback. In other words, room temperature has no effect on the operation of the controller.

TEMPERATURE CONTROL CIRCUITS

The thermostat is the basic controller in the electrical control circuits used to operate a heating and/or cooling system. In systems using gas-fired heating equipment, there are three basic control circuits:

1. Safety shutoff circuit.
2. Fan or circulator control circuit.
3. Temperature control circuit.

There are three basic types of temperature control circuits used in heating and cooling systems. These three basic circuits are:

1. Low-voltage control circuit.
2. Line voltage control circuit.
3. Millivolt control circuit.

Typical wiring diagrams for these three temperature control circuits are shown in Figs. 4-2, 4-3, and 4-4.

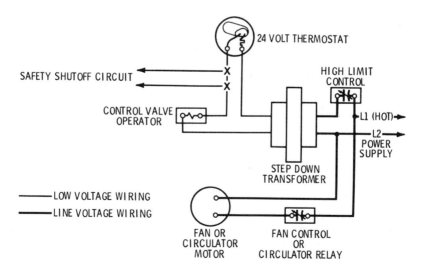

Courtesy Honeywell Tradeline Controls

Fig. 4-2. Low-voltage temperature control circuit.

105

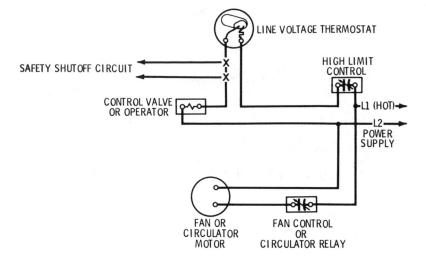

Fig. 4-3. Line voltage temperature control circuit.

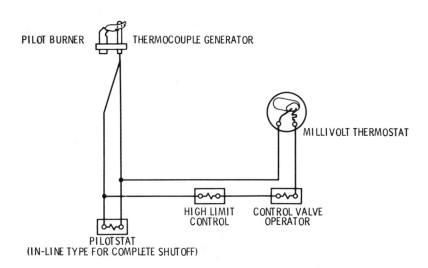

Fig. 4-4. Millivolt temperature control circuit.

106

THERMOSTATS

A *thermostat* is an automatic device designed to maintain temperature control. It accomplishes this function by reacting to temperature changes with adjustments of a controlled device such as a damper or valve motor, or the automatic firing equipment (gas burner, oil burner, or coal stoker) in space heating furnaces and boilers. Because of its specific function, a thermostat is sometimes referred to as a *temperature controller.*

Thermostats can be classified on the basis of *how* they measure (or sense) temperature changes. The devices most commonly used to measure temperature changes are:

1. Bimetallic strip sensing element.
2. Pressure-actuated sensing element.
3. Electrical resistance element.

A bimetallic strip containing two dissimilar metals is probably the most widely used of these three temperature measuring devices. Its operating principle is based on the different expansion and contraction rates of dissimilar metals. When two such metals are joined together in a bimetallic strip, the differences in expansion and contraction rates will cause a bending movement as the temperature changes. This movement is utilized to open or close an electrical circuit between the thermostat and the controlled device in the heating and/or cooling system. The bimetallic strip sensing element is used in either snap-action switch thermostats or mercury-switch thermostats.

A thermostat that operates on the positive snap-action switching principle contains movable switch contacts. One of the contacts is connected to a movable switch armature; the other is fixed in position. An auxiliary armature attached to a bimetal coil responds to temperatures induced by the expansion or contraction of a moving magnet (Fig. 4-5). The magnet, attached to the auxiliary armature, controls the movement of the switch armature. When the switch armature moves toward the magnet, it causes the contacts to close (Fig. 4-6).

On some thermostats, the switch contacts are hermetically sealed in a glass enclosure to protect them from dust or moisture.

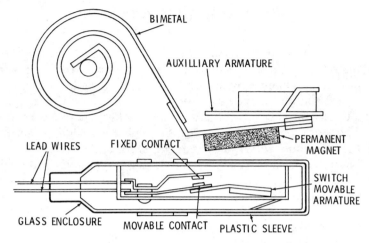

BIMETAL

AUXILLIARY ARMATURE

PERMANENT MAGNET

LEAD WIRES   FIXED CONTACT

SWITCH MOVABLE ARMATURE

GLASS ENCLOSURE   MOVABLE CONTACT   PLASTIC SLEEVE

Courtesy Robertshaw Controls Co.

Fig. 4-5. Glass-enclosed contact switch in the open position.

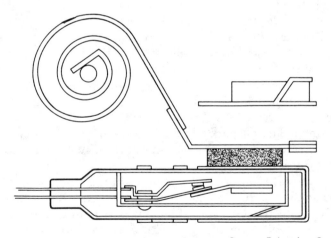

Courtesy Robertshaw Controls Co.

Fig. 4-6. Glass-enclosed contact switch in the closed position.

The thermostats illustrated in Fig. 4-5, 4-6, and 4-7 are of this design.

A mercury-switch thermostat contains fixed contacts sealed in a mercury-filled tube. The tube is attached to the end of a spiral

108

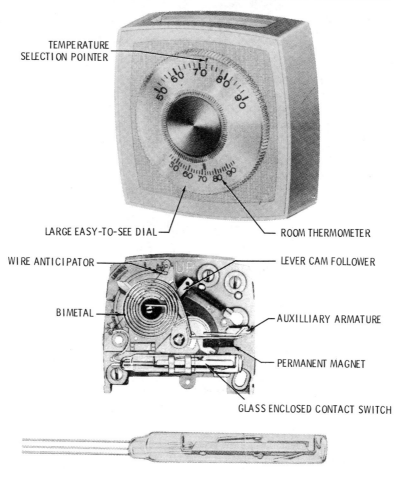

TEMPERATURE
SELECTION POINTER

LARGE EASY-TO-SEE DIAL

ROOM THERMOMETER

WIRE ANTICIPATOR

LEVER CAM FOLLOWER

BIMETAL

AUXILLIARY ARMATURE

PERMANENT MAGNET

GLASS ENCLOSED CONTACT SWITCH

Courtesy Robertshaw Controls Co.

Fig. 4-7. Thermostat with contacts in sealed glass enclosure.

bimetal element. When the temperature changes, the bimetal element tilts the tube and causes the mercury to shift its position causing a definite opening and closing of the electrical circuit.

A two-wire thermostat using the mercury tube switch method is illustrated in Fig. 4-8. A drop in temperature causes the mercury switch to complete (close) the circuit. The circuit is broken (opened) on a rise in temperature.

Fig. 4-9 illustrates the application of the mercury tube switch method in a heating and cooling thermostat. A common terminal wire runs along the bottom of the mercury tube. Mercury can make contact between either the heating or cooling terminals, but *not* both at the same time.

Sealing the contacts in a mercury-filled tube provides excellent protection against contamination; however, care must be taken to properly level the base when mounting the unit because the position of the tube determines the switching action.

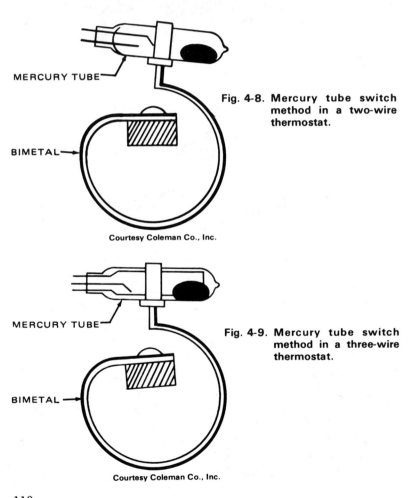

MERCURY TUBE

BIMETAL

Fig. 4-8. Mercury tube switch method in a two-wire thermostat.

Courtesy Coleman Co., Inc.

MERCURY TUBE

BIMETAL

Fig. 4-9. Mercury tube switch method in a three-wire thermostat.

Courtesy Coleman Co., Inc.

On either the snap-action or mercury-switch thermostat, the temperature setting of the thermostat can be changed by rotating the temperature dial. This device, acting through the cam, causes the bimetal coil to rotate through its mounting post to carry the temperature setting (Fig. 4-10).

Another temperature measuring device used in thermostats is the pressure-actuated sensing element. A liquid, gas, or vapor with a high coefficient of expansion is used to activate a bellows connected to a snap mechanism. A rise in temperature causes an expansion in the volume of the liquid, gas, or vapor. This expansion is transferred to the bellows, which activates the snap mechanism. See "Remote Bulb Thermostats" for additional details.

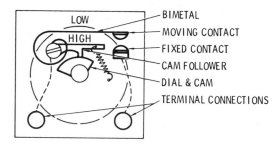

Fig. 4-10. Temperature dial and cam mechanism.

An electrical-resistance sensing element consists of a coil of wire with an electrical resistance that changes in direct proportion to temperature changes. This type of sensing element is commonly used in electronic controllers.

A thermocouple device consisting of two dissimilar electrical wires welded together at one end also serves as a temperature sensing element in some thermostats. Temperature changes at the welded juncture of the two wires cause electrical changes in the control circuit, which operates a regulatory device.

Thermostat Components

A thermostat consists of two basic parts: the base and the cover. A subbase can also be added to a thermostat to provide fan and various switch functions (Fig. 4-11).

The cover protects the internal wiring of the thermostat base from dust, lint, and other possible contaminants. It also contains the temperature setting lever, the system switching levers, and the temperature indicator (Fig. 4-12).

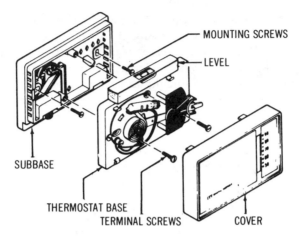

Courtesy ITT General Controls

Fig. 4-11. Thermostat cover, base, and subbase.

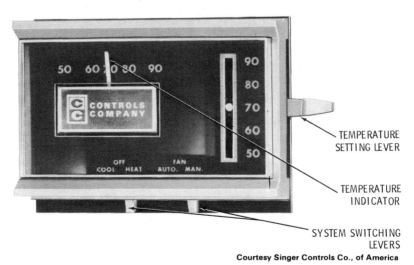

Courtesy Singer Controls Co., of America

Fig. 4-12. Low-voltage thermostat for electric heating and/or cooling system.

The cover is secured to the base either with a positive friction snap or screws. If the former is the case, the cover can be removed by grasping the base with one hand and pulling it off with the other (Fig. 4-13). Screws may require the use of an Allen wrench (Fig. 4-14), which is usually supplied by the thermostat manufacturer and shipped with the unit.

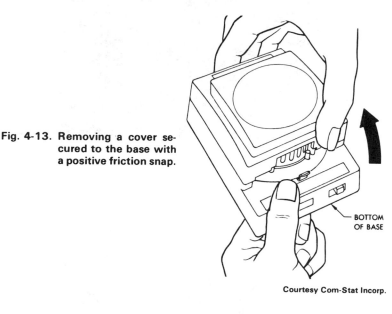

Fig. 4-13. Removing a cover secured to the base with a positive friction snap.

BOTTOM OF BASE

Courtesy Com-Stat Incorp.

Fig. 4-14. Loosening cover screws with an Allen wrench.

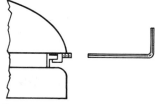

Courtesy Honeywell Tradeline Controls

The base contains the internal wiring of the thermostat. Fig. 4-15 shows the back and front views of typical base wiring for the low-voltage thermostat illustrated in Fig. 4-12. This is *not* the only possible way the base can be wired. *How* the base is wired will depend upon the particular application. A few of the possible

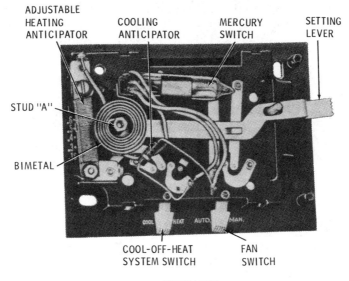

ADJUSTABLE
HEATING   COOLING   MERCURY   SETTING
ANTICIPATOR   ANTICIPATOR   SWITCH   LEVER

STUD "A"

BIMETAL

COOL-OFF-HEAT   FAN
SYSTEM SWITCH   SWITCH

FRONT VIEW

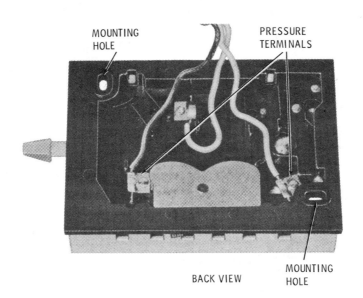

MOUNTING   PRESSURE
HOLE   TERMINALS

                               MOUNTING
BACK VIEW   HOLE

Fig. 4-15. Front and back view of Singer thermostat model 360.

variations are illustrated by the thermostat base wiring diagrams in Fig. 4-16. The many variations in wiring will depend upon which combination of the following features is required by the installation:

1. Type of switch and switching action (snap-action or mercury bulb; spst or spdt contact).
2. Number of field wires (2, 3, 4, or 5).
3. Type of anticipator (fixed heating, adjustable heating, fixed cooling).

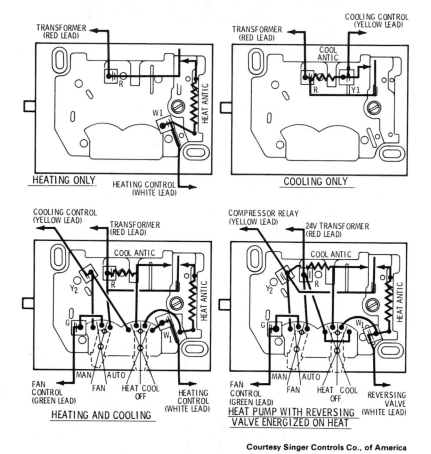

Courtesy Singer Controls Co., of America

Fig. 4-16. Thermostat base wiring.

115

4. Use of fan and system switches.
5. Type of operating voltage (low voltage, line voltage, or millivolt circuit).

Adding a subbase to a thermostat provides fan and other switching functions. The wiring diagrams of the ITT General Controls thermostats and subbases shown in Fig. 4-17 illustrate the variety of different switching combinations available.

Thermostat guards can be purchased to protect the thermostat from damage. This is particularly important in warehouses, stores, and other areas where there is a greater possibility of the thermostat being damaged. Some examples of how these guards are used are shown in Fig. 4-18. Adapter plates (wall plates) are also available from thermostat manufacturers to cover electrical utility or junction boxes (Fig. 4-19).

Thermostat Terminal Identification

The National Electrical Manufacturers Association is at present attempting to standardize thermostat markings in order to aid the installer in wiring and servicing. Thermostat terminals have been given standard identification letters that specify the function of the terminal. These identification letters are also matched, in most cases, with the color coding of the wire. A partial list of equivalent terminal markings is given in Table 4-1.

Thermostat installation literature will generally contain at least one internal view and/or wiring diagram in which the various terminals are identified by specific letter.

Thermostat Anticipators

A *thermostat anticipator* is a device used to reduce the operating differential of the heating or cooling system. It is designed to enable the thermostat to shut off the furnace or boiler slightly in advance of the actual set temperature. As a result, the thermostat shuts off the heating equipment sooner than it would if it were affected only by the room temperature, thereby compensating for heat transfer lag. A thermostat may be equipped with a heat anticipator, a cold anticipator, or both.

A *heat anticipator* is a small resistor (resistive heater) con-

116

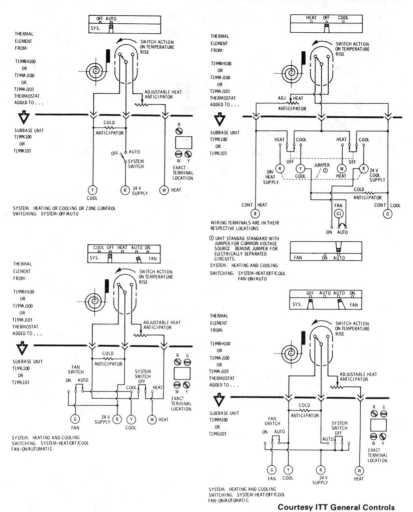

Courtesy ITT General Controls

Fig. 4-17. Thermostat switching combinations.

nected in series with the switch inside the thermostat. Heat generated by the resistor when the switch is in the *on* position heats the thermostat bimetal actuator and causes the internal temperature of the thermostat to rise faster than the surrounding room temperature. Thermostats are available with fixed anticipators, plug anticipators, or variable anticipators.

117

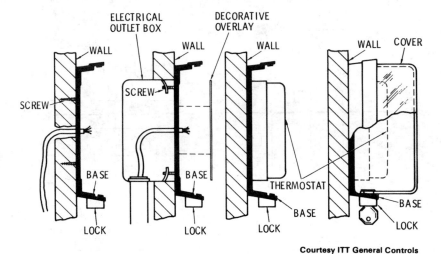

Courtesy ITT General Controls

Fig. 4-18. Thermostat guards.

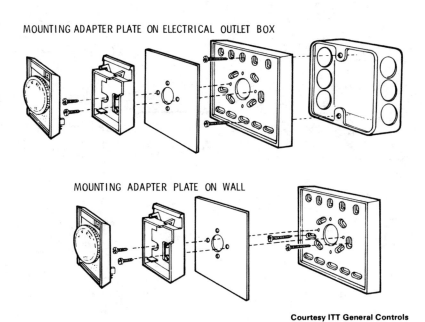

Courtesy ITT General Controls

Fig. 4-19. Adapter plate kit.

118

Table 4-1. Common Terminal Identification

Letter	Wire Color	Terminal Function
R	Red	Power supply; transformer
W	White	Heating control; heating relay or valve coil
Y	Yellow	Cooling control; cooling contactor coil
G	Green	Fan relay coil
O	Orange	Cooling damper
B	Brown	Heating damper
X	—	Malfunction light
P	—	Heat pump contactor coil
Z	—	Low-voltage fan switch

Note: The letters R_H (heating) and R_C (cooling) are used on thermostats with isolated circuits.

A fixed heat anticipator must be sized in accordance with the amperage or current draw of the operating valve in order to secure the correct degree of heat anticipation. These anticipators are generally available in the range of 0.1 to 1.5 amperes. When using a thermostat equipped with a fixed heat anticipator, check the nameplate on the valve or relay to make certain the ampere rating does not exceed the maximum amp (current) draw. For example, if the thermostat is equipped with a .40- to .60-amp fixed heat anticipator, the valve or relay should not exceed .60 amps.

The thermostat shown in Fig. 4-20 is an example of one equipped with a fixed heat anticipator. The anticipator is contained in the thermostat element along with the magnetic switch and a room temperature thermometer.

Fig. 4-21 illustrates a thermostat equipped with a plug-type heat anticipator. This design offers some degree of latitude in selecting a proper anticipator. The current (amps) drawn by the primary control or valves is first determined, and an anticipator having the proper value is selected from among those listed in Table 4-2. The heating cycle can be lengthened by selecting an anticipator one step above the proper value. Shorter cycles can be obtained by selecting an anticipator one step below the proper value.

Thermostats with adjustable (variable resistance) heat anticipators can be adjusted over a range of approximately 0.1 to 1.5

119

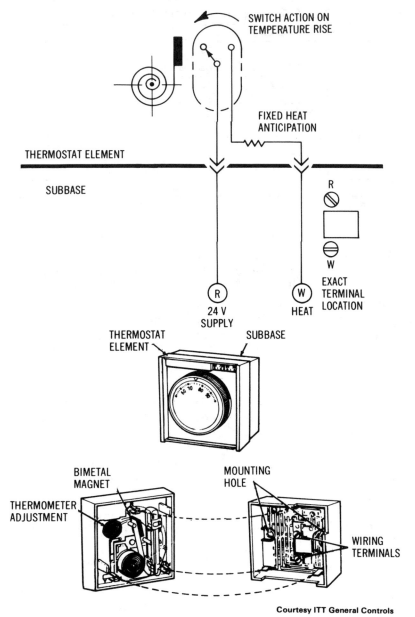

Fig. 4-20. Thermostat with fixed heat anticipator.

SNAP-ON COVER

CONNECT WIRES TO SUBBASE TERMINALS USE #19
(OR LARGER) COLOR-CODER SOLID CONDUCTOR
COPPER WIRE. DO NOT STRIP LEAD WIRE
INSULATION BELOW BOTTOM OF TERMINAL. PUSH
ACCESS WIRE BACK THROUGH SLOT IN SUBBASE.
WIRE IN ACCORDANCE WITH APPLICABLE CODES.

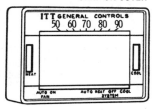

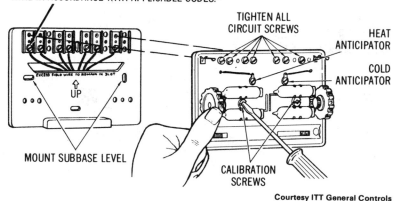

TIGHTEN ALL
CIRCUIT SCREWS

HEAT
ANTICIPATOR

COLD
ANTICIPATOR

MOUNT SUBBASE LEVEL

CALIBRATION
SCREWS

Courtesy ITT General Controls

Fig. 4-21. Thermostat with plug heat anticipator.

Table 4-2. Anticipator Values

Amperes	Color	Amperes	Color
.83–.72	Brown-Red	.33–.29	Orange-Yellow
.72–.68	Brown-Blue	.29–.25	Orange-Green
.68–.55	Orange	.25–.22	Red
.55–.48	Blue	.22–.19	Green-Blue
.48–.41	Blue-Orange	.11–.10	Orange-Red
.41–.36	Blue-Yellow	.10–.09	Red-Yellow
.36–.33	Green	.09–.08	Green-Yellow

Courtesy ITT General Controls

amps (Fig. 4-22). Before making any adjustments, you should
first read the equipment manufacturer's instructions for selecting
proper anticipator values.

Heat anticipator adjustments are made on a thermostat with an

121

SNAP-ON COVER

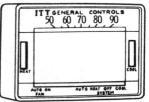

CONNECT WIRES TO SUBBASE TERMINALS USE #19 (OR LARGER) COLOR-CODER SOLID CONDUCTOR COPPER WIRE. DO NOT STRIP LEAD WIRE INSULATION BELOW BOTTOM OF TERMINAL. PUSH ACCESS WIRE BACK THROUGH SLOT IN SUBBASE. WIRE IN ACCORDANCE WITH APPLICABLE CODES.

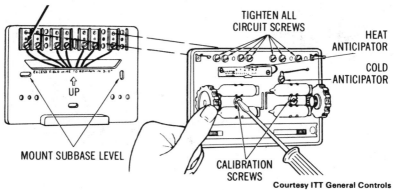

Courtesy ITT General Controls

Fig. 4-22. Thermostat with adjustable heat anticipator.

indicator or adjustment lever that moves along a scale (Fig. 4-23). Adjustments are made in accordance with the values marked along this scale.

As with the plug-type heat anticipator (see above), the current

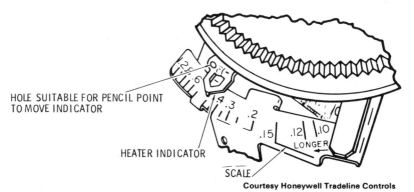

Courtesy Honeywell Tradeline Controls

Fig. 4-23. Heat anticipator adjustment lever and scale.

(amps) drawn by the primary control or valve must be determined first. In a gas-fired heating system, the heat anticipator should be set to correspond to the secondary (thermostat) current of the valve or relay. In an oil-fired heating system, the heat anticipation indicator should be set .15 amps higher than the rated secondary current of the relay.

The anticipator adjustment lever should be moved only ¼ to ½ scale division at a time. *Never* move or set the lever more than 1½ scale divisions under the valve or relay current ratings. Longer *on* periods can be obtained by setting the adjustment lever at a slightly higher amp value. For shorter *on* periods, set the lever at a slightly lower amp value.

The cold anticipator for the thermostat shown in Fig. 4-24 is not adjustable and should not be changed. The same is true of the other thermostats. Only *fixed* cold anticipation is used. Furthermore, on cooling thermostats, the cold anticipation is in parallel with the cooling contacts so that anticipation heating occurs while the cooling unit is off.

Types of Thermostats

Many different thermostats are manufactured for use in heating and cooling systems. The design differences depend largely upon the type of application.

The most common thermostat is the wall-mounted *room thermostat* used to control a heating and/or cooling system. The measuring element is contained in the thermostat unit itself. This distinguishes it from the *remote-bulb-type thermostat* used to measure temperatures in spaces separate from the location of the thermostat.

An *insertion-type thermostat* (or *duct thermostat*) is used to measure temperatures inside an air duct. The temperature measuring element is contained in an insertion device which extends into the duct. The *immersion-type thermostat* is similar in design, but is used to measure the temperature of fluids inside a pipe or tank. These thermostats are commonly used on water heaters.

A *heating-cooling thermostat* (also referred to as a *summer-winter thermostat*) is designed to be switched to either a heating or cooling application. The *day-night thermostat* (or *electric*

123

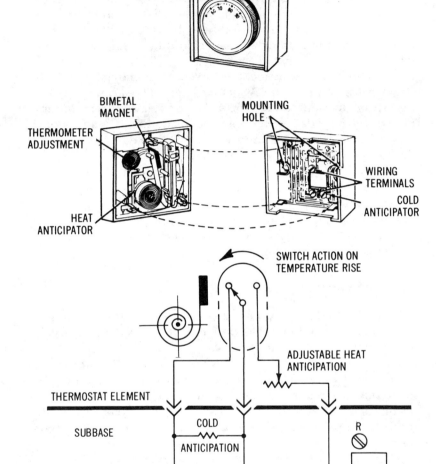

Fig. 4-24. Thermostat with fixed cool and adjustable heat anticipator.

124

clock thermostat) operates on a similar working principle except that it is designed to automatically switch from day to night operation and back again.

A *multistage thermostat* is designed to operate two or more circuits in sequence. These thermostats are used for line voltage or low-voltage temperature control of heating and cooling equipment. They are commonly used in heating and/or cooling systems where zone control is necessary.

A thermostat and humidistat can also be combined in the same control unit. These combined units sometimes also include the electronic air cleaner control.

Room Thermostats

The *room thermostat* (Fig. 4-25) is regarded as the nerve center of the heating and cooling system because it controls the operation of the furnace, boiler, or air conditioner. Ideally, it should be mounted in an area of the living or working spaces where it is not subjected to temperature or moisture extremes (see below, "Location of Room Thermostats").

Low-voltage room thermostats are recommended in preference to the line voltage types for residential heating and/or cooling systems. The low-voltage thermostats respond more quickly to temperature changes and will maintain the temperature and humidity more closely than the line voltage types. A low-voltage thermostat requires the use of a transformer to reduce the line voltage for the control circuit, but the cost of the transformer is more than offset by the lower installation cost of this thermostat.

Location of Room Thermostats

The location of the room thermostat is very important to the efficient operation of the heating and/or cooling system. If the room thermostat is improperly located, it will often call for heat or cool air when neither is necessary. It is therefore important to locate the thermostat where it will measure the actual temperature conditions of the space.

The following recommendations are offered as a guide for the proper location of a room thermostat.

125

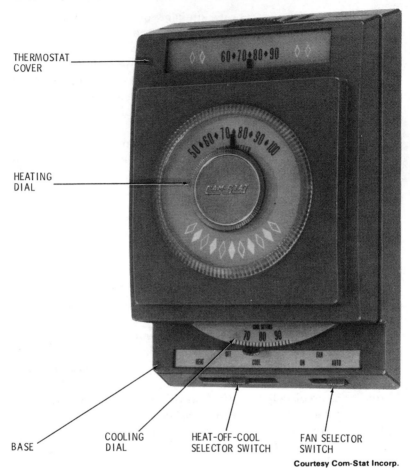

THERMOSTAT COVER

HEATING DIAL

BASE

COOLING DIAL

HEAT-OFF-COOL SELECTOR SWITCH

FAN SELECTOR SWITCH

Courtesy Com-Stat Incorp.

Fig. 4-25. Room heating and cooling thermostat.

1. *Never locate the thermostat on the interior surface of an outside wall.* Outside walls are subject to temperature extremes caused by weather changes. Always locate the thermostat on a *suitable* inside partition.

2. *Never locate the thermostat in the path of warm or cold air drafts.* The thermostat should not be placed opposite warm air outlets or near a window or an outside door.

3. *Never locate the thermostat where it will be subjected to the*

direct rays of the sun or other forms of heat radiation. Fireplaces, table lamps, and floor lamps are common sources of this type of heat.

When you select a location for the room thermostat on an inside wall, make certain that you are not placing it over a warm air duct, steam pipe, or hot-water pipe. The warmth from these ducts or pipes will interfere with the operation of the thermostat. For the most satisfactory operation, locate the room thermostat about 5 ft. above the floor on an inside wall where there is good natural air circulation and where the thermostat will be exposed to *average* room temperatures.

Installing a Room Thermostat

Before attempting to install the thermostat, read the manufacturer's installation instructions. These instructions should be followed as closely as possible to ensure efficient thermostat operation.

Always handle the thermostat carefully. Rough handling may decrease its accuracy. Inspect the thermostat carefully when you unpack it. Report any damage to the shipper if the damage was caused after it left the factory or distribution center. Their insurance should pay for any replacement. Any other damage or malfunction should be reported to the thermostat manufacturer or his field representative. The thermostat is the basic controller in any heating and/or cooling system. It must operate accurately and efficiently.

The installation instructions differ from one thermostat manufacturer to another, usually on the basis of design and construction; however, the following points are common to all:

1. Mercury-switch thermostats are usually shipped with some form of protection around the mercury tube to prevent breakage. This *must* be removed before operating the thermostat.
2. Disconnect the power supply before connecting the wiring to the thermostat in order to prevent electrical shock and/or equipment damage.
3. All wiring must be done in accordance with local electrical

codes and ordinances. Be sure to follow the thermostat manufacturer's wiring diagram when making the connections.

4. Use a plumb line or spirit level to accurately level the sub-base or wall plate when mounting it on the wall (Fig. 4-26). Thermostat control deviations are often caused by inaccurate leveling.

5. Follow the manufacturer's instructions for making any internal wiring connections (e.g., connecting lead wires to terminal screws), connecting a heating thermostat to a cooling base, etc.

6. Check out the installation to make sure the thermostat is operating correctly.

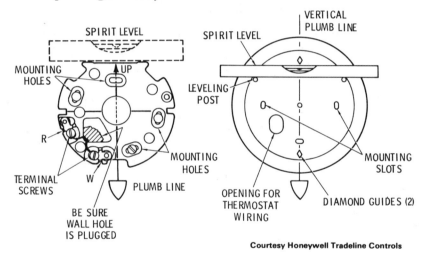

Courtesy Honeywell Tradeline Controls

Fig. 4-26. Using a plumb line and level.

Thermostat Calibration

Thermostats are accurately calibrated at the factory under controlled conditions and should not require recalibration. Sometimes a thermostat will *appear* to require recalibration when the problem is actually a quite different one. For example, a thermostat that is not level or one subjected to a high degree of radiant heat from the sun, radiators, convectors, or other heat sources often fails to function properly. Before jumping to the conclusion that it needs to be recalibrated, check out the possibility that some external cause may be the source of the difficulty. If you are certain the problem is in the thermostat, then you should call a trained serviceman to recalibrate it. Do not attempt to recalibrate it yourself unless you have the necessary experience.

A new thermostat is generally shipped with complete installation literature. This literature also usually contains instructions for recalibrating the thermostat. By way of example, the instructions for recalibrating a Honeywell T87 Room Thermostat include the following steps:

1. Remove the thermostat cover ring (Fig. 4-27). The locking cover will require the use of an Allen wrench to loosen the screws securing the cover.
2. Set the thermostat below the room temperature, and allow it to remain in an *off* position for approximately 10 minutes.
3. Slowly raise the setting until the switch just makes contact. If the thermostat pointer and setting indicator do *not* read

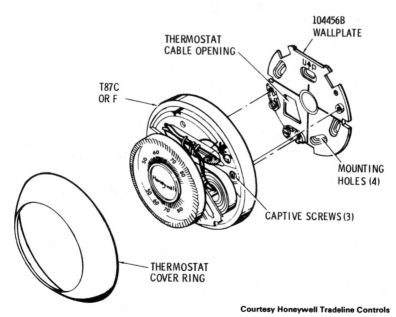

Fig. 4-27. Removing the cover ring.

129

the same the instant the switch makes contact, then the thermostat requires recalibration.

4. Turn the setting dial a few degrees above room temperature.

5. Slip the calibration wrench onto the calibration (hex) nut under the bimetal coil (Fig. 4-28).

6. If the thermostat has a stationary pointer, hold the dial firmly and turn the calibration nut counterclockwise until the mercury breaks contact. If the thermostat has a movable pointer, turn the calibration nut *clockwise*.

7. Turn the thermostat dial to a low setting, and wait approximately 5 minutes.

8. Slowly turn the dial until the pointers read the same.

9. Hold the dial firmly, and turn the calibration in the opposite direction from the one in Step 6 until the mercury switch slips to the heating contact end of the tube.

10. Recheck the calibration, select the desired temperature, and replace the cover.

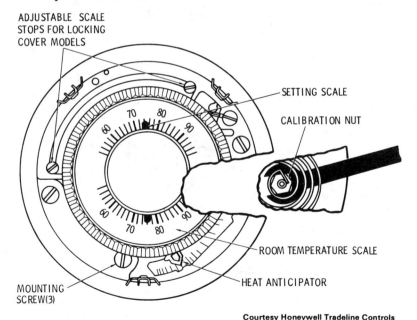

Fig. 4-28. Calibration nut under bimetal coil.

Day-Night Thermostats

A *day-night* (or *twin-type*) *thermostat* is an assembly of two thermostats mounted on a single base operating in conjunction with a timer or clock. The electric clock can be set to throw the temperature controls from one thermostat for the daytime onto the other for night (or vice versa) at a predetermined time setting on the clock. This conveniently permits a low temperature at night and normal temperature during the day. Fig. 4-29 shows a wiring diagram of a typical twin-type thermostat and illustrates the connections between the clock and primary control.

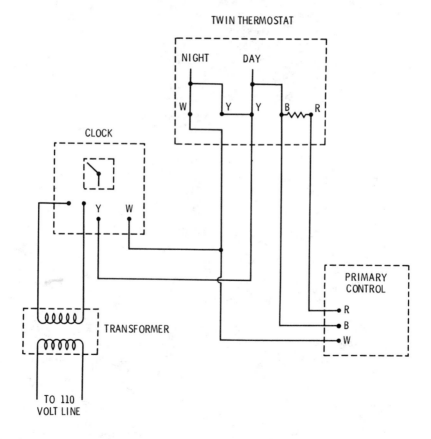

Fig. 4-29. Twin-type thermostat wiring diagram.

131

Day-night thermostats designed to provide automatic temperature switching control only for a heating system can be modified to provide system and fan switching. The Honeywell T882 Chronotherm clock thermostat provides these functions for a heating and cooling installation when used with the Honeywell Q611A thermostat subbase (Figs. 4-30 and 4-31).

Insertion Thermostats

An *insertion thermostat* is used primarily to measure the air temperature in a duct. The thermostat is mounted on the outside

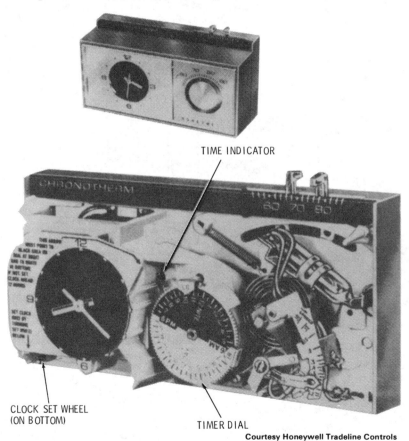

TIME INDICATOR

CLOCK SET WHEEL
(ON BOTTOM)

TIMER DIAL

Courtesy Honeywell Tradeline Controls

Fig. 4-30. Honeywell T882 Chronotherm clock thermostat.

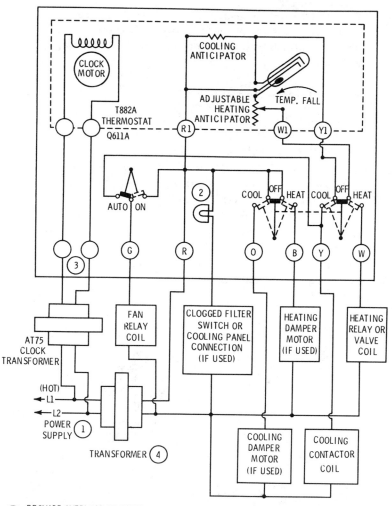

PROVIDE OVERLOAD PROTECTION AND DISCONNECTION MEANS AS REQUIRED

FILTER MALFUNCTION LIGHT (OPTIONAL)

CLOCK TERMINAL. CONNECT CLOCK TO AT75 TRANSFORMER ONLY - NO OTHER POWER SOURCE

USE HEATING TRANSFORMER IF ADEQUATE. OTHERWISE REPLACE.

Courtesy Honeywell Tradeline Controls

Fig. 4-31. Wiring diagram of the T882 thermostat and Q611A subbase.

of the air duct with the sensing element extending inside. Because of this application, it is sometimes referred to as a *duct thermostat*. Remote-bulb-type thermostats are also used to measure the air temperature in ducts.

Immersion Thermostats

An *immersion thermostat* is commonly used for water temperature control in an automatic gas-fired water heater (Fig. 4-32). This is usually a direct, snap-action, bimetallic thermostat in which contraction of the thermal element immersed in the stored hot water causes the main gas valve to open. This occurs when there is a drop in the temperature of the water in the storage tank. Expansion of this element serves to close the main gas valve when the tank water attains the selected predetermined temperature.

These thermostats normally operate at a temperature differential of approximately 12°F. In other words, if the thermostat is set

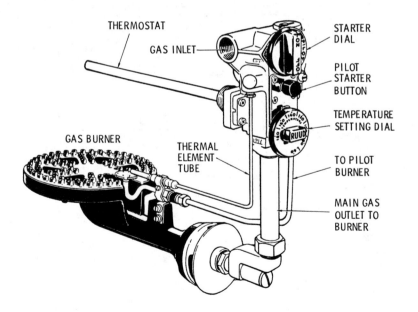

Fig. 4-32. Immersion thermostat.

134

to shut off the gas to the main burner when the tank water temperature reaches 140°F, it will react to open the valve when the temperature drops to 128°F.

Quite often, an immersion water heater thermostat will be included in a combination control. These combination controls used in water heaters are described in Chapter 4 of Volume 3 (Water Heaters and Other Appliances).

Remote-Bulb Thermostats

A *remote-bulb thermostat* is distinguished from other thermostats by having a sensing element (usually enclosed in a bulb-type device) located at some distance from the thermostat controller body.

In residential heating and/or cooling applications, remote-bulb thermostats are used for space temperature control of room and vented-recess heaters, room air conditioning units, or radiator valves. Outdoor thermostats also operate on the remote-bulb principle.

Sometimes these thermostats are combined with other controls to provide a number of different control functions in one package. Two examples of this thermostatic combination control are the Robertshaw Unitrol 110SR and 7000SR controls.

The Unitrol 110SR is used as combination control for small gas-fired room heaters (Fig. 4-33). The control contains a thermostatic valve, a sensing bulb, temperature (thermostat) dial, gas cock, automatic pilot valve, and a pressure regulator (Fig. 4-34).

The sensing bulb is located in the return air stream at the bottom of the heater. The temperature of the return air is sensed and the valve is actuated to open and close by the hydraulic system in the control. The gas cock and automatic pilot mechanism provide safe lighting of the heater. If the pilot light should go out, the main gas and pilot gas supplies are shut off by the automatic pilot valve. Main burner gas pressure regulation is provided by the pressure regulator incorporated in the Unitrol 110SR control.

Fig. 4-35 illustrates the principal components of a Unitrol 7000SR control. This is a diaphragm valve that operates through the center of a thermostatic bleed valve in an internal bleed line. The Unitrol 7000SR combines into a single package a diaphragm

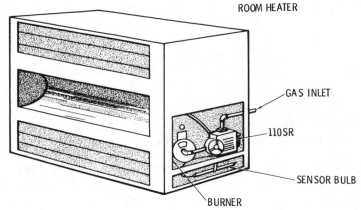

ROOM HEATER

GAS INLET

110SR

SENSOR BULB

BURNER

Courtesy Robertshaw Controls Co.

Fig. 4-33. Unitrol 110SR combination control on a gas-fired room heater.

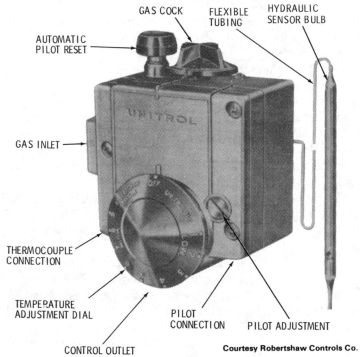

GAS COCK

FLEXIBLE TUBING

HYDRAULIC SENSOR BULB

AUTOMATIC PILOT RESET

GAS INLET

THERMOCOUPLE CONNECTION

TEMPERATURE ADJUSTMENT DIAL

PILOT CONNECTION

PILOT ADJUSTMENT

CONTROL OUTLET

Courtesy Robertshaw Controls Co.

Fig. 4-34. Principal components of a Unitrol 110SR combination gas valve.

136

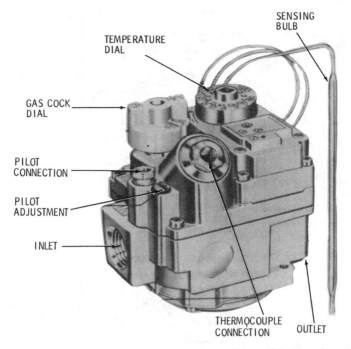

SENSING
BULB

TEMPERATURE
DIAL

GAS COCK
DIAL

PILOT
CONNECTION

PILOT
ADJUSTMENT

INLET

THERMOCOUPLE
CONNECTION   OUTLET

Courtesy Robertshaw Controls Co.

Fig. 4-35. Principal components of a Unitrol 7000SR combination gas valve.

valve, thermostat valve, temperature (thermostat) dial, gas cock, automatic pilot valve, and a pressure regulator.

Application of a Unitrol 7000SR on a gas-fired vented-recess heater is illustrated in Fig. 4-36. The sensing bulb is placed in the return air opening at the bottom of the furnace.

Both the Unitrol 110SR and Unitrol 7000SR thermostatic space heater controls use a closed hydraulic sensing and actuating device consisting of a bulb, capillary tube, and a bellows or diastat.

The cross section of a typical hydraulic sensor and actuator is illustrated in Fig. 4-37. The bulb, capillary tube, and actuator are filled with a liquid that has a high coefficient of expansion. When the bulb senses a rise in temperature, it causes the volume of the liquid to expand. This expansion is transferred through the capillary tube to expand the hydraulic bellows. The bellows is spring

137

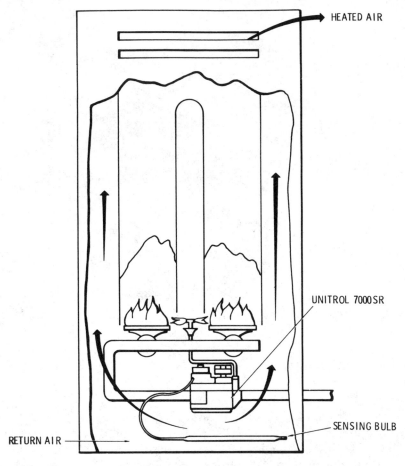

Fig. 4-36. Unitrol 7000SR combination gas valve on a gas-fired vented recess heater.

loaded to operate a snap mechanism to a valve *open* condition (Fig. 4-38). As the temperature rises, the expansion of the liquid opposes the spring loading to secure a valve *closed* (i.e., *off*) condition when the required temperature is reached.

Proportional Thermostats

Some thermostats are designed to provide proportional control for valve and damper motors in heating or cooling systems. This

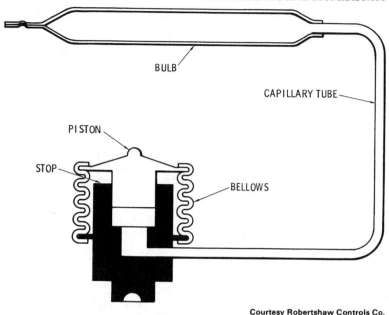

BULB

CAPILLARY TUBE

PISTON

STOP

BELLOWS

Courtesy Robertshaw Controls Co.

Fig. 4-37. Cross section of a typical hydraulic sensor and activator.

type of controller is generally referred to as a *proportional thermostat*.

The Honeywell T92 proportional thermostat, shown in Fig. 4-39, contains a bellows that adjusts one or two potentiometers in proportion to temperature changes. These potentiometer adjustments regulate the power supplied to the controlled device. This particular thermostat is designed to provide 24- to 30-volt proportional control for valve and damper motors in the system. Those models equipped with two potentiometers are capable of unison or sequence control.

Outdoor Thermostats

An *outdoor thermostat* is designed to maintain the proper balance between the temperature of the heating medium inside the structure and the outdoor temperature.

The outdoor thermostat, illustrated in Fig. 4-40, can be used as an operating controller for a hot-water or warm-air heating sys-

139

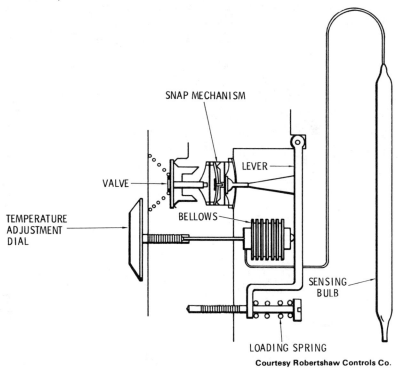

SNAP MECHANISM

LEVER

VALVE

TEMPERATURE
ADJUSTMENT
DIAL

BELLOWS

SENSING
BULB

LOADING SPRING

Courtesy Robertshaw Controls Co.

Fig. 4-38. Schematic of a hydraulic sensor and actuating system used for
snap-action control.

tem. This is a remote-bulb thermostat suitable for line voltage,
low-voltage, or millivolt switching. It is designed to automatically raise the heating medium control point as the outdoor
temperature falls.

Outdoor thermostats are frequently used as controller in hot-
water heating systems. A simple on-off control is possible, but
this usually involves stopping the circulation of the water in the
system during those periods when there is no call for heat. Anticipating control systems are preferred to the simple on-off types
because there is no noticeable lag time between cold temperatures and a call for heat. An anticipating control system provides
continuous circulation of the water temperature in direct ratio to
changes in outdoor temperature.

140

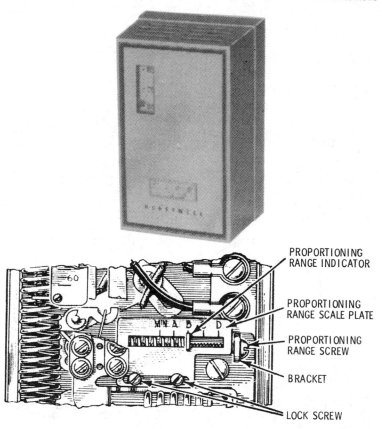

PROPORTIONING
RANGE INDICATOR

PROPORTIONING
RANGE SCALE PLATE

PROPORTIONING
RANGE SCREW

BRACKET

LOCK SCREW

Courtesy Honeywell Tradeline Controls

Fig. 4-39. Honeywell T92 proportional thermostat.

TROUBLESHOOTING THERMOSTATS

Thermostat problems are sometimes the result of poor wiring connections. Check the wiring first. You should also make certain that the fan and system switches and the temperature set point are properly set.

The following troubleshooting chart includes many of the more common symptoms and possible causes of operating problems associated with thermostats.

141

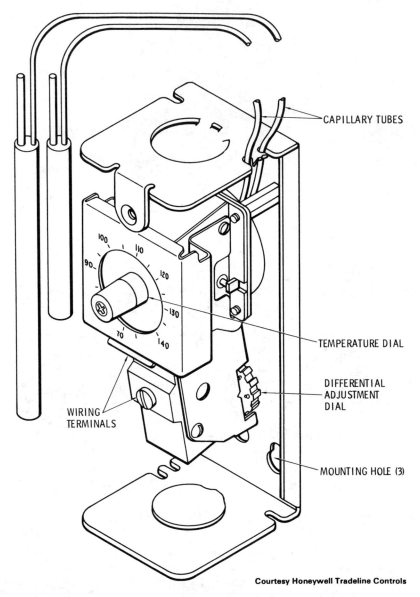

CAPILLARY TUBES

TEMPERATURE DIAL

DIFFERENTIAL
ADJUSTMENT
DIAL

WIRING
TERMINALS

MOUNTING HOLE (3)

Fig. 4-40. Outdoor reset control for hot-water or warm-air heating systems.

Symptom and Possible Cause   *Possible Remedy*

Room temperature overshoots thermostat setting (too cold)

1. Thermostat not mounted level (mercury-switch types).
2. Thermostat not properly calibrated.
3. Thermostat exposed to heat source.
4. Thermostat set point too low.
5. System sized improperly.

1. Remount thermostat in level position.
2. Recalibrate or replace.
3. Move thermostat to better location.
4. Reset.
5. Determine correct sizing and make system adjustments.

Room thermostat does not reach setting (too warm)

1. Thermostat subject to draft.

2. Thermostat not mounted level (mercury-switch types).
3. Thermostat not properly calibrated.
4. Thermostat set point too high.
5. System sized improperly.

6. Thermostat damaged.

1. Wiring hole may not be plugged. Move thermostat to better position.
2. Remount thermostat in level position.
3. Recalibrate or replace.
4. Reset.
5. Determine correct sizing and make system adjustments.
6. Replace thermostat.

System cycles too often

1. Thermostat exposed to heat source.

1. Relocate thermostat.

143

Symptom and Possible Cause   *Possible Remedy*

2. Thermostat differential too small.
3. Thermostat heating element improperly set.
4. Thermostat subject to vibrations.

5. Thermostat exposed to cold draft.

2. Reset or replace thermostat.
3. Reset or replace thermostat.
4. Remount thermostat in location free from vibrations.
5. Remount in better location.

System does not cycle often enough (burner operates too long)

1. Thermostat not exposed to circulating air.
2. Contacts dirty.
3. System sized improperly.

4. Thermostat differential too great.
5. Thermostat heating element improperly set.
6. Thermostat set too high.

1. Remount in better location.
2. Clean or replace.
3. Determine correct sizing and make system adjustments.
4. Reset or replace thermostat.
5. Reset or replace thermostat.
6. Reset.

Room temperature swings excessively

1. Thermostat not exposed to circulating air.
2. Thermostat exposed to heat source.
3. System sized improperly.

1. Remount in better position.
2. Remount in better position.
3. Determine correct sizing and make system adjustments.

144

Symptom and Possible Cause *Possible Remedy*

Thermostat jumpered (system works)

1. Thermostat contacts dirty.	1. Clean or replace.
2. Thermostat set point too high.	2. Reset or replace thermostat.
3. Thermostat damaged.	3. Replace thermostat.
4. Break in thermostat circuit.	4. Locate and correct.

Burner fails to stop

1. Thermostat in cold location.	1. Relocate in better location.
2. Thermostat set too high.	2. Reset.
3. Defective thermostat.	3. Replace thermostat.
4. Thermostat out of adjustment.	4. Recalibrate or replace thermostat.
5. Thermostat contacts stuck.	5. Correct.

HUMIDISTATS

A *humidistat* is an automatic device designed to control the degree of humidity in a confined space. It is also sometimes referred to as a *humidity controller* or *hygrostat*. The latter term is derived from the use of hygroscopic materials such as human hair, paper, wood, or a similar material in the construction of the sensing or control element (Fig. 4-41). Hygroscopic materials are those affected by the moisture present in the air. When the material absorbs moisture from the air, expansion takes place. Contraction occurs when the material gives up moisture to the air. This expansion and contraction of the control element in the humidistat opens or closes the electrical circuit controlling the humidifier or dehumidifier.

145

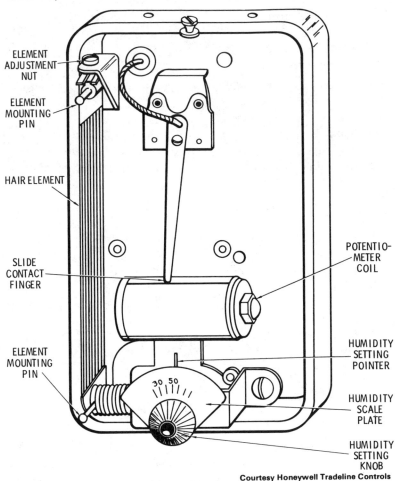

ELEMENT
ADJUSTMENT
NUT

ELEMENT
MOUNTING
PIN

HAIR ELEMENT

SLIDE
CONTACT
FINGER

ELEMENT
MOUNTING
PIN

POTENTIO-
METER
COIL

HUMIDITY
SETTING
POINTER

HUMIDITY
SCALE
PLATE

HUMIDITY
SETTING
KNOB

Courtesy Honeywell Tradeline Controls

Fig. 4-41. Components of a typical humidstat.

Location of Room Humidistats

The proper location of the room humidistat is important to its efficient operation. The following recommendations are offered as a guide to locating the humidistat:

1. *Never locate the humidistat in an area where there are heavy concentrations of moisture.* A kitchen, bathroom, or

146

laundry room will frequently have high levels of moisture in the air.

2. *Never install the humidistat on the inside surface of an outside wall.* Exterior walls are subject to temperature extremes caused by weather changes.
3. *Never locate the humidistat where the air circulation is restricted.*
4. *Never locate the humidistat where it can be affected by a nearby heat source.* Heat sources such as sunlight, lamps, television sets, fireplaces, warm-air outlets, or heat producing appliances will interfere with the efficient operation of the humidistat.

CHAPTER 5

Gas and Oil Controls

Gas-fired and oil-fired heating equipment use certain types of controls to govern the flow of gas or oil to the burner or burner assembly. These controls function together as a control circuit within the heating system.

The two principal functions of the control system are to start or stop the heating equipment in response to a signal from the thermostat, and to shut off the heating equipment if an unsafe condition such as pilot failure or overheating occurs. The individual controls can be divided into the following four basic categories:

1. Operating controls.
2. Limit controls.
3. Primary controls.
4. Auxiliary controls.

The room thermostat is an example of an operating control or controller. This device is the "nerve center" of the heating system. The thermostat signals the primary control to turn the burner or burner assembly on or off in response to temperature changes in the space or structure. Thermostats and other types of operating controls are described in Chapter 4 (Thermostats and Humidistats) and in appropriate sections of other chapters.

Limit controls are designed to prevent excessive and unsafe temperatures or pressures from developing in the heating equipment or system. Fan and limit controls, air switches, and similar types of controls are described in Chapter 6 (Other Automatic Controls).

The primary control not only governs the on-off operation of the burner, but also monitors the burner flame. Thus, its principal function is to ensure *safe* starting and operating conditions. Many auxiliary controls (valves, time delay devices, relays, etc.) operate in conjunction with the primary control to provide additional safeguards against the possibility of a breakdown somewhere in the heating system, particularly when the breakdown could result in an unsafe operating condition.

A complete description of all the controls used to govern the operation of a furnace, boiler, or water heater would be too extensive to include in a single chapter. For that reason, operating controls, limit controls, and related safety and control devices were given separate chapters. This chapter is primarily concerned with a description of those controls that directly govern the flow of gas or oil to the burner or burner assembly.

GAS CONTROLS

In addition to the room thermostat and the fan and limit control, the basic components of the control system of a gas-fired furnace, boiler, or water heater will generally consist of the following controls:

1. Gas-pressure regulators.
2. Main gas shutoff valve.
3. Pilot gas cock.
4. Automatic gas control valve.
5. Automatic pilot valve.
6. Gas pilot assembly.

The gas pilot assembly includes the pilot burner and the thermocouple or pilot generator. The system may also include a gas primary control, transformer, or safety pilot relay. Some controls may be eliminated, depending on the design and requirements of the system. For example, a safety pilot relay is not necessary if a nonelectric pilot safety valve is used in the gas line.

Gas Control Circuits

The gas control circuits used to operate modern gas-fired heating equipment can be divided into the following three basic types:

1. Low-voltage control circuits.
2. Line voltage control circuits.
3. Millivolt control circuit.

A low-voltage temperature control circuit (Fig. 5-1) uses a stepdown transformer to reduce the higher line voltage to approximately 24 to 30 volts. A 24-volt thermostat is used as the controller in most installations.

The line voltage temperature control circuit shown in Fig. 5-2 is a 120-volt system. Because the voltage is not reduced, a line voltage thermostat or controller and a line voltage operator must be used in the system.

A millivolt control circuit (Fig. 5-3) operates on the thermocouple principle. A single thermocouple automatically generates approximately 30 millivolts without the aid of an outside source of electricity. A number of thermocouples used together can generate up to as high as 750 millivolts. This combination is variously referred to as a *generator, pilot generator, thermopile generator, thermopile system,* or *powerpile system.*

Each of the three temperature control circuits described in the preceding paragraphs is also wired into a pilot safety shutoff circuit, generally via a switch-type pilot safety shutoff device. An inline pilot safety shutoff device is also located in each safety shutoff circuit, and these provide *complete* gas shutoff.

Gas Burner Primary Control

The primary control shown in Fig. 5-4 is a solid-state electronic relay used on gas, oil, or combination gas-oil burners. It is

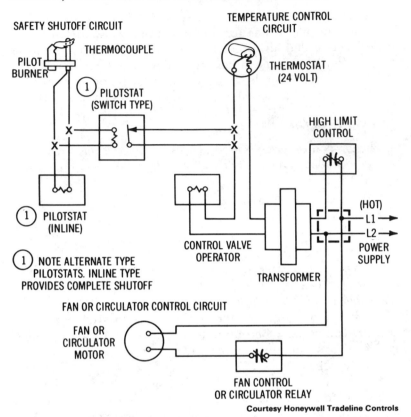

SAFETY SHUTOFF CIRCUIT

TEMPERATURE CONTROL CIRCUIT

THERMOCOUPLE

PILOT BURNER

THERMOSTAT (24 VOLT)

① PILOTSTAT (SWITCH TYPE)

HIGH LIMIT CONTROL

① PILOTSTAT (INLINE)

CONTROL VALVE OPERATOR

(HOT) L1 →
L2 →
POWER SUPPLY

① NOTE ALTERNATE TYPE PILOTSTATS. INLINE TYPE PROVIDES COMPLETE SHUTOFF

TRANSFORMER

FAN OR CIRCULATOR CONTROL CIRCUIT

FAN OR CIRCULATOR MOTOR

FAN CONTROL OR CIRCULATOR RELAY

Courtesy Honeywell Tradeline Controls

Fig. 5-1. Low-voltage control circuit.

designed to provide operational control of the burner in response to the room thermostat and limit controls, and to instantly shut off the burner in the event of flame failure. Fig. 5-5 illustrates a typical wiring diagram for a Honeywell RA890F Protectorelay primary control used in a control circuit for a gas-fired boiler.

This primary control is used with rectification-type flame detectors to sense the presence or absence of flame. The heart of the flame detector circuit in a gas-fired system is an electrode inserted in the pilot flame. In the wiring diagram, shown in Fig. 5-6, the flame rod is connected to terminal *(F)* of the primary control.

152

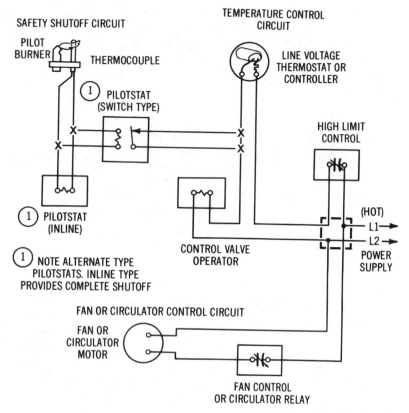

Fig. 5-2. Line voltage control circuit.

In the event of pilot flame failure, the flame detector circuit responds to control the gas valve in the manifold. When there is an absence of flame, the primary control shuts off the supply of gas by closing the gas valve or by keeping it closed if it is not already open.

These units are designed to fail safe. Abnormal conditions in the flame detector circuit, such as an open circuit, short circuit, or current leakage to ground, simulate absence of flame and cause the system to shut down. Safety controls such as temperature or pressure limits, or low-water cutoffs are connected *ahead* of the

153

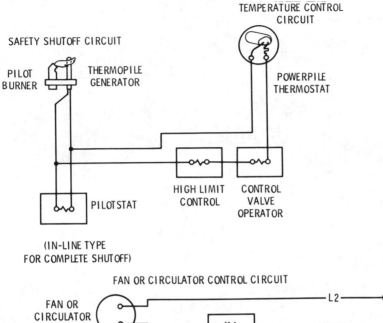

Courtesy Honeywell Tradeline Controls

Fig. 5-3. Millivolt control circuit.

switching terminals of the relay so that shutdown of the burner occurs even in spite of relay malfunction such as fused contacts.

The main valve circuit is deenergized $\frac{8}{10}$ of a second after flame failure occurs. On starting up or after flame failure, a trial for ignition period of approximately 45 seconds maximum occurs. During this ignition period, only pilot gas is allowed to flow to the burner. If the flame circuit is not completed within this time period, safety lockout of the relay occurs, causing a total shutdown of the system. Manual reset is then required to restart.

Two sets of relays are contained in a Honeywell RA890F Protectorelay primary control. The load relay (left hand) supplies current to the No. 3 terminal to control the blower motor. The flame relay (right hand) responds to the load relay but only if

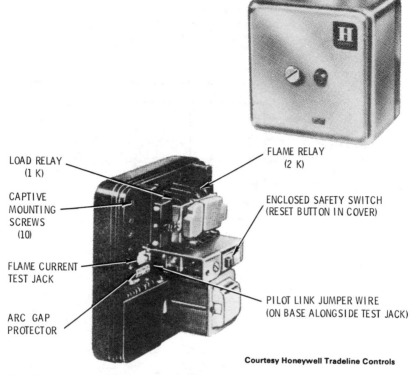

LOAD RELAY
(1 K)

CAPTIVE
MOUNTING
SCREWS
(10)

FLAME CURRENT
TEST JACK

ARC GAP
PROTECTOR

FLAME RELAY
(2 K)

ENCLOSED SAFETY SWITCH
(RESET BUTTON IN COVER)

PILOT LINK JUMPER WIRE
(ON BASE ALONGSIDE TEST JACK)

Courtesy Honeywell Tradeline Controls

Fig. 5-4. Honeywell RA890F Protectorelay primary control.

allowed by the flame detecting electronic network of the relay. The flame relay can supply current to the No. 5 terminal controlling the gas valve only if the load relay has also "pulled in."

The load relay is responsive to the thermostat or other operating control connected to the T/T terminal provided that safety controls, located in the 1-6 circuit, indicate safe conditions exist for main burner operation.

On an interruption of power to the No. 1 and No. 2 terminals of the primary control, the relay returns to the standby position. When power is restored, normal operation is resumed except that the starting cycle is maintained longer than usual while the vacuum tube is warming up. Relay positions and their effect on the burner are listed in Table 5-1.

155

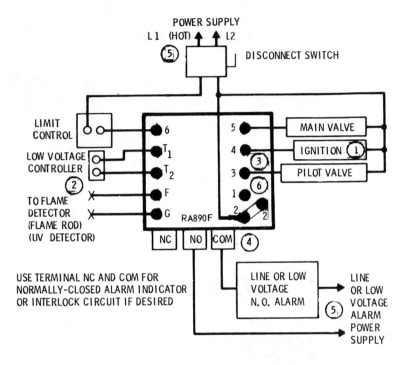

POWER SUPPLY

L1 (HOT)   L2

DISCONNECT SWITCH

LIMIT CONTROL

LOW VOLTAGE CONTROLLER

TO FLAME DETECTOR (FLAME ROD) (UV DETECTOR)

6

T$_1$

T$_2$

F

G

RA890F

5   MAIN VALVE

4   IGNITION

3   PILOT VALVE

1

2

2

NC   NO   COM

USE TERMINAL NC AND COM FOR NORMALLY-CLOSED ALARM INDICATOR OR INTERLOCK CIRCUIT IF DESIRED

LINE OR LOW VOLTAGE N.O. ALARM

LINE OR LOW VOLTAGE ALARM POWER SUPPLY

(1) FOR INTERMITTENT IGNITION, CONNECT TO TERMINAL 3.

(2) IF LINE VOLTAGE CONTROLLER IS USED CONNECT IT BETWEEN THE LIMIT CONTROL AND TERMINAL 6. JUMPER T1 AND T2.

(3) HOOKUP FOR STANDING PILOT IS THE SAME AS FOR INTERRUPTED IGNITION EXCEPT IGNITION AND PILOT VALVE CONNECTIONS ARE NOT MADE.

(4) SPDT ALARM TERMINALS OPTIONAL. IF LINE VOLTAGE ALARM IS USED, RA890F MUST BE MOUNTED IN SUITABLE ENCLOSURE.

(5) POWER SUPPLY-PROVIDE OVERLOAD PROTECTION AND DISCONNECT MEANS AS REQUIRED.

(6) FOR RA890E REPLACEMENT, LEAVE POWER CONNECTED TO TERMINAL 1. SUBBASE WIRING CHANGES ARE NOT REQUIRED.

Fig. 5-5. Wiring diagram of the RA890F control.

Table 5-1. Relay Position and Effect on Gas Burner

Relay Position	Description	Effect on Burner
Standby	Load and flame relays both "out."	Motor and gas valve both deenergized.
Starting	Load relay "in," Flame relay "out."	Motor energized, gas valve deenergized, trial for ignition period, safety lockout occurs after 45 seconds.
Running	Load and flame relays both "in."	Motor and gas valve both energized.
Abnormal conditions due to flame simulatimg failure	Load relay "out," Flame relay "in."	Motor and gas valve both deenergized. Safety lockout occurs after 45 seconds.

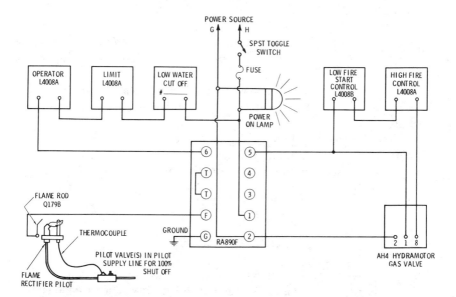

Fig. 5-6. Wiring diagram showing connections between a RA890F control and other components in a gas control circuit.

157

Servicing a Gas Burner Primary Control

Access to the wiring terminals of the primary control illustrated in Fig. 5-4 is obtained by loosening the screws that secure the chassis to the base. When remounting the chassis, be sure to tighten all mounting screws because they also serve as electrical connections.

No attempt should be made to repair a primary control except for tube replacement. Vacuum tubes are used in Honeywell primary controls. *Never* replace them with radio tubes. If a primary control is defective, the entire chassis should be replaced with a good one.

Operating controls located in the T/T circuit (see Fig. 5-6) should be of the low-voltage, two-wire type. A low-voltage transformer for this purpose is built into a Honeywell Protectorelay. Safety controls located in 1-6 terminals must be two-wire, line voltage type. With the exception of the line switch, no controls should ever be placed in the line ahead of the 1-2 terminals of the primary control.

Before assuming that the primary control is defective, be sure to check the pilot, pilot adjustment, flame detector circuit, and all operating and safety controls—proper operation is also dependent on these external factors. The flame circuit can be more accurately checked by the use of a microammeter to read "flame current." Normal operation requires a current of 2 microamps or more.

Never push relays in manually because it can result in accidental opening of the main diaphragm valve. Be sure to turn off the electrical power before removing the primary control chassis from the base.

GAS VALVES

The valves used to control the flow of gas through a gas-fired furnace, boiler, or water heater can be divided into two basic categories: (1) manually operated valves and (2) power operated valves.

The two manually operated valves (gas cocks) used on gas-

158

fired heating equipment provide a backup safety function in case the automatic gas valves fail to operate. One of these valves is located in either the main gas supply riser or the manifold. The other one is located on the pilot gas line.

The manual gas valve installed on the main gas supply line (riser) or manifold is variously referred to as the *main gas shutoff valve, manual shutoff plug cock,* or simply the *gas cock* (Fig. 5-7). This valve provides manual control of the gas flow to the main gas burners. It is *not* used to control the gas supply to the pilot burner, the latter being provided with its own separate shutoff valves.

The manual valve located on the pilot gas line is called the *pilot shutoff cock* or the *pilot gas cock.* It is usually the first controlling device on the pilot line (Fig. 5-7). It provides complete gas shutoff whenever it is necessary to remove and service other controls on the pilot line, such as the pilot gas regulator or the pilot solenoid valve.

Power operated or automatic valves are actuated by some form of auxiliary power such as hydraulic pressure, pneumatic pressure, electricity, or a combination of these sources. The principal types of power operated valves used on gas-fired heating equipment are:

1. Solenoid valves.
2. Direct-acting heat motor valves.
3. Diaphragm valves.

SOLENOID GAS VALVES

The solenoid gas valve is commonly used on gas-fired heating equipment to provide on-off control of the flow of gas.

The primary function of a solenoid valve is to provide direct valve operation. The power to operate the valve is obtained from the magnetic flux developed in a solenoid coil. A valve disc in the valve body is connected by a rod to the core of an electromagnet. When the room thermostat or power switch directs an electrical current to the solenoid, it pulls the rod (plunger) to the top of the plunger tube and lifts the attached valve disc. Gas then flows through the main valve port until the electrical circuit is inter-

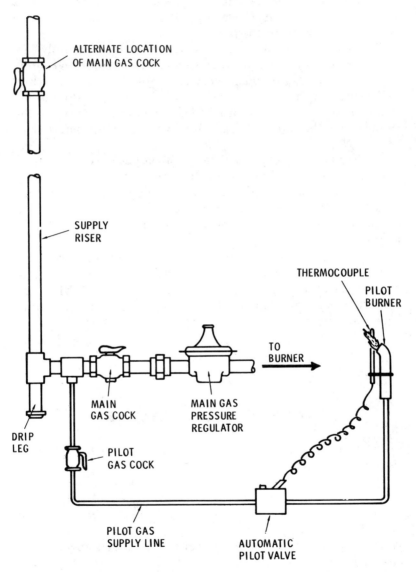

Fig. 5-7. Typical arrangement of gas cocks and main gas-pressure regulator.

rupted by the controller. This action releases the rod, which falls and shuts off the valve. The weight of the rod, seat assembly, and gas pressure on top of the valve seat ensures a tight shutoff. The ITT General Controls K3 Series gas valve shown in Fig. 5-8 is an example of a direct-operated solenoid gas valve.

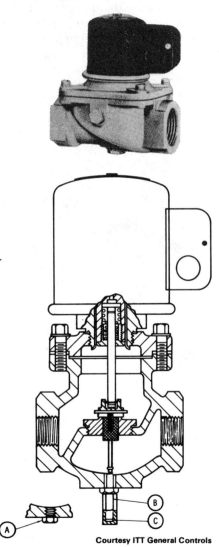

Fig. 5-8. Direct-operated solenoid gas valve.

Courtesy ITT General Controls

161

Some solenoid valves use a balanced diaphragm to control the flow of gas (Fig. 5-9). When the solenoid coil is energized, it lifts the rod or plunger just enough to open a bleed valve (or so-called pilot valve). Gas then bleeds from the area above the diaphragm faster than it can be replaced. This eventually results in the pressure above the diaphragm being the same as the pressure below the seat disc. This is referred to as a balanced or unloaded condition. The solenoid coil lifts the complete interior assembly to full

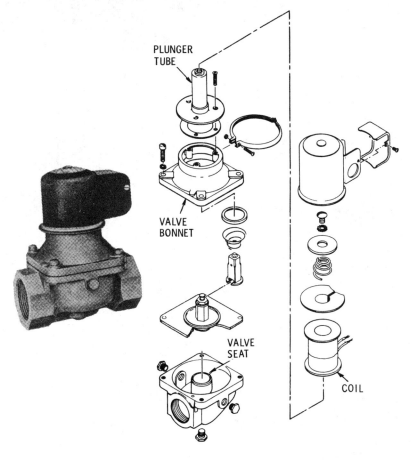

PLUNGER
TUBE

VALVE
BONNET

VALVE
SEAT

COIL

Fig. 5-9. Balanced diaphragm solenoid gas valve.

open position. When the solenoid is deenergized, the pressure recovers above the diaphragm. The weight of the interior assembly and the gas pressure across the seat disc are sufficient to hold the valve closed. In this type of valve, the pressure of the gas is used to control its operation.

A third type of solenoid valve consists of a solenoid operated (i.e. magnetically operated) puff bleed three-way valve and a diaphragm valve in a single unit (Fig. 5-10). The combined unit provides on-off control of the gas to the gas-fired heating equipment.

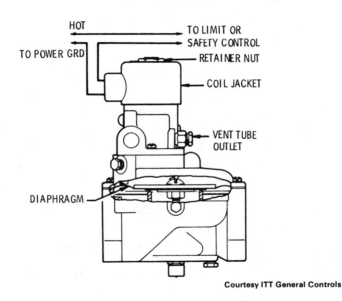

Courtesy ITT General Controls

Fig. 5-10. Electric diaphragm gas valve.

The three-way valve (also referred to as a *pilot valve*), responding to electrical signals from the limit or safety controls, opens or closes the gas valve by controlling the gas-pressure bleed-off above the diaphragm in the main valve body. In the normally closed position, inlet gas pressure above the diaphragm prevents the valve from opening. In the open (energized) position, the three-way or pilot valve closes off the inlet gas pressure and allows the gas pressure above the diaphragm to bleed off so

163

that gas pressure below the diaphragm forces the diaphragm up to open the valve.

Dual solenoid valves are designed for three-stage control (high-low-off) of the flow of gas (Fig. 5-11). Both a high-fire solenoid and a low-fire solenoid are used to accomplish this purpose. Low-fire adjustments can be made by turning the adjustment screw clockwise (to decrease low fire) or counterclockwise (to increase it).

Fig. 5-12 shows a schematic wiring diagram of a two-stage control containing a dual solenoid valve. If both solenoids are to be energized at one time, the circuit requires a 40-volt transformer.

SOLENOID COILS

Several different solenoid coils are available from manufacturers, and the type selected for use will depend on the specific application. For example, most standard applications will require a moisture-resistant coil for normal usage of gas or fluid up to 175°F. Special applications include those with especially high ambient and fluid temperatures, high voltage, or high steam pressure. A solenoid coil may also be specifically required for moisture or water applications. Under these circumstances the soil should be both waterproof and fungus proof. Some examples of solenoid coils are shown in Fig. 5-13.

A principal cause of coil malfunction is excessive heat. If the valve is subjected to temperatures above the coil rating, it will probably fail. A missing part, a damaged plunger tube or tube sleeve, or improper assembly may also be a cause of excess heat. The applied voltage must be at the coil rated frequency and voltage.

Always turn off the electrical power to the solenoid valve before attempting to replace the coil. Then, having turned off the electrical power, disconnect the coil leads. Figs. 5-14 and 5-15 are schematics of some typical solenoid valves. The numbers in the illustrations refer to the valve components in their order of disassembly and are identified as follows:

164

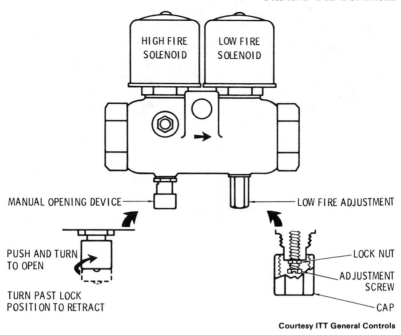

MANUAL OPENING DEVICE

LOW FIRE ADJUSTMENT

PUSH AND TURN
TO OPEN

TURN PAST LOCK
POSITION TO RETRACT

LOCK NUT

ADJUSTMENT
SCREW

CAP

Courtesy ITT General Controls

Fig. 5-11. Dual solenoid valve.

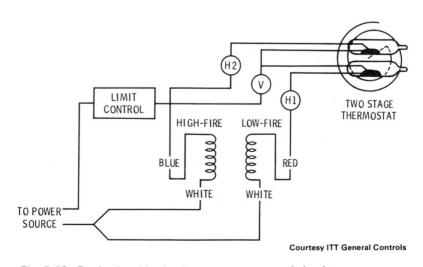

Courtesy ITT General Controls

Fig. 5-12. Dual solenoid valve in a two-stage control circuit.

165

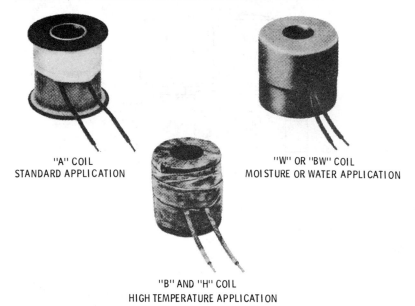

"A" COIL
STANDARD APPLICATION

"W" OR "BW" COIL
MOISTURE OR WATER APPLICATION

"B" AND "H" COIL
HIGH TEMPERATURE APPLICATION

Courtesy ITT General Controls

Fig. 5-13. Various types of solenoid coils.

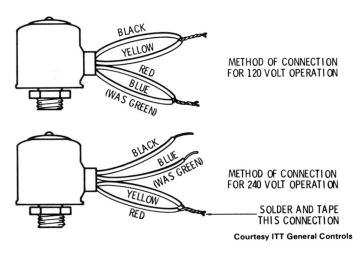

BLACK
YELLOW
RED
BLUE
(WAS GREEN)

METHOD OF CONNECTION
FOR 120 VOLT OPERATION

BLACK
BLUE
(WAS GREEN)
YELLOW
RED

METHOD OF CONNECTION
FOR 240 VOLT OPERATION

SOLDER AND TAPE
THIS CONNECTION

Courtesy ITT General Controls

Fig. 5-14. Wiring diagrams for dual voltage coils.

166

1. Jacket retaining nut or screw assembly.
2. Elbow for coil leads.
3. Valve O-Ring.
4. Coil jacket or coil assembly.
5. Nut or screw.
6. Spring retainer.
7. Plunger tube spring.
8. Screw assembly spacer.
9. Top washer and/or sleeve assembly.
10. Solenoid coil.
11. Bottom washer and/or sleeve assembly.
12. Plunger tube.

The solenoid coil should be reassembled in reverse order, but with the following precautions:

1. Be *very* careful to reassemble the top washer and/or sleeve assembly (part 9 above) exactly as it had been assembled. Improper assembly will cause the solenoid coil to burn out.
2. Be sure to align the top washer and/or sleeve assembly so the coil leads have an unobstructed passage out of the solenoid.
3. Properly align all slots in the bottom washer and/or sleeve assembly.

DIRECT-ACTING HEAT MOTOR VALVES

A *direct-acting heat motor valve* depends on the heat-induced expansion and contraction of a rod-type element to provide the movement and force for its operation.

The heat is generated by the passage of an electrical current through a resistance coil wound around a metal rod. One end of the rod is secured in place. The other end rests against a flexible snap mechanism (Fig. 5-16). When the room thermostat calls for heat, the electrical current flows through the coil and heats the metal rod. The heat generated by the resistance of the coil causes the rod to expand against the snap mechanism. When enough force is applied to the rod, the snap mechanism snaps over center

167

and opens the valve. When the rod cools and contracts, the snap mechanism returns to its original position, and the valve closes.

Some combination gas valves utilize the direct-acting heat motor principle of operation. The Robertshaw Unitrol 1000E is an example of this type of valve (Fig. 5-17). It combines in one unit a gas cock, an automatic pilot, a pilot gas filtration device,

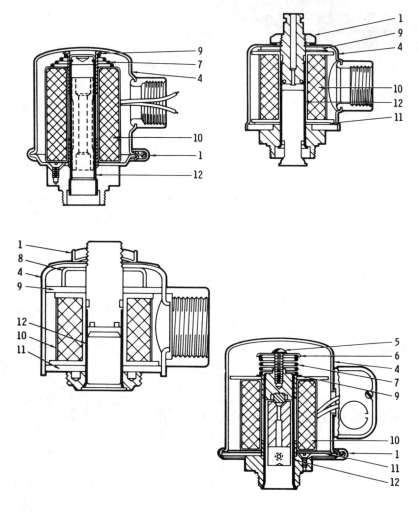

Fig. 5-15. Examples of typical

and a heat-motor-actuated automatic valve. A main gas-pressure regulator can be added as an option. As shown in Fig. 5-17, a manual opener or bypass selector is a common feature on these valves. When the bypass selector is in *on* position and the room thermostat is in *off* position, a bypass rate is provided to the burner for minimum input conditions.

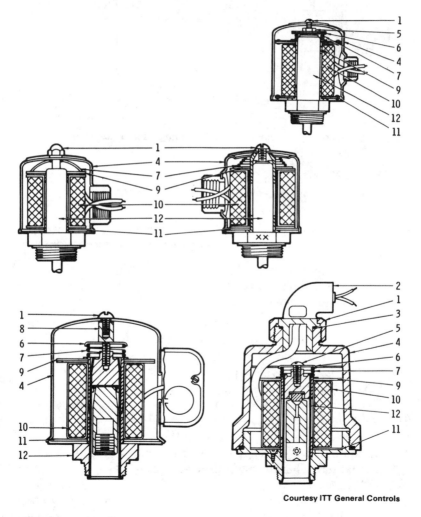

Courtesy ITT General Controls

solenoid construction.

169

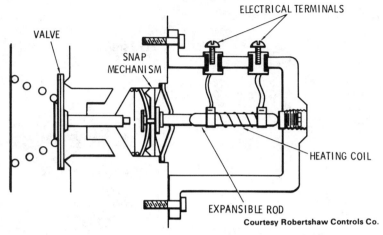

Fig. 5-16. Operating principle of a direct-acting heat motor valve.

DIAPHRAGM VALVES

There are three principal diaphragm valves, each distinguished by the kind of power used to actuate them. These three types are:

1. Hydraulic-actuated valves.
2. Solenoid-actuated valves.
3. Heat-motor-actuated valves.

A hydraulic-actuated valve utilizes a hydraulic element to provide both the thermostatic sensing means and the power for valve operation. The closed hydraulic sensing and actuating device consists of a bulb, capillary tube, and a bellows or diastat (Fig. 5-18). Temperature changes cause the liquid in the remote-bulb sensing device to expand or contract. This expansion or contraction of the liquid operates the valve by controlling the pressure exerted against a bellows in the valve body. Additional information about this type of valve is contained in the section "Remote-Bulb Thermostats" in Chapter 4 (Thermostats and Humidistats).

Both solenoid-actuated diaphragm valves and heat-motor-actuated diaphragm valves are described elsewhere in this chap-

MANUAL OPENER
OR BY-PASS
SELECTOR

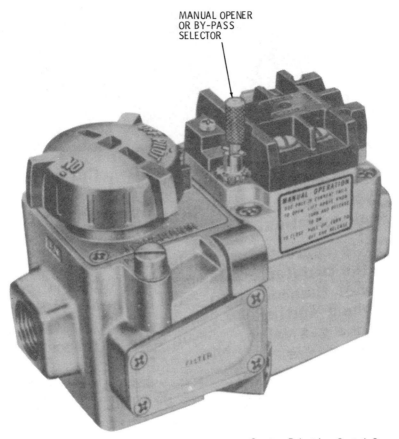

Courtesy Robertshaw Controls Co.

Fig. 5-17. Heat-motor-actuated combination gas valve.

ter (see "Solenoid Gas Valves," "Oil Valves," and "Direct-Acting Heat Motor Valves").

The term "diaphragm valve" can be confusing because valves that use a diaphragm are usually referred to by the power used to actuate them (e.g., hydraulic-actuated valves, solenoid-actuated valves) or their specific function (e.g., electric gas valves, oil burner valves).

A *diaphragm valve* is *any* valve that contains a diaphragm; its purpose is to respond to pressure variations. Because this is an

171

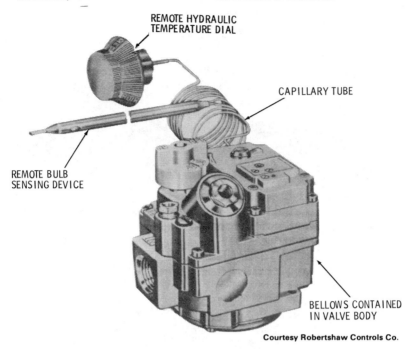

REMOTE HYDRAULIC
TEMPERATURE DIAL

CAPILLARY TUBE

REMOTE BULB
SENSING DEVICE

BELLOWS CONTAINED
IN VALVE BODY

Courtesy Robertshaw Controls Co.

Fig. 5-18. Unitrol 7000SR-1H hydraulic actuated valve.

essential feature of pressure regulators, the operating principles of a diaphragm valve are described in the sections "Gas-Pressure Regulators" and "Combination Gas Controls."

GAS-PRESSURE REGULATORS

Natural gas is distributed through the city mains at pressures of 7 in. water gauge or higher. Normally this gas will be at a higher pressure than the heating equipment or appliance can properly use. Furthermore, the gas pressure in the mains (and in the building supply lines) will often fluctuate because of load demand variations. Excessively high gas pressure and gas-pressure variations are detrimental to the operating efficiency and safety of a gas-fired furnace, boiler, or water heater. Hence, they must be brought under control before the gas enters the burners.

A *gas-pressure regulator* (or *manifold pressure regulator* as it is also called) is a regulating device used to control manifold gas pressure. Gas is delivered to the burners from the outlet orifice of the regulator at a single, nonfluctuating constant pressure regardless of inlet pressure changes.

A regulator must sense all changes in gas pressure and be able to adjust the gas flow as required. The sensing device by which this is accomplished is a diaphragm and spring arrangement attached to a valve ball or disc used to restrict gas flow through the seat. These and other components are illustrated in the cutaway of the low-pressure regulator shown in Fig. 5-19.

A pressure regulator uses the available gas pressure as the primary force to open or close the valve. Outlet gas pressure presses against the diaphragm and spring. If the gas pressure is

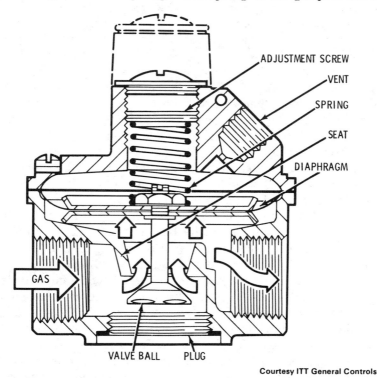

Courtesy ITT General Controls

Fig. 5-19. Principal components of a low-pressure regulator.

too little to overcome the force of the spring, then the attached ball or disc is pushed away from its seat. This enlarged opening allows more gas to flow. If the outlet pressure against the diaphragm is greater than the spring setting, then the valve ball or disc is brought toward its seat, narrowing the opening and restricting flow. As the gas pressure against the diaphragm equals the force exerted by the spring, the valve ball or disc is so positioned from the orifice to maintain a steady downstream pressure. This principle of operation is basic to all diaphragm valves.

The spring-loaded side of the diaphragm must be vented or the movement of the diaphragm will be restricted. The most elementary form of venting is shown in Fig. 5-20. This is simply an orifice installed in a vent hole on the spring-loaded side of the diaphragm. A more complex method of venting involves connecting a tube to a tapping in the vent. This represents either the internal or external bleed system of venting gas. The principal difference between the two systems lies in how and where the vented gas is disposed.

Proper venting allows the valve diaphragm to move freely in either direction. Installing an orifice in the vent hole slows the diaphragm action, thereby providing a smoother operating control (Fig. 5-20). The use of a vent hole orifice also prevents the rapid and potentially dangerous escape of gas in the event of a diaphragm rupture.

Both internal and external bleed systems are used to vent the gas from the spring-loaded side of the diaphragm. In an *internal bleed system*, the bleed gas is routed to the burner or pilot where it is burned, or to the burner manifold where it is mixed with the main gas supply and eventually burned. In an *external bleed system*, the bleed gas is vented by a tube extending to the outdoors. A variation of the external bleed system is to place the outlet end of a tube in the heat exchanger to vent the gas outside.

A simple diaphragm valve functions only to open or close the valve and is generally of the external bleed type (Fig. 5-21). The operator mechanism for a restricting internal bleed orifice diaphragm valve is basically as shown in Fig. 5-22.

A gas-pressure regulator can be either an independent control on a gas manifold or it can be a part of a combination gas control. The obvious advantage to using a combination control is the

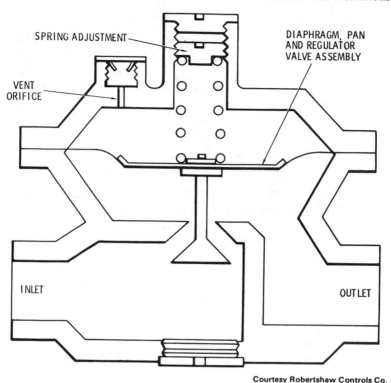

SPRING ADJUSTMENT

DIAPHRAGM, PAN
AND REGULATOR
VALVE ASSEMBLY

VENT
ORIFICE

INLET

OUTLET

Fig. 5-20. Pressure regulator with vent orifice.

simplification of appliance assembly and the saving of space obtained when compared to the use of separate components. Less obvious are the operational advantages.

When the pressure regulator is *not* an integral part of the combination gas control, it either precedes or follows the location of the latter in the manifold. If the separate pressure regulator precedes the combination control, both the main burner gas and the pilot gas are regulated by the same regulator in most installations. If this is the case, a problem will sometimes occur with pilot outage. As the main gas valve opens to provide gas to the main burners, a temporary starving of the pilot gas can occur due to regulator response delay. This condition can be avoided by installing a separate pilot gas regulator for pilot gas only or by

175

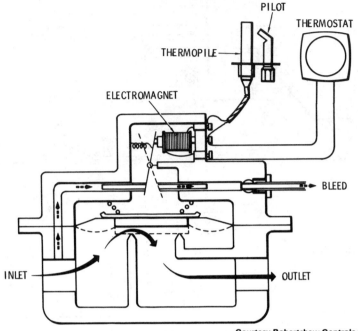

Courtesy Robertshaw Controls Co.

Fig. 5-21. Schematic drawing of a millivolt diaphragm bleed valve.

using a regulator equipped combination gas control with the proper sequencing of operation.

Installing a separate pressure regulator *after* the combination gas control (i.e., downstream from it) can sometimes result in overgassing the main burners. This occurs because the regulator remains in a wide open position when the main gas valve remains closed. When the gas valve opens, overgassing can result from the delay of the regulator valve in resuming regulation. A combination gas control can eliminate this problem.

Some gas-fired water heaters use a so-called *balanced* pressure regulator. This type of pressure regulator uses two internal diaphragms to control gas pressure. The operating principle is quite different from that described for the regulators used on gas-fired heating equipment. For additional information, read the section "Balanced Pressure Regulators" in Chapter 4 of Volume 3 (Water Heaters and Other Appliances).

PRESSURE SWITCHES

A *pressure switch* is a safety device used in positive-pressure or differential-pressure systems to sense gas- or air-pressure changes. A typical arrangement of gas-pressure switches on a gas manifold is shown in Fig. 5-23. The wiring diagram in Fig. 5-24 illustrates the connections between the high and low gas-pressure switches on a gas-fired boiler manifold and the primary control.

Gas-pressure switches are available in two basic types:

1. Falling-pressure switches.
2. Rising-pressure switches.

In the falling-pressure switch, decreased pressure on the diaphragm actuates the device. The switch is designed to lock out when the pressure falls to the set point (minus differential). As the pressure increases, the diaphragm rises and the switch is deactuated (except on manual reset models). An adjustable spring-loaded diaphragm determines the amount of pressure to actuate the switch.

In a rising-pressure switch, the switch is actuated by increased pressure on the diaphragm. As the pressure falls, the diaphragm

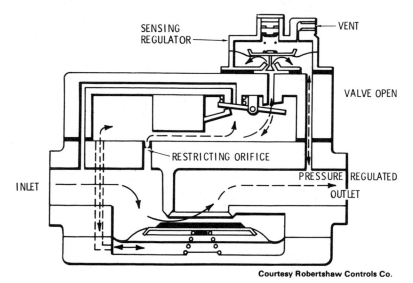

Courtesy Robertshaw Controls Co.

Fig. 5-22. Restricting internal bleed orifice diaphragm valve.

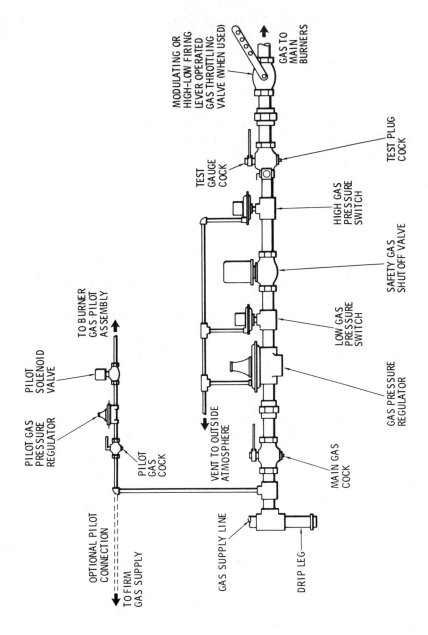

Fig. 5-23. Typical arrangement of gas-pressure switches on a manifold.

GAS AND OIL CONTROLS

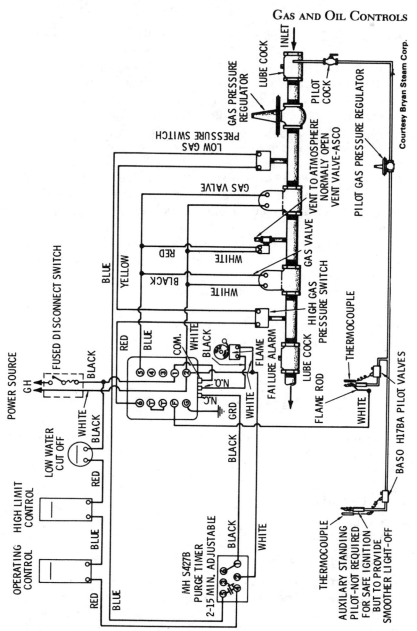

Fig. 5-24. Wiring diagram and gas train for a Bryan atmospheric gas-fired boiler showing the wiring connections for the high and low gas-pressure switches.

179

lowers and the switch is deactuated (again, except on manual reset models). An adjustable spring-loaded diaphragm determines the amount of pressure to actuate the switch.

The pressure required to move the diaphragm in these switches is adjustable within the pressure range stamped on the switch nameplate. The pressure switch shown in Fig. 5-25 can be adjusted by removing the cover and turning the adjustment screws clockwise. This action raises the actuation point of the switch. Turning the screws counterclockwise lowers the actuation point. The range scale plate in the switch is marked for four relative pressure settings. Setting A corresponds to a minimum range, D to a maximum range, and both B and C to intermediate ranges.

A typical wiring diagram for a single-pole double-throw SPDT switch is shown in Fig. 5-26. An SPDT switch may be wired to open or close the circuit on pressure rise.

On switches equipped with manual reset (Fig. 5-25), the switch contacts open on pressure rise or pressure drop (depending on which model is used) and remains open regardless of pressure change. The reset button is pushed to close the switch *after* the pressure has returned to an acceptable level.

AUTOMATIC PILOT SAFETY VALVE

The automatic pilot safety valve is a device used to shut off the gas supply when the pilot flame is extinguished or fails to light.

There are a variety of different pilot safety controls available on the market, but all are based on one of the following three operating principles:

1. Thermocouple.
2. Metal expansion.
3. Liquid pressure.

Pilot safety controls based on the thermocouple operating principle are probably the most common. The schematic in Fig. 5-27 shows the principal components of an automatic pilot in which a thermocouple is used. The constant-burning pilot of the system illustrated here provides not only burner gas ignition but

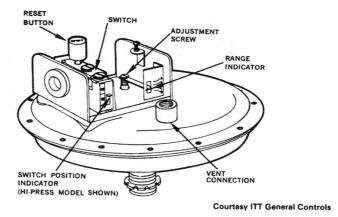

Courtesy ITT General Controls

Fig. 5-25. Gas-pressure switch with manual reset.

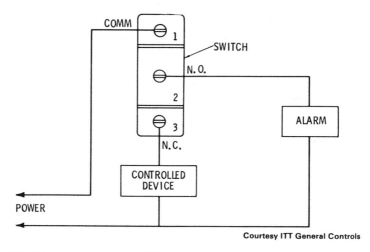

Courtesy ITT General Controls

Fig. 5-26. Wiring diagram of SPDT gas-pressure switch.

also heat for the hot junction of the thermocouple. As shown in Fig. 5-28, the hot junction of a thermocouple is positioned so that it is directly in the path of the pilot flame.

The thermocouple itself is actually a miniature generator that can convert heat (from the pilot flame) into millivolts of electricity. It consists of two dissimilar metals joined together at their extremities. When one of the junctions (the hot junction) remains

181

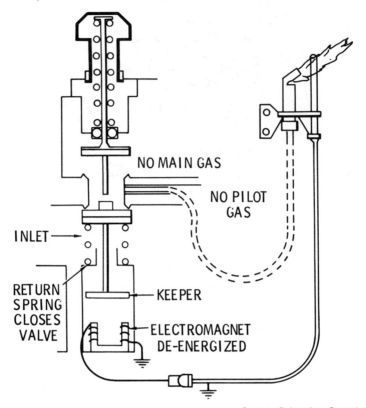

NO MAIN GAS

NO PILOT GAS

INLET →

RETURN SPRING CLOSES VALVE

KEEPER

ELECTROMAGNET DE-ENERGIZED

Courtesy Robertshaw Controls Co.

Fig. 5-27. Thermocouple used in conjunction with an automatic pilot valve.

cold, electrical energy is generated. The amount of energy generated by the thermocouple is directly proportional to the temperature difference between the hot and cold junctions (Fig. 5-28). One thermocouple or junction will deliver approximately 25 millivolts. Several thermocouples or junctions can be wired in series to produce a higher voltage (see "Pilot Generators"). The electrical energy generated by the thermocouple energizes an electromagnet which operates the automatic pilot valve.

The automatic pilot safety valve may be an individual control but is more commonly a part of a combination control, which

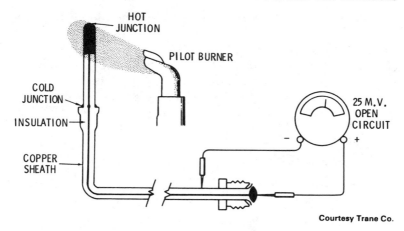

HOT
JUNCTION

PILOT BURNER

COLD
JUNCTION

INSULATION

25 M.V.
OPEN
CIRCUIT

COPPER
SHEATH

Courtesy Trane Co.

Fig. 5-28. Common thermocouple construction.

usually combines manual valve and thermostatic or automatic valve functions. These valves are described in this chapter (see "Combination Gas Valves").

Some automatic pilot valves are designed to shut off both the main gas (i.e., the gas to the main burners) and the pilot gas. This is called 100 percent safety automatic pilot or a 100 percent safety shutoff. They are recommended for use with either natural or LP gases.

A 90 percent safety automatic pilot (or 90 percent safety shutoff) shuts off the gas supply to the main burners but allows gas to continue flowing to the pilot burner. This type of gas pilot may be used safely with natural gas (and is permitted by some local codes), but it should *never* be used with LP gas. LP gas is heavier than air and will not vent.

The procedure for lighting the pilot in a system containing a thermocouple (or pilot generator) and an automatic pilot safety valve is fairly simple. The reset button on the automatic pilot valve must be pushed down while the pilot is being lit (Figs. 5-29 and 5-30). While the reset button is depressed, an auxiliary valve causes the main gas porting to close. At the same time, the spring-loaded automatic pilot valve is opened, allowing gas to flow only to the pilot burner, which can now be lit.

The same action that opened the spring-loaded automatic

183

valve has also forced the keeper (Fig. 5-29) against the pole faces of the electromagnet. Eventually the heat of the pilot flame causes the thermocouple to generate an electrical current, which causes the electromagnet to become magnetized. As a result, the keeper is held against the magnet and the automatic pilot valve is maintained in an open position. The pilot should be allowed to burn for at least 60 seconds before releasing the button. This

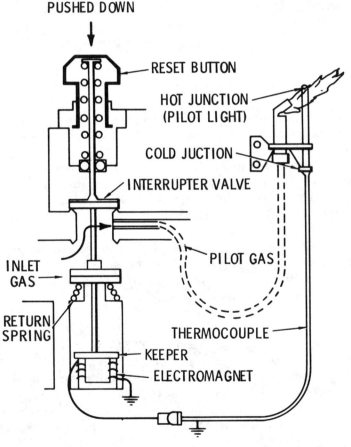

Fig. 5-29. Reset button depressed while pilot is being lit.

permits enough current to be built up by the thermocouple or pilot generator to hold the valve open.

Once the pilot has been established, the reset button can be released; and the main gas and pilot gas will be free to move past the automatic pilot valve (Fig. 5-30). When the pilot flame is extinguished, the thermocouple hot junction cools and breaks the electrical current. The electromagnet is demagnetized as a result of this loss of current, and the keeper is released, causing the automatic pilot valve to close and shut off the flow of gas. Before relighting the pilot, allow at least 5 minutes for all the gas to clear the system.

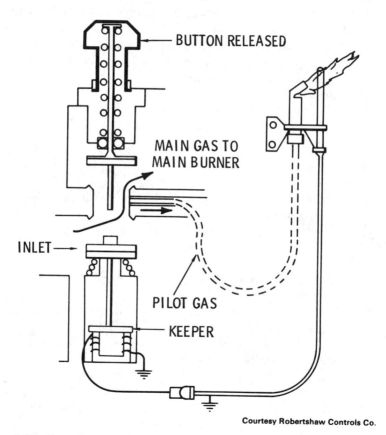

BUTTON RELEASED

MAIN GAS TO
MAIN BURNER

INLET

PILOT GAS

KEEPER

Courtesy Robertshaw Controls Co.

Fig. 5-30. Reset button released after pilot flame is established.

THERMOPILOT VALVES

A *thermopilot valve* is a 100 percent safety shutoff control for gas-fired heating equipment and appliances. It operates on current supplied by a thermocouple or pilot generator. This current energizes a thermomagnet, which holds the valve open after manual reset. When there is an unstable or low pilot flame or no flame at all, the current to the thermomagnet is lost. As a result, the valve closes, shutting off the gas flow.

An example of a thermopilot valve operated by a thermocouple is the ITT General Controls MR-2G valve. Using it in conjunction with a solenoid gas valve provides 100 percent safety shutoff control (Fig. 5-31).

The thermomagnet of an ITT General Controls MR-2YA valve is energized by a pilot generator. It must be used with a pilot-operated diaphragm valve to provide 100 percent safety shutoff control (Fig. 5-32).

The following suggestions should be carefully observed when installing ITT thermopilot valves in order to ensure efficient valve operation:

1. Locate the valve so that it is easily accessible and where the ambient temperature is below 200°F.
2. Make sure the piping is clean (blow out all particles and other impurities).
3. Apply thread seal sparingly to male threads only.
4. Install valve with gas flow in the same direction as the arrow in the valve body is pointing.
5. Use the pipe wrench on the valve body flats at the end being connected.
6. Avoid stress on the valve body by aligning the inlet and outlet pipe connections.
7. Check all pipe connections for gas leaks with a soap and water solution. *Never* use a flame.
8. Check the valve with a millivolt meter (see below).
9. All wiring connections should be clean and tight.
10. Pilot burner should be installed on main burner so that the ignition flame will light the main burner with the pilot turned down low.

186

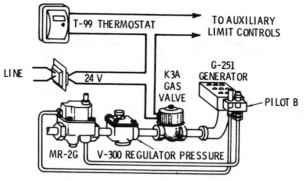

Courtesy ITT General Controls

Fig. 5-31. ITT control valve model MR-2G.

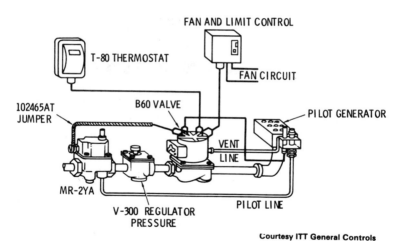

Courtesy ITT General Controls

Fig. 5-32. ITT control valve model MR-2YA.

Testing an ITT General Controls MR-2G Thermopilot valve with a millivolt meter requires the use of a special adapter. The adapter and thermocouple are connected to the valve as shown in Fig. 5-33. Attach the meter clips as shown. Fig. 5-34 illustrates the testing of an ITT MR-2YA valve. In either case, the valve should be replaced if the meter reading is above that shown on the applicable scale *and* the valve fails to open after following the lighting instructions.

187

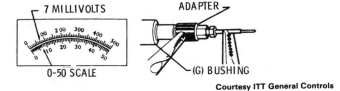

0-50 SCALE

Courtesy ITT General Controls

Fig. 5-33. Testing control valve MR-2G.

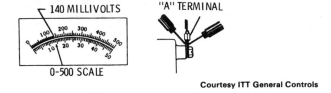

0-500 SCALE

Courtesy ITT General Controls

Fig. 5-34. Testing control of valve MR-2YA.

PILOT GENERATORS

The amount of electrical energy generated by a thermocouple having a single hot junction and a single cold junction (i.e., approximately 25 millivolts) is considered adequate for most residential heating equipment. However, a few residential furnaces and boilers and most commercial types require a higher voltage. An increase in voltage can be obtained by using a number of thermocouples wired in series (Fig. 5-35). A series of thermocouples in one unit connected in series is called a *pilot generator,*

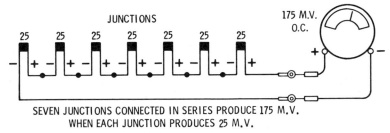

SEVEN JUNCTIONS CONNECTED IN SERIES PRODUCE 175 M.V.
WHEN EACH JUNCTION PRODUCES 25 M.V.

Courtesy Trane Co.

Fig. 5-35. Thermocouple construction and series wiring.

188

thermopile, or *thermopile generator.* A pilot generator forms a part of a millivolt or self-energizing control circuit.

An example of a pilot generator used in gas-fired heating equipment and appliances is shown in Fig. 5-36. This particular unit produces approximately 320 millivolts in an open circuit for gas valve control. It is available with an adjustable ignition port and interchangeable orifices for use with any gas.

A pilot generator that produces an even higher voltage is the ITT General Controls PG9 (Fig. 5-37). This pilot generator provides a pilot flame for gas burner ignition and generates electricity from the heat of the pilot flame to operate millivolt gas valves and relays. The replaceable generator cartridge in the generator contains many thermocouples connected in series. The top $\frac{3}{8}$ to $\frac{1}{2}$ in. of the cartridge is heated by the pilot flame which produces approximately 500 to 750 millivolts ($\frac{1}{2}$ to $\frac{3}{4}$ of 1 volt) open circuit.

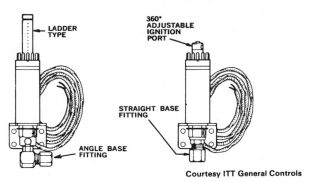

Courtesy ITT General Controls

Fig. 5-36. ITT pilot generator model PG-1.

Fig. 5-37. ITT pilot generator model PG-9A with cartridge.

Courtesy ITT General Controls

189

A millivolt meter must be used to test a pilot generator. The meter leads are attached to the valve or relay terminals to which the wires of the pilot generator are also attached. The thermostat must be calling for heat and the pilot burning during a millivolt meter test.

PILOT-OPERATED DIAPHRAGM VALVES

A pilot-operated diaphragm valve is used to provide automatic shutoff of main line gas when there is an unstable pilot flame or no flame at all. These valves are operated by electrical energy (millivoltage) produced by the pilot generator. Their operation is controlled by the room thermostat, limit devices, and other operating controls.

The ITT General Controls B60 gas valve (Fig. 5-38) is an example of a pilot-operated diaphragm valve commonly used with gas-fired heating equipment. Where 100 percent shutoff is required, it should be used in conjunction with a thermopilot valve (Fig. 5-39).

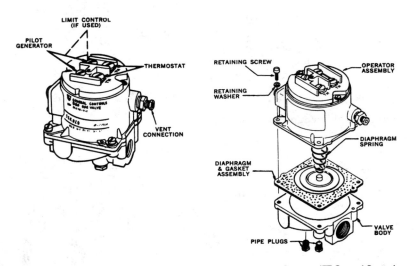

Courtesy ITT General Controls

Fig. 5-38. Pilot-operated diaphragm valve.

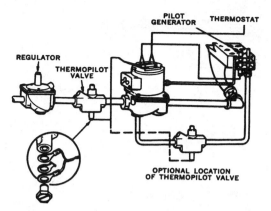

PILOT
GENERATOR   THERMOSTAT

REGULATOR
THERMOPILOT
VALVE

OPTIONAL LOCATION
OF THERMOPILOT VALVE

Courtesy ITT General Controls

Fig. 5-39. ITT valve model B60Y used in conjunction with a thermopilot valve.

THERMOCOUPLES

Thermocouples are made from flame-resistant metal alloys and are available in standard lengths of 13.5 to 48 in. (Fig. 5-40). Industry standard "G" bushing or "R" bushing threads are used for connecting the thermocouple to the valve. Thermocouples from the various manufacturers can also be interchanged. The thermocouples manufactured by ITT General Controls, Robert-shaw-Grayson, Baso, Honeywell and others are designed for universal installation when used with the proper adapter and retainer (where required) (Figs. 5-41 and 5-42).

Always test the thermocouple before replacing it. The problem with the pilot flame or safety control may be caused by something other than a malfunctioning thermocouple.

Manufacturers of thermocouples generally provide a special adapter for testing thermomagnet valves. The adapter is screwed into the gas valve; the thermocouple (or generator) leads into the adapter; and the test is run as follows:

1. Attach the meter clips as shown in Fig. 5-43. Reverse the meter clips if the needle moves to the left of zero on the millivoltmeter.

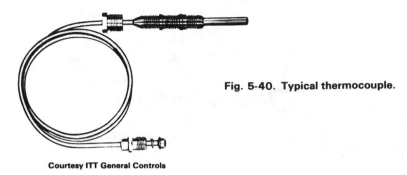

Fig. 5-40. Typical thermocouple.

Courtesy ITT General Controls

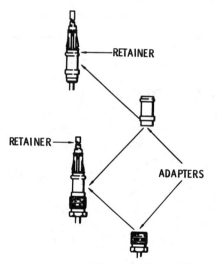

RETAINER

Fig. 5-41. Thermocouple adapters and retainers.

RETAINER

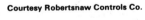

ADAPTERS

Courtesy Robertsnaw Controls Co.

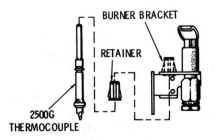

BURNER BRACKET

RETAINER

Fig. 5-42. Retainer used to hold thermocouple in burner bracket.

2500G
THERMOCOUPLE

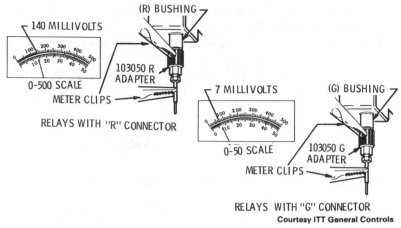

Fig. 5-43. Relay bushing connections.

2. A meter reading of less than 7 millivolts for thermocouples or less than 140 millivolts for pilot generators indicates the orifice and primary air holes on the pilot burner need to be cleaned.

Sometimes a thermocouple will have a terminal block connected to a high-energy cutoff switch. The millivoltage should be checked with meter probes as shown in Fig. 5-44. The various readings suggest the following courses of action:

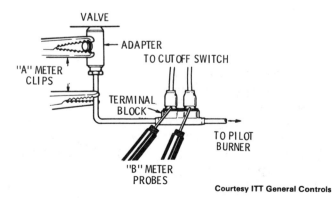

Fig. 5-44. Checking a terminal block connected to a high-energy cutoff switch.

193

1. Replace the high-energy cutoff switch if the reading exceeds 4 millivolts.
2. Replace the thermocouple if the reading is still less than 7 millivolts.
3. Replace (or repair) the valve if it fails to hold open with a meter reading of more than 7 millivolts.

TROUBLESHOOTING THERMOCOUPLES

The following suggestions are offered as possible remedies for a number of different operating problems associated with thermocouples.

Symptom and Possible Cause   *Possible Remedy*

Pilot flame lit but safety control fails to function

1. Thermocouple not hot enough to generate current.	1. Wait at least 1 minute for thermocouple to become hot enough.
2. Drafts deflecting flame away from thermocouple.	2. Eliminate source of draft.
3. Pilot flame too small or yellow in color due to restricted pilot line or dirt in primary air opening or burner head.	3. Disconnect, clean thoroughly, and reconnect. Change orifice if necessary.
4. Loose or dirty electrical connections.	4. Disconnect, clean, reconnect, and tighten.
5. Thermocouple tip too low in pilot flame.	5. Check installation to make sure thermocouple is properly mounted in bracket.

Symptom and Possible Cause   *Possible Remedy*

Safety control operates but fails when main burner has been on a short time.

1. Restriction in pilot or main gas tubing.
2. Draft-deflecting flame couple.

1. Eliminate restriction. Provide normal pressure.
2. Eliminate draft or baffle.

COMBINATION GAS VALVES

A *combination gas valve* (or *combination gas control*) combines in a single unit all manual and automatic control functions required for the operation of gas-fired heating equipment. In other words, a single valve replaces the various individual pilot line and main line gas controls. A gas-pressure regulator is usually optional.

Many manufacturers of gas controls offer a complete line of combination gas valves; each valve is designed for a different kind of installation or application. Usually these valves will differ on the basis of the controller voltage or voltage source, valve application or function, required Btu capacity for the installation, and type of gas used.

Honeywell manufactures a line of standardized and interchangeable gas control components. A complete preassembled combination gas control can be ordered from the factory or one can be assembled in the field from a variety of different standardized components in order to meet the needs of a particular installation. This add-on feature also allows field replacement of a defective component without removing the complete valve from the installation. A number of possible combinations are illustrated in Figs. 5-45 and 5-46.

A typical Honeywell combination gas control consists of the following *basic* components (Fig. 5-47):

195

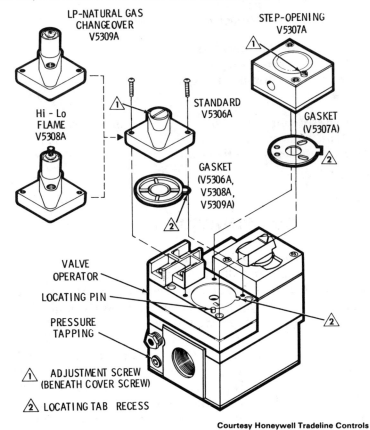

Courtesy Honeywell Tradeline Controls

Fig. 5-45. Various gas-pressure regulators available for use on a combination gas control equipped with a valve operator.

1. Main valve body.
2. Valve operator.
3. Pressure regulator.

As shown in the schematic diagram of the control (Fig. 5-48), the main valve body (or manifold control) contains a valve diaphragm (5) and disc (3). This portion of the combination gas control operates as a conventional diaphragm valve. The valve opens and closes in response to the presence or absence of gas in the pressure chamber (4). This gas is called the "working gas"

196

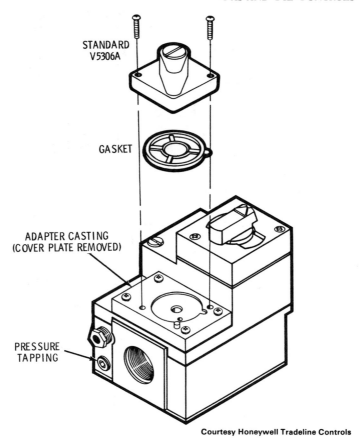

STANDARD
V5306A

GASKET

ADAPTER CASTING
(COVER PLATE REMOVED)

PRESSURE
TAPPING

Fig. 5-46. Standard gas-pressure regulators installed on manual manifold control not equipped with valve operator.

because it provides the lifting force necessary to raise the valve disc off its seat.

The valve operator controls the flow of working gas by means of an electrically actuated lever (1). This lever is opened or closed by the temperature control circuit. When the burner is off and there is no call for heat from the room thermostat, the lever is in the position shown in Fig. 5-48. Note that while the lever is in this position it blocks the admission of gas into the valve. Furthermore, the working gas can escape through the working channel

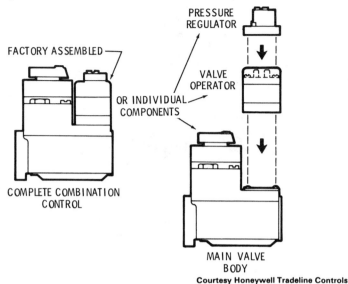

Courtesy Honeywell Tradeline Controls

Fig. 5-47. Basic components of a typical Honeywell combination gas control.

(2), resulting in a reduction of the gas pressure in the pressure chamber. The main gas valve closes with this loss of working gas pressure.

A call for heat from the room thermostat energizes the valve operator and causes the lever to open the inlet port (Fig. 5-49). The working gas then flows into the pressure chamber (4) and pushes the diaphragm (5) up against the valve disc assembly (3) to allow the flow of gas through the valve to the burner. At the same time, gas also flows into the pressure regulator chamber (8) and through the evacuation gas channel (6) into the combination gas control outlet.

Changes in the outlet pressure of the gas cause changes in the position of the regulator diaphragm (9). If the outlet pressure rises, the regulator valve (7) opens slightly to allow more working gas into the evacuation gas channel (6). This discharge of working gas causes the main valve diaphragm (5) to drop and allows the main valve disc (3) to move downward on its seat. This action reduces the flow of main burner gas through the control to correct the rise in outlet pressure.

A drop in the outlet pressure has the opposite effect. The regulator valve (7) closes slightly and reduces the amount of working gas entering the evacuation gas channel (6). This action increases the gas pressure in the main valve pressure chamber (4) and forces the main valve disc upward away from its seat. As a result, the flow of gas through the control is increased enough to correct the fall in outlet pressure.

LIGHTING INSTRUCTIONS

Manufacturers provide detailed instructions for lighting a pilot in an installation equipped with a combination gas valve. These instructions should be read carefully before attempting to light the pilot burner. If there are no instructions with the equipment, the following procedure is suggested:

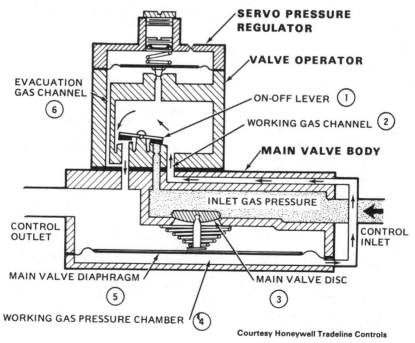

Courtesy Honeywell Tradeline Controls

Fig. 5-48. Schematic diagram of combination gas control in the burner *off* position.

199

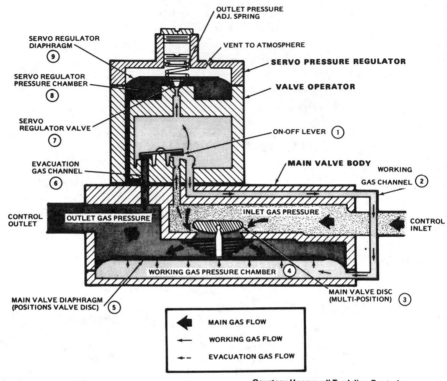

Courtesy Honeywell Tradeline Controls

Fig. 5-49. Wiring diagram of a combination gas control in the burner *on* position.

1. Turn the wall thermostat to *off* or its lowest setting.
2. Depress the gas cock and rotate it to the *off* position (Fig. 5-50).
3. Allow 5 minutes for any gas in the burner compartment to escape.
4. Turn the gas cock dial to the pilot position.
5. Depress and hold the gas cock dial down while lighting the gas pilot.
6. Allow the pilot to burn 60 seconds before releasing the gas cock dial.
7. Turn the gas cock dial to *on* position after the pilot burner flame has been established.

8. Set the room thermostat to the desired temperature position.

The pilot position on a combination gas valve is used for temporary or seasonal shutdown. The *off* position is used when complete shutdown is necessary.

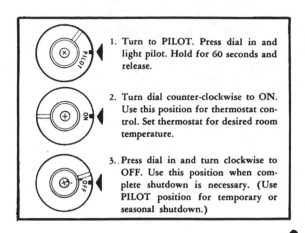

1. Turn to PILOT. Press dial in and light pilot. Hold for 60 seconds and release.

2. Turn dial counter-clockwise to ON. Use this position for thermostat control. Set thermostat for desired room temperature.

3. Press dial in and turn clockwise to OFF. Use this position when complete shutdown is necessary. (Use PILOT position for temporary or seasonal shutdown.)

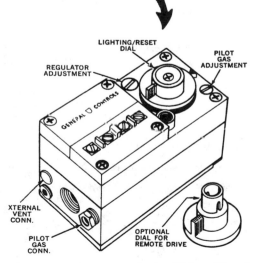

Courtesy ITT General Controls

Fig. 5-50. Lighting procedure for a combination gas control.

201

PILOT BURNERS

A *pilot burner* is a device used in a gas-fired appliance to light the main gas burners and generate sufficient millivoltage to operate a thermocouple or thermopile pilot safety shutoff device (see "Automatic Pilot Valves" in this chapter). Modern pilot burners are designed to burn continually in order to provide an automatic ignition of the burners when the main gas supply is turned on. The pilot burners in common use today can be divided into either aerated or nonaerated types.

An *aerated pilot* (Fig. 5-51) is one that injects primary air through an air intake opening into the gas stream. The air and gas are mixed before burning.

An aerated pilot burner produces a very stable flame. For this reason, these pilots are often used where the pilot location is particularly inaccessible. Although the flame produced by an aerated pilot burner is more stable than one produced by a nonaerated type, an aerated pilot burner does have some disadvantages. An important one to remember is the tendency for the small primary air openings to clog with lint and dirt. Frequent cleaning is required, particularly when these pilots are used in areas having a large amount of foreign material in the air.

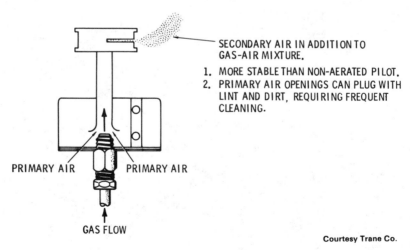

SECONDARY AIR IN ADDITION TO GAS-AIR MIXTURE.

1. MORE STABLE THAN NON-AERATED PILOT.
2. PRIMARY AIR OPENINGS CAN PLUG WITH LINT AND DIRT, REQUIRING FREQUENT CLEANING.

PRIMARY AIR   PRIMARY AIR

GAS FLOW

Courtesy Trane Co.

Fig. 5-51. Aerated pilot.

A *nonaerated* pilot (Fig. 5-52) does not inject primary air. As a result, the air and gas are not premixed, and the combustion process must be completed with secondary air only. This results in a less stable flame than the one produced by an aerated pilot. On the credit side, a nonaerated pilot requires less maintenance than an aerated pilot. For this reason, the nonaerated pilot burner is usually preferred by most utility companies.

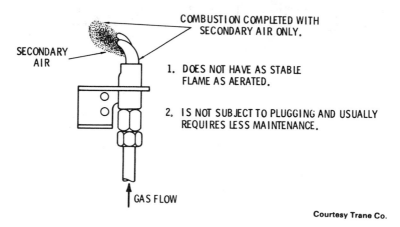

Courtesy Trane Co.

Fig. 5-52. Nonaerated pilot.

A pilot burner assembly consists of the pilot bracket, pilot orifice, primary air intake, lint screen, mixing chamber, pilot ports, and pilot hood.

The *pilot bracket* is a device used to mount the pilot in a fixed relationship to the burner. Some pilot brackets also contain means for mounting the thermocouple or pilot generator so that the hot junction is located directly in the path of the pilot flame (Fig. 5-53).

The pilot ports are the openings through which the gas (in nonaerated pilot burners) or the gas and air mixture (in aerated pilot burners) pass before burning. The gas and air are premixed in the *mixing* chamber of an aerated pilot. The air is injected into the mixing chamber of an aerated pilot through a hole or opening called the *primary air intake*. The amount of primary air can be

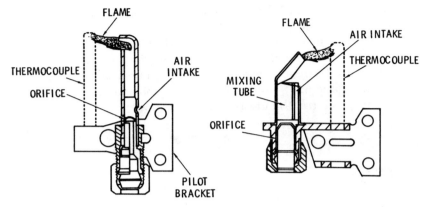

Fig. 5-53. Pilot bracket with means for mounting thermocouple.

controlled by adjusting an *air shutter* that covers the primary air intake opening.

A *lint screen* is generally used to remove lint, dirt, and other contaminants from the primary air before it enters the primary air opening. In some pilots, the lint and other particles are burned before entering the primary air intake. A so-called *incinerated pilot* is an aerated pilot in which the primary air passes adjacent to the flame area where the particles are burned out of the air before it enters the primary air intake.

Impurities are also found in the pilot gas. These impurities can be removed by installing a pilot gas *filter* in the line upstream from the pilot adjustment means in the control. The pilot gas filter is expected to operate several years without service. Clogging of the filter will be indicated by shortened pilot flames, which will result in improper pilot operation. Shortened pilot flames can also be caused by pilot tube stoppage or a dirty pilot orifice. These possibilities should be checked before removing the filter. If the filter should become clogged, replace the *entire* filter rather than just the filtering medium.

The *pilot orifice* is a removable component in the pilot that contains precisely sized openings that control the admission of gas to the pilot. Pilot orifices are either of the spud or insert type. A *spud orifice* screws into the bottom of the pilot burner. It is

204

both an orifice and a fitting combined into a single unit with threads at either end (Fig. 5-54). An *insert orifice* must be held in position by a separate fitting (Fig. 5-55). The pilot gas line (tubing) is connected to the bottom of the pilot orifice.

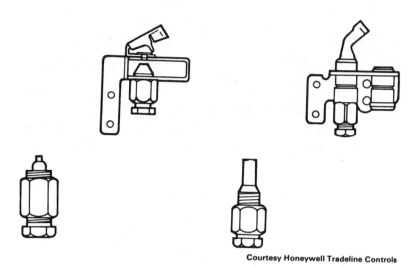

Courtesy Honeywell Tradeline Controls

Fig. 5-54. Spud orifice.

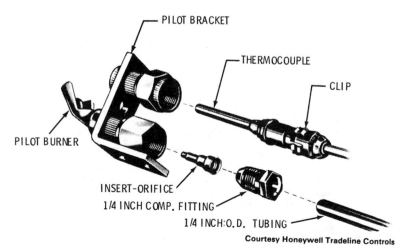

PILOT BRACKET

THERMOCOUPLE

CLIP

PILOT BURNER

INSERT-ORIFICE
1/4 INCH COMP. FITTING
1/4 INCH:O.D. TUBING

Courtesy Honeywell Tradeline Controls

Fig. 5-55. Insert orifice.

205

INSTALLING A PILOT BURNER

Always follow the manufacturer's instructions when installing a new pilot burner. Read these instructions very carefully before beginning any work.

If no installation instructions are available, the following location and mounting requirements should be carefully observed:

1. Choose a location for the pilot burner that provides easy access, observation, and lighting.
2. Rigidly affix the pilot burner to the main burner. Other mounting surfaces should not be used (Fig. 5-56).
3. Mount the pilot burner so that the flame is properly positioned with respect to the main burner flame (Fig. 5-57).

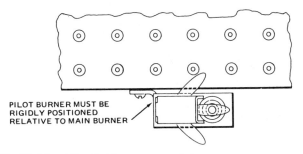

PILOT BURNER MUST BE
RIGIDLY POSITIONED
RELATIVE TO MAIN BURNER

Fig. 5-56. Rigidly affix pilot burner to main burner.

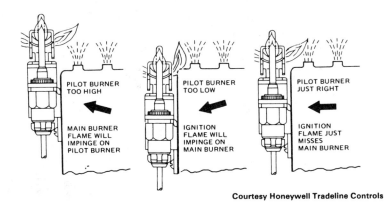

PILOT BURNER
TOO HIGH

MAIN BURNER
FLAME WILL
IMPINGE ON
PILOT BURNER

PILOT BURNER
TOO LOW

IGNITION
FLAME WILL
IMPINGE ON
MAIN BURNER

PILOT BURNER
JUST RIGHT

IGNITION
FLAME JUST
MISSES
MAIN BURNER

Courtesy Honeywell Tradeline Controls

Fig. 5-57. Proper positioning of the pilot burner flame.

206

A pilot flame should never be exposed to falling scale, which could impair the ignition of the main burners. Furthermore, the flame should not be exposed to draft conditions or to sudden puffs of air sometimes caused by igniting or extinguishing the main burner. Always provide an ample air supply free of contaminating products of combustion.

REPLACING THE PILOT BURNER ORIFICE

The procedure for replacing a pilot burner orifice depends on whether it is a spud or insert orifice. For a Honeywell spud orifice, the procedure is as follows:

1. Disconnect the gas supply tubing from the pilot burner.
2. Unscrew the spud orifice and throw it away.
3. Remove any burrs from the tubing and square off the ends.
4. Remove the one-piece nut and ferrule from the new assembly and slip them over the tubing.
5. Install the new assembly in the pilot burner and tighten securely.
6. Push the tubing (along with the nut and ferrule) into the burner as far as it will go and engage the nut.
7. Tighten the nut by hand until it will turn no further. Use a wrench to make one final turn.

The connection in Step 7 should be tight enough to prevent any gas leakage. *Do not tighten* it too much or you will run the risk of stripping the threads. Try not to bend the tubing near the fitting after the nut has been tightened.

As shown in Fig. 5-55, the procedure for replacing an insert orifice presents no serious problems. The gas supply tubing must first be disconnected from the pilot burner by unscrewing the compression fitting. The small insert orifice can then be removed. Sometimes a light tap on the pilot burner bracket will be required to dislodge the orifice.

Place the new orifice on the end of the gas tubing and insert both the orifice and tubing into the pilot burner. Tighten the compression nut until it is secure. Use the same procedure described for tightening a spud orifice.

LIGHTING THE PILOT

Read the appliance manufacturer's lighting instructions before attempting to light the pilot burner. The basic procedure for lighting a pilot is as follows:

1. Turn the room thermostat to its lowest setting.
2. Shut off the main gas supply to the main burner and the pilot burner.
3. Allow at least 5 minutes for the unburned gas to vent.
4. Light the pilot burner in accordance with the appliance manufacturer's lighting instructions.

Venting the unburned gas (Step 3) is very important. This is especially true for LP gas because it is heavier than air and will not vent upward naturally. *Every* precaution should be taken to insure that the appliance is properly venting any unburned gas.

Appliance and pilot burner manufacturers provide very detailed lighting instructions for their equipment. Moreover, the development of various types of combination gas controls has simplified the lighting procedure and increased the safety factor (see "Combination Gas Valves" in this chapter).

PILOT FLAME ADJUSTMENT

The pilot flame must be adjusted for proper color and size. Appliance manufacturers refer to this procedure as pilot *flame* adjustment or pilot *gas* adjustment.

The appliance manufacturer will provide instructions for making pilot flame adjustments. On combination gas valves, this involves the removal of a pilot adjustment cap and turning the adjustment screw until ,the desired flame characteristics are obtained. The best pilot flame is a steady, nonblowing blue flame that envelops the upper $\frac{3}{8}$ to $\frac{1}{2}$ in. of the thermocouple or generator (Fig. 5-58).

208

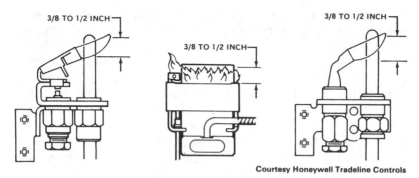

Courtesy Honeywell Tradeline Controls

Fig. 5-58. Pilot flame should generally envelop ⅜ to ½ in. of the thermo-couple or generator tip.

MAIN BURNER IGNITION

The pilot burner should ignite the main burner quietly and reliably under all operating conditions, including low gas supply pressure. The ignition of the main burner gas should occur within 4 seconds from the time that gas is admitted to the main burner. This should occur when the pilot gas supply is reduced to an amount just above the point of pilot flame extinction.

A main burner ignition test should be performed *after* the main burner gas input and primary air adjustments have been made. The type of test used to check main burner ignition will depend upon the type of pilot used in the gas-fired appliance. For example, in a pilot generator system, main burner ignition is checked with the pilot flame adjusted to the minimum millivoltage required to open the main valve. The manufacturer's installation literature for the appliance will probably include instructions for testing main burner ignition. If no literature is available, check with the local gas company.

PILOT-PRESSURE SWITCH

A *pilot-pressure switch* operates on the same principle as the gas-pressure switch used in the manifold of the gas-fired furnace or boiler. Its function is to prevent the premature failure of the spark igniter or glow coil should a prolonged gas interruption occur while the thermostat is calling for heat.

The pilot-pressure switch is installed in the pilot gas line between the pilot and pilot regulator. Installing a pilot switch requires that the pilot gas line be disconnected *downstream* from the pilot gas regulator. The gas line is then connected to the pilot-pressure switch by cutting the existing tubing to the regulator and connecting into the tee provided for switch mounting. A typical wiring of a pilot-pressure switch is shown in Fig. 5-59.

FLAME RECTIFIER PILOTS

The *flame rectifier* pilot consists of a rectification flame rod and pilot burner combined in a single unit (Fig. 5-60). When used with a suitable electronic relay, it forms a part of the flame detector circuit in a gas-fired appliance.

OIL CONTROLS

The principal functions of the oil controls are: (1) to turn the oil burner on and off in response to temperature changes in the space or spaces being heated and (2) to stop the system if an unsafe condition develops. The controls necessary to perform these functions are:

1. Thermostat.
2. Limit controls.
3. Primary control.
4. Oil valves.
5. Time-delay controls.
6. Circulator or fan control.
7. Other auxiliary controls.

This chapter is concerned with a description of the oil burner primary control, oil valves, and time-delay controls. The remaining controls found in an oil burner control system are described

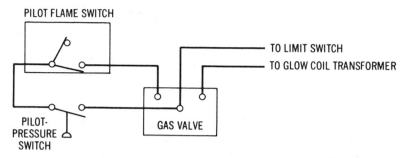

PILOT FLAME SWITCH

TO LIMIT SWITCH
TO GLOW COIL TRANSFORMER

GAS VALVE

PILOT-
PRESSURE
SWITCH

Fig. 5-59. Typical wiring of a pilot-pressure switch.

Fig. 5-60. Rectification flame rod and pilot burner combined in a single unit.

in Chapter 4 (Thermostats and Humidistats) and Chapter 6 (Other Automatic Controls).

OIL VALVES

Oil valves are used to provide on-off control of the flow of oil to the oil burner. These are normally closed solenoid valves that open when energized and close immediately when deenergized. They are variously referred to as *solenoid oil valves, magnetic oil valves,* or *oil burner valves* and are available in either immediate discharge or delayed discharge models.

An immediate discharge oil valve discharges oil as soon as it is energized. A delayed discharge valve is equipped with an integral thermistor to delay the valve opening for about 3 to 15

211

seconds (the length of time will vary depending on the manufacturer). This delay allows the burner fan to reach operating speed and establish sufficient draft before the oil is discharged.

A solenoid oil valve will make an audible click when it is opening and closing properly. If the valve fails to open after the room thermostat calls for heat, the following conditions may be responsible:

1. Inadequate fuel pressure available at the valve.
2. An obstructed bleed line.
3. No voltage indicated at valve.

Check the voltage at the coil lead terminals against the voltage shown in the nameplate. Also check the inlet pressure against the rating on the nameplate. If none of these conditions is causing the problem, the failure of the valve to open is probably due to a malfunctioning solenoid coil. The position of the coil is shown in the exploded view of the valve in Fig. 5-61. The steps for replacing the solenoid are as follows:

1. Remove the nut on top of the valve by turning it counterclockwise.
2. Remove the powerhead assembly from the spindle.
3. Disconnect and remove the solenoid coil.
4. Connect the replacement coil and reassemble.

Examples of delayed discharge valves are shown in Figs. 5-62 and 5-63. In both valves, the timing delay is governed by a thermistor attached to the solenoid coil. In these valves, the timing delay will vary with ambient temperature, voltage level, and other factors during normal operation. If the timing is *significantly* off, it may be necessary to replace the thermistor. Because the thermistor is attached to the solenoid coil, the coil must also be replaced in order to replace the thermistor.

Delayed valve opening can also be obtained by using an electronic time delay wired in series with the oil valve (Fig. 5-64). Unlike the thermistor, the timing of this device is not affected by ambient temperature. On a call for heat, the valve opening is delayed for approximately 5 seconds.

212

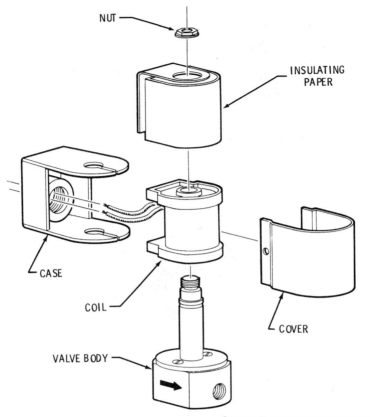

Fig. 5-61. Position of the coil in a solenoid oil valve.

Fig. 5-62. Magnetic valve used in controlling oil flow to the oil burner.

213

Fig. 5-63. Magnetic oil valve.

Courtesy Honeywell Tradeline Controls

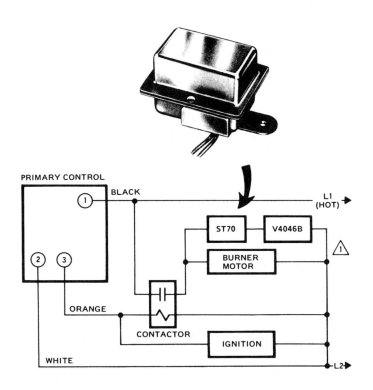

⚠ POWER SUPPLY. PROVIDE OVERLOAD PROTECTION AND DISCONNECT
MEANS AS REQUIRED.

Courtesy Honeywell Tradeline Controls

Fig. 5-64. Electronic time delay wired in series with the oil valve.

OIL BURNER PRIMARY CONTROL

The oil burner *primary control* is an automatic safety device designed to turn off the oil burner motor should ignition or flame failure occur.

Each primary control operates in conjunction with a sensor by which the burner flame is monitored throughout the burner *on* cycle. The method used to sense the burner flame determines the type of primary control used and its location in the heating system.

The two types of primary controls commonly used in oil burner control systems are: (1) the cadmium cell primary control and (2) the stack detector primary control. The cadmium cell primary control is burner mounted and uses a light-sensitive cad cell flame detector (sensor). The stack detector primary control relies on a thermal sensor to detect flame or ignition failure. This type of primary control assembly is available with the thermal sensor mounted in the stack and the primary control mounted on the burner, or with both the primary control and thermal sensor mounted in the stack as a single unit.

CADMIUM CELL PRIMARY CONTROLS

A cadmium cell primary control consists of a primary control assembly operating in conjunction with a cadmium detection cell.

The cadmium detection cell is considered the most effective sensor used to monitor the burner flame. It consists of a light-sensitive photocell, a holder, and a cord assembly (Fig. 5-65). The surface of the detection cell is coated with cadmium sulfide and overlaid with a conductive grid. Electrodes attached to the detection cell are used to transmit an electrical signal to the primary control. The variable resistance of this surface to the presence of light (i.e., the burner flame) is used to actuate the flame detection circuit. When light is present (in the form of the burner flame), the resistance of the cadmium sulfide surface to the passage of electrical current is very low. Consequently, as long as the burner flame lasts, an electrical current will pass between the

215

Fig. 5-65. Cadmium detector cell.

Courtesy Honeywell Tradeline Controls

cadmium detection cell and the primary control unit, and the burner motor *on* cycle will continue operating. If the burner flame should fail or if ignition should fail to occur, the resistance of the cadmium sulfide surface to the passage of electrical current will be very high. This will interrupt the passage of the electrical current to the primary control and will cause the latter to shut off the burner motor.

The detection cell is mounted inside the burner air tube so that it faces the burner flame (Fig. 5-66). The exact location of the detection cell is determined by the oil burner manufacturer and dictated by the design of the oil burner. In any event, the detection cell must be placed so that it views the burner flame directly.

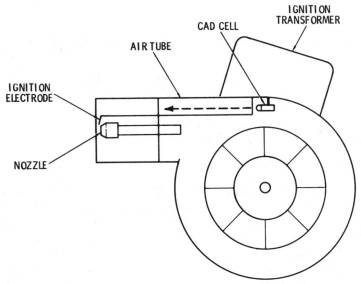

Fig. 5-66. Cadmium detection cell mounted inside burner air tube and facing burner flame.

The fact that the detection cell responds to *any* light source means that it must be located where its surface will be shielded from any form of direct or reflected external light. Moreover, the ambient temperature should be kept below 140°F because excessive temperatures can also cause the detection cell to malfunction.

Sometimes a malfunctioning oil burner will cause a heavy layer of soot to accumulate on the cell surface. The cell surface should be carefully wiped to remove the soot and restore full view of the oil flame. A damaged detection cell should be replaced.

The *type* of primary control used with a cadmium detection cell will depend upon the type of controller voltage, the type of ignition system, and the length of safety switch timing required by the installation.

The Honeywell R8184G Protectorelay primary control shown in Fig. 5-67 has a transformer included in the unit to supply 24-volt power to the control circuit. This is a low-voltage primary, and it requires a 24-volt thermostat. Other models are available that require a line voltage controller (Fig. 5-68).

The primaries illustrated in Figs. 5-67 and 5-68 are designed for use with nonrecycling constant ignition oil burners. Automatic

Courtesy Honeywell Tradeline Controls

Fig. 5-67. Honeywell model R8184G protectorelay primary control.

217

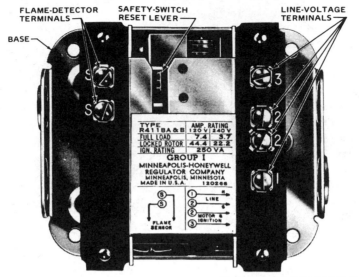

Fig. 5-68. Primary control used with a line voltage thermostat.

recycling control of an *intermittent* ignition oil burner can be obtained by using the primary shown in Fig. 5-69. The basic differences between the constant ignition and intermittent ignition systems can best be illustrated by the wiring diagrams shown in Figs. 5-70 and 5-71. An intermittent ignition system contains the same components as does a constant ignition system *plus* the following:

1. A interlock contact in the ignition circuit (T_1).
2. An ignition timer heater (T).
3. An interlock contact (T_2) in the circuit between the safety switch heater (SS) and the ignition timer heater (T).

Safety switch timing can be 15, 30, 45, or 80 seconds depending on the manufacturer and model.

STACK DETECTOR PRIMARY CONTROL

Stack-mounted oil burner primary controls employ thermal sensors to detect ignition or flame failure. A typical stack detec-

218

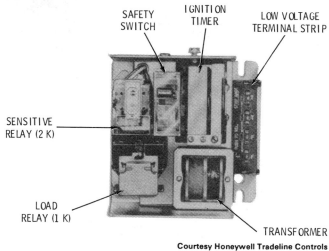

SAFETY SWITCH

IGNITION TIMER

LOW VOLTAGE TERMINAL STRIP

SENSITIVE RELAY (2 K)

LOAD RELAY (1 K)

TRANSFORMER

Courtesy Honeywell Tradeline Controls

Fig. 5-69. Honeywell R8185E protectorelay primary control.

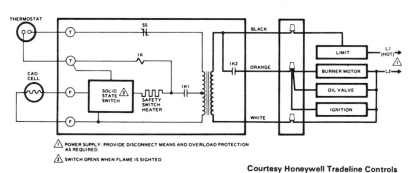

Courtesy Honeywell Tradeline Controls

Fig. 5-70. Schematic diagram of Honeywell model R8184G primary control.

219

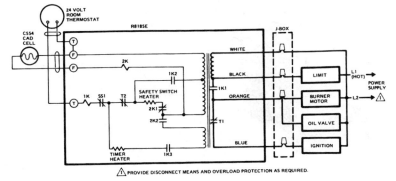

Courtesy Honeywell Tradeline Controls

Fig. 5-71. Schematic diagram of Honeywell model R8185E primary control.

tor thermal sensor consists of a bimetal element (Fig. 5-72) inserted into the stack (Fig. 5-73). The thermal sensor (combustion thermostat) is usually located on the stack where the element will be exposed to the most rapid temperature changes. The thermal sensor should always be mounted ahead of any draft regulator. If installed on an elbow, it should be mounted on the outside curve of the elbow.

The stack-mounted primary control illustrated in Fig. 5-74

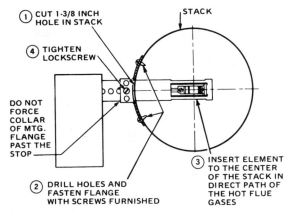

Courtesy Honeywell Tradeline Controls

Fig. 5-72. Typical stack detector thermal sensor.

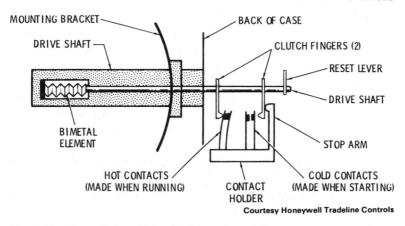

MOUNTING BRACKET

DRIVE SHAFT

BACK OF CASE

CLUTCH FINGERS (2)

RESET LEVER

DRIVE SHAFT

BIMETAL
ELEMENT

STOP ARM

HOT CONTACTS
(MADE WHEN RUNNING)

CONTACT
HOLDER

COLD CONTACTS
(MADE WHEN STARTING)

Courtesy Honeywell Tradeline Controls

Fig. 5-73. Bimetal element inserted into the stack.

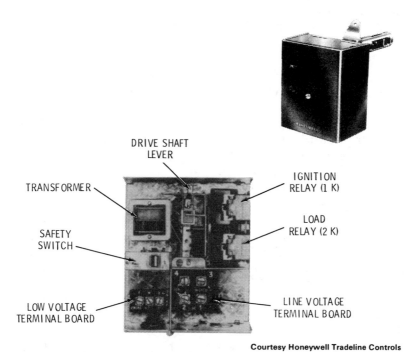

DRIVE SHAFT
LEVER

TRANSFORMER

IGNITION
RELAY (1 K)

LOAD
RELAY (2 K)

SAFETY
SWITCH

LOW VOLTAGE
TERMINAL BOARD

LINE VOLTAGE
TERMINAL BOARD

Courtesy Honeywell Tradeline Controls

Fig. 5-74. Honeywell RA117A protectorelay primary control for stack-mounted installation.

221

combines a Honeywell RA117A Protectorelay control for burner cycling control and a thermal detector for sensing temperature changes of the flue gases (as high as 1000°F maximum temperature). The safety switch shown on the center left of the unit is designed to lock out if the flame is not properly established. If the flame goes out during the burner *on* cycle, the primary control will make one attempt to restart. If the attempt is unsuccessful, the safety switch will lock out. A manual reset is then required in order to restart the burner. The primary control shown in Fig. 5-74 is used with a 2-wire or 3-wire primary controller.

A stack-mounted combination line voltage primary control and flame detector is shown in Fig. 5-75. This type of primary control is used with constant ignition oil burners and is designed for flange mounting on a stack, flue pipe, or combustion chamber door of a furnace or boiler. It must be used with a line voltage thermostat or controller.

COMBINATION PRIMARY CONTROL AND AQUASTAT

The combination primary control and aquastat is designed for use with a *constant* ignition oil burner in a hydronic heating system. The purpose of this unit is to supervise the operation of the oil burner and provide both water temperature and circulator control. A remote sensor (cadmium detection cell) is used to detect any irregularities in the oil burner flame.

Figs. 5-76 and 5-77 illustrate a number of different combination primary control and aquastat units used in hydronic heating systems. In operation, the high-limit switch (SPST) will automatically turn off the burner if the boiler overheats. The low-limit circulator switch (SPDT) is used to maintain water temperature for the domestic hot-water supply. It will also prevent the circulator from operating if the water temperature is too low (i.e., below the set point).

On the units shown in Figs. 5-76 and 5-77, a call for heat from the room thermostat pulls in relays 1K and 2K to turn on the oil burner and start heating the safety switch. Under normal operat-

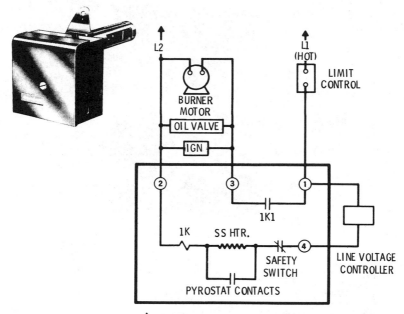

POWER SUPPLY. PROVIDE DISCONNECT MEANS
AND OVERLOAD PROTECTION AS REQUIRED.

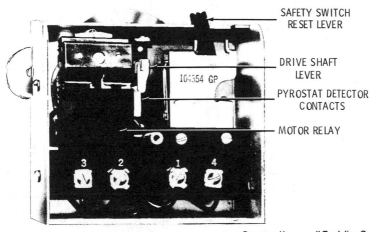

Courtesy Honeywell Tradeline Controls

Fig. 5-75. Stack-mounted combination line voltage primary control and flame detector.

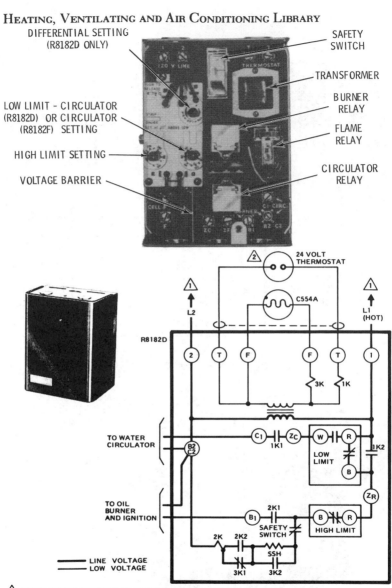

DIFFERENTIAL SETTING
(R8182D ONLY)

SAFETY
SWITCH

TRANSFORMER

BURNER
RELAY

LOW LIMIT - CIRCULATOR
(R8182D) OR CIRCULATOR
(R8182F) SETTING

FLAME
RELAY

HIGH LIMIT SETTING

CIRCULATOR
RELAY

VOLTAGE BARRIER

LINE VOLTAGE
LOW VOLTAGE

1. POWER SUPPLY, 120 VOLTS AC. PROVIDE DISCONNECT MEANS AND OVER-
LOAD PROTECTION AS REQUIRED.

2. CONTROL WIRES CAN BE RUN WITH LINE VOLTAGE WIRES IN CONDUIT
BUT THEN MUST HAVE NEC CLASS 1 INSULATION.

Courtesy Honeywell Tradeline Controls

Fig. 5-76. Model R8182H protectorelay primary control. High-limit/low-limit aquastat switching with remote-bulb sensor.

224

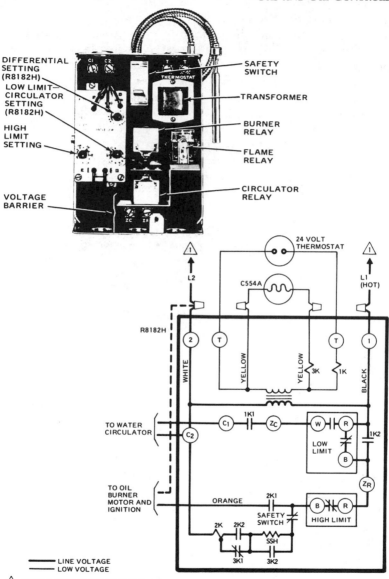

Fig. 5-77. Model R8182D protectorelay primary control. High-limit/low-limit circulator aquastat switching.

ing conditions, the burner should ignite within safety switch timing. If such is the case, the cadmium cell detects the flame and relay 3K pulls in to deenergize the safety switch heater. The oil burner then continues to operate until the call for heat is satisfied.

The circulator (pump) in the heating system operates when relay 1K pulls in *only* if the R to W contact on the aquastat control is made (Fig. 5-77). A drop in water temperature will cause the R to B (low limit) contact to be made. This acts as a call for heat, pulling in relay 2K to turn on the oil burner.

Combination primary control and aquastat units are also used on water heaters (see Chapter 4 of Vol. 3, Water Heaters and Other Appliances).

TROUBLESHOOTING THE OIL BURNER PRIMARY CONTROL

The following suggestions are offered as possible remedies to a number of different operating problems associated with oil burner primary controls. Before checking the primary control, examine the following parts of the oil burner and ignition systems:

1. Main power supply and burner motor fuse.
2. Ignition transformer.
3. Electrode gap and position.
4. Contacts between ignition transformer and electrode.

Other system components that should be checked before examining the primary control are the oil piping to the tank, the oil filter, oil pump pressure, oil nozzle, and oil supply.

Symptom and Possible Cause   *Possible Remedy*

Repeated safety shutdown

1. Slow combustion thermostat response.	1. Move combustion thermostat to better location. Adjust for more efficient burner flame. Clean surface of cad cell.

GAS AND OIL CONTROLS

Symptom and Possible Cause   *Possible Remedy*

2. Low line voltage.

3. High resistance in combustion thermostat circuit.

4. High resistance in thermostat or operating control circuit.

5. Short cycling of burner.

6. Short circuit in combustion thermostat cable.

2. Check wiring and rewire if necessary. Contact local power company.

3. Replace combustion thermostat.

4. Check circuit and correct cause.

5. Clean filters. Reset or replace differential of auxiliary controls. Repair or replace faulty auxiliary control. Set thermostat heat anticipation at higher amp value. Clean holding circuit contacts.

6. Repair cable or replace combustion thermostat.

Relay will not pull in

1. No power. Open power circuit.

2. Open thermostat circuit.

3. Combustion thermostat open.

4. Ignition timer contacts open.

5. Open circuit in relay coil.

1. Repair, replace, or reset fuses, line switch, limit control, auxiliary controls.

2. With power to relay, momentarily short thermostat terminals on relay. If burner starts, check wiring.

3. Repair or replace combustion thermostat.

4. Clean magnet.

5. Replace relay.

227

CHAPTER 6

Other Automatic
Controls

Modern heating and cooling systems contain electrical control circuits that are interconnected and interlocked with the various system components by a series of switches and relays. Most of these components, particularly the heating and cooling equipment (furnaces, boilers, compressors, condensers, etc.) and most system controls have been described in other chapters of the book. This chapter is reserved for a description of the fan and limit controls, the various electrical control circuit switches and relays, transformers, and a number of different control devices used in cooling systems.

FAN CONTROLS

A number of different devices are available for controlling the operation of fans in heating and/or cooling installations. Most of these devices function as fan safety controls; a few of them serve as fan primary controllers. The fan controls described in this chapter are:

1. Fan controller.
2. Air switch.
3. Fan relays.
4. Fan center.
5. Fan timer switch.
6. Fan safety cutoff switch.

FAN CONTROLLER

A *fan controller* (or *fan control*) is a device used to turn the system fan on and off in response to air temperature changes in the furnace plenum. This fan controller is frequently combined with a limit controller in one unit (see "Combination Fan and Limit Controller" in this chapter).

In the operation of the furnace, the burner or burner assembly starts first and heats the air, which rises through the heat exchanger to the furnace plenum. The fan controller is located in the plenum and is present for a specific cut-in temperature. When the temperature of the rising air reaches the cut-in temperature setting on the fan controller, the fan is automatically turned on and warm air is moved through the distribution ducts. After a period of time, the room thermostat will no longer call for heat and will shut off the burner. The air in the plenum then begins to cool. When the air temperature drops below the cut-in temperature of the fan controller, the fan is automatically shut off.

In most forced warm-air heating systems, the fan control is usually a line voltage device wired in the *hot* lead (L1) of the power supply to the fan motor (Fig. 6-1). If a stepdown transformer is used to provide a low-voltage control circuit for the room thermostat, the fan motor and fan controller will connect at the line side of the transformer (Fig. 6-2).

The National Warm Air Heating and Air Conditioning Association in its CAC (Constant Air Circulation) program recommends the following procedure for setting a fan control:

1. Allow the burner or burner assembly to operate for a normal running period.
2. Lower the thermostat setting so that the burner(s) will not operate during the fan control setting procedure.
3. Place a thermometer in the furnace plenum, bonnet, or in one of the farm air ducts near the furnace.
4. Set the fan adjustment lever and the fan differential adjustment (when used) to their lowest or coldest position so that the fan will run continuously.
5. Watch the thermometer until the temperature drops to about 5°F (3°C) above the temperature normally maintained in the rooms being heated.
6. As soon as the temperature on the thermometer has reached the appropriate level (see Step 5), slowly move the fan temperature adjustment lever up to a point where it will stop the fan.

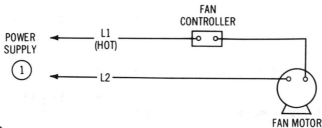

Fig. 6-1. Fan controller wired in the hot lead.

When setting a fan according to the procedure described above, the speed of the fan must be set so that the *average* temperature rise through the furnace is about 90°F (50°C).

In some installations, a drafty condition may result from the location or types of warm-air outlets. This condition can be corrected by the following procedure recommended by the National Warm Air Heating and Air Conditioning Association (see Steps 1–6 above) for setting a fan control, but with the following adjustments in the procedure:

1. Place the thermometer in front of the return air grille in the room or space most difficult to heat.
2. When the air *leaving* the room begins to feel cool, slowly move the fan adjustment up till the fan just stops.

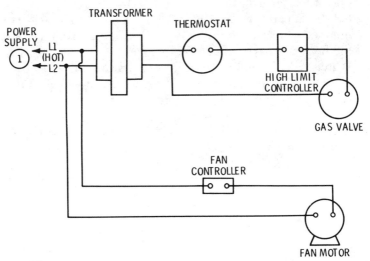

PROVIDE DISCONNECT MEANS AND OVERLOAD PROTECTION AS REQUIRED.

Courtesy Honeywell Tradeline Controls

Fig. 6-2. Typical warm-air fan control circuit.

AIR SWITCH

An *air switch* is a device designed to control a two-speed fan in response to air temperature changes in the furnace plenum (Fig. 6-3). When the temperature in the plenum rises to the set point, the air switch will change fan operation from low to high. In other words, the air will be removed from the plenum at a higher rate of speed in order to lower the temperature of the air in the furnace. Because this device is frequently used in conjunction with a fan controller or a combination fan and limit controller, it is sometimes referred to as an *upper fan control.*

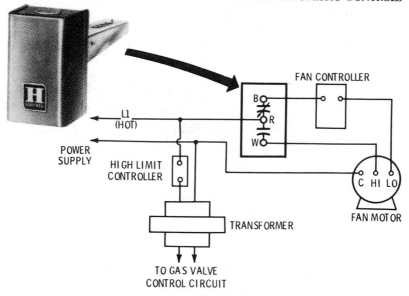

FAN CONTROLLER

B

L1
(HOT)

R

W

POWER
SUPPLY

HIGH LIMIT
CONTROLLER

C   HI LO

FAN MOTOR

TRANSFORMER

TO GAS VALVE
CONTROL CIRCUIT

TYPICAL WIRING WIRING CONNECTIONS FOR L6068 USED TO O CONTROL A 2-SPEED
FAN MOTOR. R-W MAKES, R-B BREAKS ON TEMPERATURE RISE TO SET POINT

Courtesy Honeywell Tradeline Controls

Fig. 6-3. Model L6068A air switch and wiring diagram.

Air switches are generally available with fixed temperature settings (e.g., 125°, 135°, 165°, or 200°F) or with an adjustable temperature range (125° to 165°F or 160° to 200°F). Temperature setting adjustments can be made as shown in Fig. 6-4.

An air switch can also be used to shut off the burner and turn on the fan when the temperature between the filter and heat exchanger rises to the set point. In this manner, it functions as a limit control (see "Secondary High-Limit Switch" in this chapter).

FAN RELAYS

A *fan relay* is a primary controller designed to provide 24-volt circuit control of line voltage fan motors and auxiliary circuits in heating and/or cooling systems (Figs. 6-5 and 6-6). It also pro-

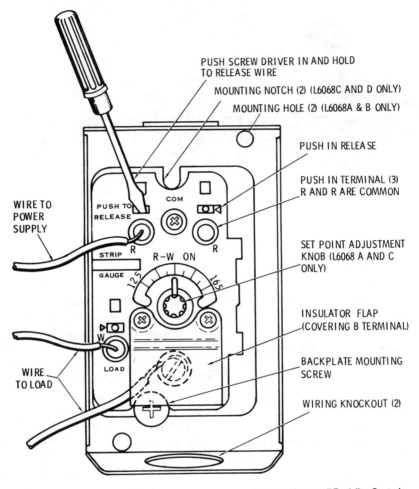

PUSH SCREW DRIVER IN AND HOLD
TO RELEASE WIRE

MOUNTING NOTCH (2) (L6068C AND D ONLY)

MOUNTING HOLE (2) (L6068A & B ONLY)

PUSH IN RELEASE

PUSH IN TERMINAL (3)
R AND R ARE COMMON

WIRE TO
POWER
SUPPLY

PUSH TO
RELEASE

COM

SET POINT ADJUSTMENT
KNOB (L6068 A AND C
ONLY)

STRIP R   R-W ON   R

GAUGE

INSULATOR FLAP
(COVERING B TERMINAL)

BACKPLATE MOUNTING
SCREW

WIRE
TO LOAD

LOAD

WIRING KNOCKOUT (2)

Fig. 6-4. Adjustment and connection points on a model **L6068** air switch.

vides manual fan operation at any time by using the manual fan
switch of the thermostat base.

Fan relays are often used with multiple-speed fans to provide
low-speed fan operation during the heating cycle and high-speed
operation during the cooling cycle.

A 24-volt room thermostat is generally used to switch the fan

234

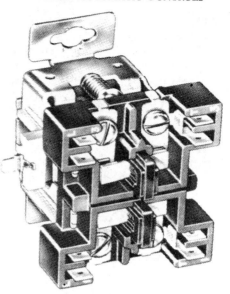

Fig. 6-5. Model R851 fan relay (contactor) provides 24-volt control of single- or two-speed fan motors up to ½ hp.

Courtesy Honeywell Tradeline Controls

Courtesy Honeywell Tradeline Controls

Fig. 6-6. Fan relay used for low-voltage control of line voltage fan motors and auxiliary circuits.

235

relay controlling the indoor fan (120- or 240-volt AC power). In cooling systems, a second fan relay must be added if switching control of the condenser fan motor is desired.

The operation of a fan relay can be checked by applying power to the coil and listening for the click of the contacts closing or by testing for electrical continuity. If the fan relay does not operate, check the voltage to the coil.

Sometimes a fan will operate at low speed but not at high speed. Check the fan relay first. If it is not defective and is receiving proper voltage, the failure of the fan to operate at high speed may be caused by loose wiring or dirty contacts. The method used for cleaning relay contacts is described elsewhere in this chapter (see "Cleaning Contactors").

A defective fan relay is also the *occasional* cause of compressor short cycling; however, this operating problem is more commonly traced to dirty air filters and other air movement restrictions on the low side of the compressor. These possible causes should be checked first.

FAN CENTER

A *fan center* is a primary controller designed to provide automatic low-voltage control of line voltage fan motors and auxiliary circuits in heating, cooling, and heating/cooling circuits. A typical wiring hookup with a two-speed fan motor, electronic air cleaner, and humidifier is illustrated in Fig. 6-7.

In addition to providing the same switching functions as a fan relay (see above), a fan center includes an integral low-voltage transformer and a terminal board for low-voltage system wiring. In air conditioning systems, a thermal delay relay is often added to a fan center to prevent short cycling of the compressor motor.

FAN TIMER SWITCH

A *fan timer switch* is a device that provides *timed* fan operation for forced warm-air furnaces and unit heaters when they are wired in parallel with a furnace or heater controller. The switch

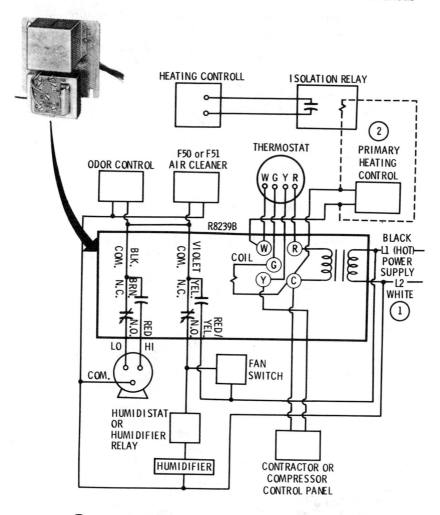

PROVIDE DISCONNECT MEANS AND OVERLOAD PROTECTION AS REQUIRED.

USE OPTIONAL HOOKUP WITH ISOLATING RELAY (DASHED LINE) IF HEATING CONTROL HAS A SEPARATE POWER SUPPLY. ISOLATION OF THE POWER SUPPLIES MAY ALSO BE ACCOMPLISHED BY USING SPECIAL THERMOSTAT SUBBASE COMBINATIONS WITH ISOLATED CIRCUITS.

Courtesy Honeywell Tradeline Controls

Fig. 6-7. Fan control center and wiring diagram.

237

operation is independent of furnace plenum temperature changes. This factor ensures fan operation and eliminates unnecessary recycling at the beginning and end of burner operation. The use of a fan switch is particularly recommended for horizontal and downflow furnaces.

The Honeywell S876 fan timer switch shown in Fig. 6-8 contains a heater-actuated, SPST bimetal snap switch that turns the fan on after the burner starts and off after the burner stops. Typical wiring connections for a timer switch are shown in Figs. 6-9 and 6-10.

FAN SAFETY CUTOFF SWITCH

A *fan safety cutoff switch* can be installed in any heating, ventilating, or air conditioning system to control fan motor operation

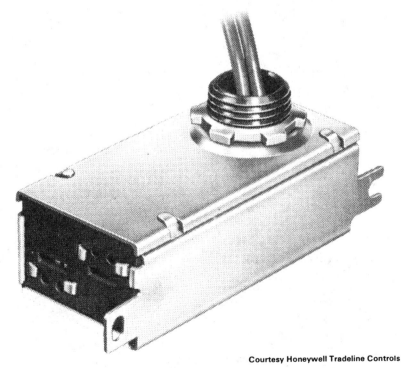

Courtesy Honeywell Tradeline Controls

Fig. 6-8. Model S876 fan timer switch.

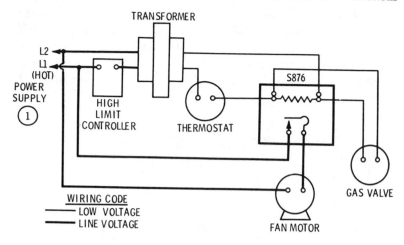

① PROVIDE DISCONNECT MEANS AND OVERLOAD PROTECTION AS REQUIRED.

Courtesy Honeywell Tradeline Controls

Fig. 6-9. Fan motor controlled by a time switch.

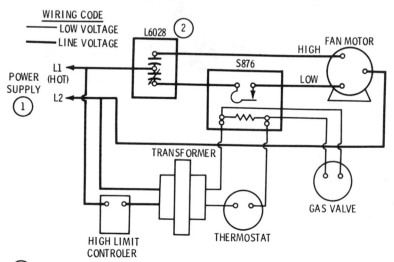

① ADD DISCONNECT MEANS AND OVERLOAD PROTECTION AS REQUIRED.

② BREAKS ONE CIRCUIT, MAKES ANOTHER ON TEMPERATURE RISE.

Courtesy Honeywell Tradeline Controls

Fig. 6-10. Two-speed fan control circuit with low fan speed controlled by a fan timer switch.

239

(Fig. 6-11). These are manually reset mercury switches that automatically break the fan motor circuit on temperature rise to the set point. The set point is established by adjusting the temperature setting screw on the unit. This switch is designed to lock out to prevent the return of fan operation until manually reset.

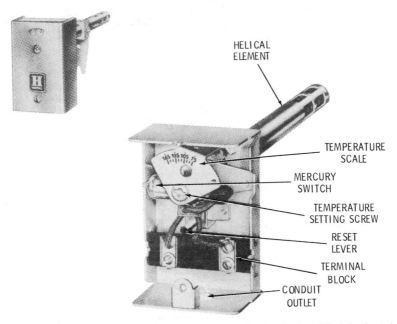

Courtesy Honeywell Tradeline Controls

Fig. 6-11. Fan safety shutoff switch.

LIMIT CONTROLS

Limit controls are also used to prevent the buildup of excessive and dangerous high temperatures in the furnace plenums. They accomplish this task by shutting off the burner when the maximum temperature setting on the control is reached and by turning it on again when the air temperature returns to normal.

The limit controls described in this chapter are:

OTHER AUTOMATIC CONTROLS

1. Limit controller.
2. Secondary high-limit switch.

Limit Controller

A *limit controller* (or *limit control*) is a device designed to provide high-limit protection for a forced warm-air furnace (Fig. 6-12). It controls the operation of a burner or burner assembly in response to air temperature changes in the furnace plenum. If the air temperature in the plenum becomes excessively high, the limit controller shuts off the burner or burner assembly until the air temperature returns to normal.

Limit controllers are available in models suitable for use in low-voltage, line voltage, and self-energizing (millivolt) systems. Typical wiring hookups for these different systems are shown in

Courtesy Honeywell Tradeline Controls

Fig. 6-12. Limit controller.

241

Figs. 6-13 and 6-14. These limit controllers have so-called univer-sal contacts in the limit switch, which makes them suitable for all voltages from millivolt to line voltage.

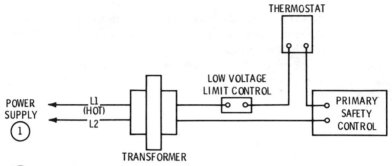

PROVIDE DISCONNECT MEANS AND OVERLOAD PROTECTION AS REQUIRED.

Courtesy Honeywell Tradeline Controls

Fig. 6-13. Limit controller in a low-voltage circuit.

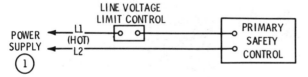

PROVIDE DISCONNECT MEANS AND OVERLOAD PROTECTION AS REQUIRED.

Courtesy Honeywell Tradeline Controls

Fig. 6-14. Limit controller in a line voltage circuit.

A limit controller contains a snap-acting switch operated by either a fluid-filled or bimetallic sensing element.

Fluid-filled sensing elements are connected to the controller by a length of capillary tube, which is available in lengths up to 72 in. The tube is filled with a temperature-sensitive liquid. A tempera-ture change causes the liquid to expand against a diaphragm that operates a snap-action switch (Fig. 6-15).

Bimetal sensing elements are available in helical, flat-blade, and spiral types (Figs. 6-16, 6-17, and 6-18). The bimetal sensing element is connected directly to the switch operator.

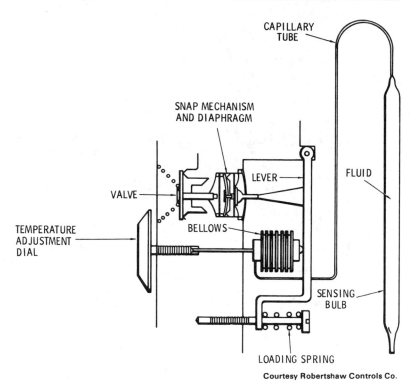

Courtesy Robertshaw Controls Co.

Fig. 6-15. Fluid-filled sensing bulb, diaphragm, and snap mechanism.

Fig. 6-16. Helical bimetal sensing element.

The temperature setting of the limit controller should be high enough not to interfere with the normal operation of the furnace, but low enough to shut off the burner or burner assembly before air temperatures in the furnace plenum reach the danger point. After the temperature in the furnace plenum has cooled and dropped below the setting on the limit controller, the limit switch closes and starts the burner or burner assembly again.

243

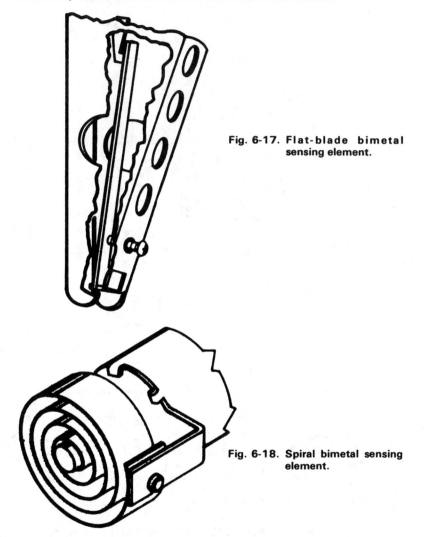

Fig. 6-17. Flat-blade bimetal
sensing element.

Fig. 6-18. Spiral bimetal sensing
element.

SECONDARY HIGH-LIMIT SWITCH

The same type of air switch used to provide two-speed control of fan motors can also serve as a *secondary high-limit switch* (or *upper-limit control*) on downflow or horizontal warm-air furnaces (Fig. 6-19).

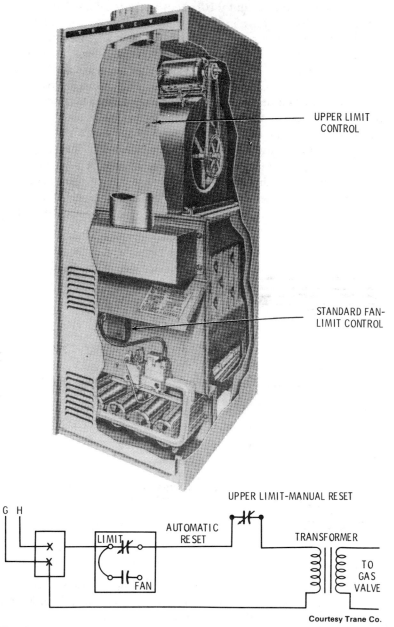

UPPER LIMIT
CONTROL

STANDARD FAN-
LIMIT CONTROL

Courtesy Trane Co.

Fig. 6-19. Secondary high-limit switch.

Downflow and horizontal furnaces are sometimes subject to a reverse air circulation condition that can result in a dangerous buildup of temperatures. This condition is usually caused by fan failure or clogged filters. The secondary high-limit switch is a safety device used as a backup system for the regular high-limit controller (see Fig. 6-3). It is particularly important to have a secondary high-limit switch on a furnace if there is a possibility that the location of the regular high-limit controller may fail to detect a fan malfunction.

The secondary high-limit switch is located between the filter and the furnace fan (Figs. 6-20 and 6-21). When the air temperature exceeds a certain setting, the switch opens and shuts off the burner or burner assembly, and turns on the fan. Secondary high-limit switches are either automatic or manually reset types. The

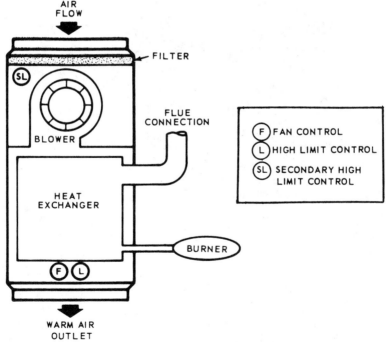

Courtesy Honeywell Tradeline Controls

Fig. 6-20. Approximate location of a secondary high-limit control on a downflow warm-air furnace.

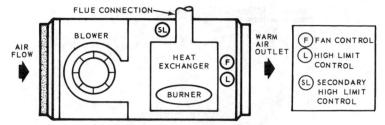

Fig. 6-21. Approximate location of the secondary high-limit control in a horizontal warm-air furnace.

automatic type is a single-pole, double-throw (SPDT) switch that turns on the burner when the limit cuts out. The manually reset switch must be reset before the burner will operate again.

When the air temperature between the filter and the heat exchanger reaches the setting (i.e., set point) in the air switch, one internal switch closes (R to W) and another opens (R to B). This operation shuts off the burner and starts the fan.

COMBINATION FAN AND LIMIT CONTROLLER

A *combination fan and limit controller* combines the functions of a fan controller and a limit controller in a single unit. One sensing element (either bimetal or fluid filled) is used for both controls (Figs. 6-22 and 6-23).

Combination controllers are wired in much the same way as the individual controls. Examples of some typical wiring hook-ups are shown in Figs. 6-24, 6-25, 6-26, and 6-27. These combined controls can be used in line voltage, low-voltage, or self-energizing millivolt systems.

The combination fan and limit controller should be located where it will provide the best possible operating characteristics.

Limit switch terminals are on the left side of the control; fan switch terminals on the right. This arrangement is true of combination fan and limit controllers as well as single-purpose types (Figs. 6-28, 6-29, 6-30, and 6-31).

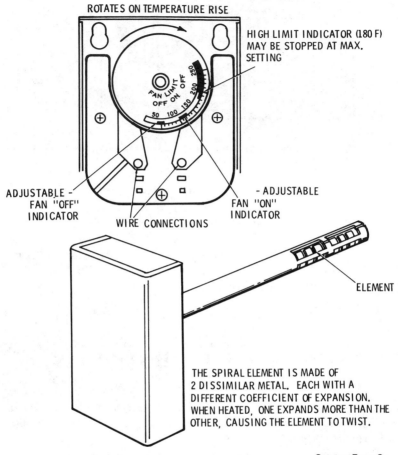

ROTATES ON TEMPERATURE RISE

HIGH LIMIT INDICATOR (180 F)
MAY BE STOPPED AT MAX.
SETTING

ADJUSTABLE -
FAN "OFF"
INDICATOR

- ADJUSTABLE
FAN "ON"
INDICATOR

WIRE CONNECTIONS

ELEMENT

THE SPIRAL ELEMENT IS MADE OF
2 DISSIMILAR METAL, EACH WITH A
DIFFERENT COEFFICIENT OF EXPANSION.
WHEN HEATED, ONE EXPANDS MORE THAN THE
OTHER, CAUSING THE ELEMENT TO TWIST.

Courtesy Trane Co.

Fig. 6-22. Typical combination fan and limit control.

On the controllers shown in Figs. 6-28, 6-29, 6-30, and 6-31, temperature settings can be changed by moving the temperature setting pointers. Temperature settings are interlocked to prevent the limit *off* from being set as low as the fan *on* pointer. Sometimes the limit *off* setting will be factory locked to a specific setting. If this is the case, do *not* attempt to adjust this setting. A safety interlock prevents the fan *on* pointer being set as high as the limit *off* pointer.

248

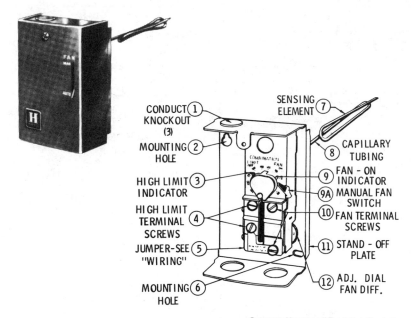

CONDUCT ① KNOCKOUT (3)

SENSING ⑦ ELEMENT

MOUNTING ② HOLE

CAPILLARY ⑧ TUBING

HIGH LIMIT ③ INDICATOR

FAN - ON ⑨ INDICATOR

⑨A MANUAL FAN SWITCH

HIGH LIMIT TERMINAL ④ SCREWS

⑩ FAN TERMINAL SCREWS

JUMPER-SEE ⑤ "WIRING"

⑪ STAND - OFF PLATE

MOUNTING ⑥ HOLE

⑫ ADJ. DIAL FAN DIFF.

Courtesy Honeywell Tradeline Controls

Fig. 6-23. Model L4017 combination fan and limit controller.

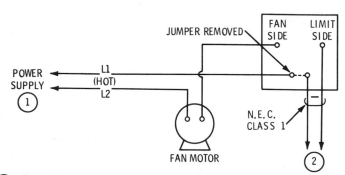

JUMPER REMOVED

FAN SIDE   LIMIT SIDE

POWER SUPPLY ①

L1 (HOT)

L2

N.E.C. CLASS 1

FAN MOTOR

②

① ADD DISCONNECTING MEANS AND OVERLOAD PROTECTION AS REQUIRED.

② TO CONTROLLED LOW-VOLTAGE EQUIPMENT.

Courtesy Honeywell Tradeline Controls

Fig. 6-24. Combination fan and limit controller wiring diagram with limit controller in the low-voltage circuit.

249

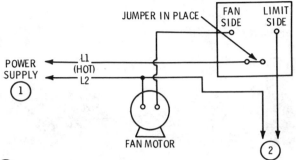

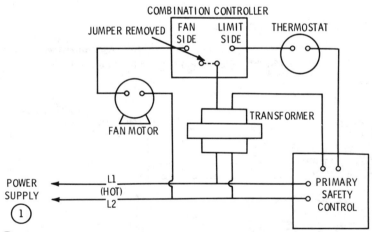

(1) ADD DISCONNECTING MEANS AND OVERLOAD PROTECTION AS REQUIRED.

(2) TO CONTROLLED LOW-VOLTAGE EQUIPMENT.

Courtesy Honeywell Tradeline Controls

Fig. 6-25. Combination fan and limit controller wiring diagram with limit controller in the line voltage circuit.

(1) PROVIDE DISCONNECT MEANS AND OVERLOAD PROTECTION AS REQUIRED.

Courtesy Honeywell Tradeline Controls

Fig. 6-26. Combination fan and limit controller in warm-air heating control circuit with low-voltage limit.

Some combination fan and limit controllers are equipped with a manual summer fan switch to provide continuous fan operation for summer ventilation. To operate the blower during summer weather *without* the burner in operation, move the switch lever

250

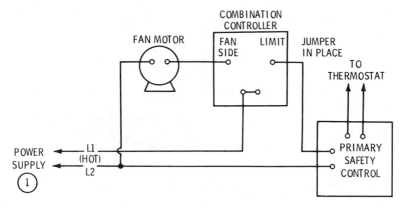

Fig. 6-27. Combination fan and limit controller in warm-air heating control circuit with line voltage limit.

on the manual summer fan switch from *auto* to *on* position. This will provide continuous fan operation until the lever is moved to the *auto* position.

As shown in Fig. 6-32, the metal tabs on the temperature dial of the Robertshaw combination fan and limit controller can be bent back to serve as stops for the fan and limit setting pointers. The dial is held to prevent rotation, and the pointer is pressed in and rotates to the proper temperature setting. The first tab *above* the pointer is bent back 90° toward the sensing element.

A certain amount of caution must be exercised when installing a combination fan and limit controller. These controllers can be located either in the furnace plenum or mounted directly on a panel, *but* the sensing element *must* be located in the path of free-flowing air. Never mount the control near the cool air intake, and keep the sensing element away from any hot metal surfaces. Furthermore, make sure you mount the control where it is accessible for making temperature adjustments.

SWITCHING RELAYS

A *switching relay* is a device used to increase system switching capabilities, to isolate electrical circuits, and to provide electrical

251

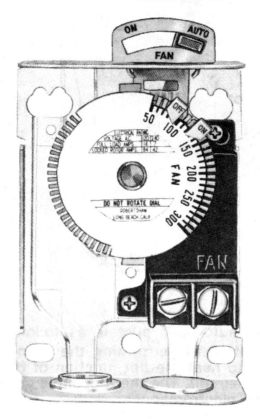

Fig. 6-28. Fan controller with adjustable fan differential and summer fan switch.

interlocks in a heating and cooling system. These devices are especially useful in systems where the heating and cooling equipment have separate supplies.

A typical switching relay contains an integral transformer and a magnetic relay with contacts designed to make or break an electrical circuit. These contacts will either be normally open or normally closed, depending on the design of the relay and its purpose in the heating and/or cooling system.

The 24-volt switching relay illustrated by the wiring diagram in Fig. 6-33 is designed to control 115-volt and 24-volt or millivolt

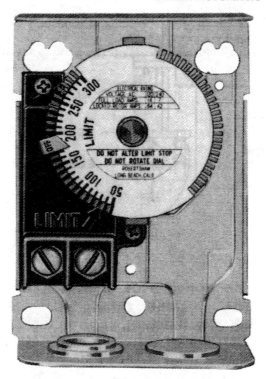

Fig. 6-29. Limit controller without summer fan switch.

circuits. It incorporates a 20 VA 115V/24V transformer and a 24V/60 Hz .2 amp magnetic relay with two normally open contacts. One set of the relay contacts is line voltage rated for the switching of a circulator or other device. The other set of contacts is used for the switching of a self-energized (millivolt) or 24-volt circuit. A terminal board is located on top of the relay cover with screw terminals for connecting a thermostat (terminals T1 and T2) and a gas valve or oil burner control (terminals X1 and X2).

The switching relay illustrated in Fig. 6-33 is shown as used in a gas-fired, forced hot-water heating system. In operation, a thermostat or some other switching accessory (e.g., an aquastat or zone valve) connected to the T1 and T2 terminals starts the

253

circulator and boiler simultaneously by energizing and closing the two-pole, normally open relay. When the relay is activated by the thermostat, it closes a circuit from the integral transformer to a magnetic switch, which causes high voltage to be fed to the circulator to start the system pump. At the same time, the second pole of the relay switches 24-volt power to the gas valve or oil burner control, thereby starting the boiler.

A switching relay can also be used as a pilot duty relay to power a contactor and control a crankcase heater for the com-

Courtesy Robertshaw Controls Co.

Fig. 6-30. Combination fan and limit controller with nonadjustable fan differential and summer fan switch.

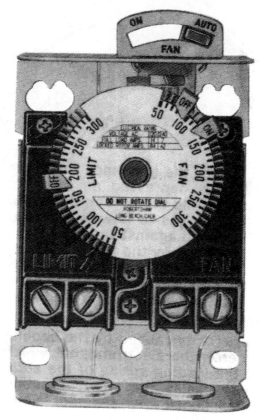

Courtesy Robertshaw Controls Co.

Fig. 6-31. Combination fan and limit controller with adjustable fan differential and summer switch.

pressor motor. This type of switching relay has normally closed contacts that complete an electrical circuit to the heater until the thermostat calls for cooling. When this occurs, the relay switches to break the heater circuit and power the compressor motor circuit.

In installations where a cooling system has been added to a self-energizing (millivolt) heating system, a switching relay may be used to isolate the cooling and heating power supplies (Fig. 6-34). When the room thermostat calls for heat, an isolating relay is used to switch the heating equipment directly.

255

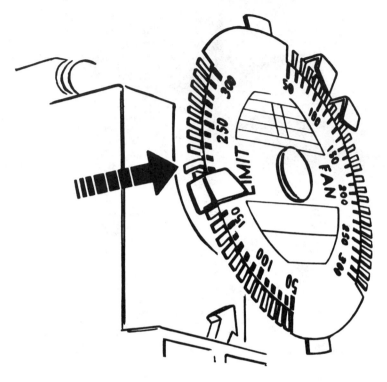

Courtesy Robertshaw Controls Co.

Fig. 6-32. Temperature-dial metal tabs bent to form a stop.

Heavy-duty switching relays are used for control of high-current loads such as cooling compressors or electric heating where sudden high-current demands are not unusual. The relay is wired to break both sides of the circuit with DPST switching (Fig. 6-35).

IMPEDANCE RELAYS

An *impedance relay* (Fig. 6-36) is used to provide lockout and remote reset in refrigeration, air conditioning, and other systems. A typical wiring diagram for a low-voltage impedance relay is shown in Fig. 6-37. A low-voltage relay requires the use of a

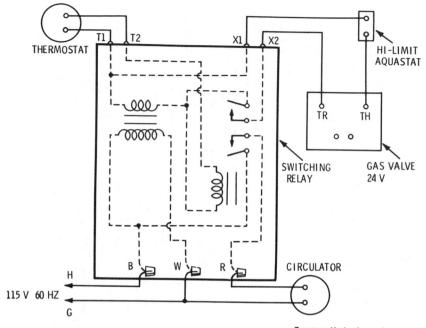

Fig. 6-33. Switching relay used with 24-volt power supply.

transformer with an open circuit secondary of between 24 and 27 volts AC.

As shown in Fig. 6-37, one pair of contacts is normally open and the other pair is normally closed. During normal operation, the normally closed contacts of the pressure controls (i.e., the low-pressure and high-pressure cutout switches) and the motor overloads short out the impedance relay coil so that the compressor contactor pulls in. If one of the pressure controls or overloads opens, the impedance relay coil is energized in series with the contactor coil and most of the available voltage is used by the high impedance of the relay coil. Because insufficient voltage remains to operate the contactor coil, the contactor drops out and compressor operation stops.

As the impedance relay pulls in, its normally closed contacts open to keep the contactor out, even though the pressure control or overload (automatic reset) remakes. The system can be reset

257

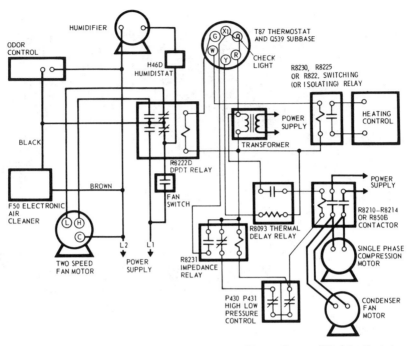

Fig. 6-34. Switching relay used to isolate cooling and heating power supply.

by breaking the contactor circuit to allow the impedance relay to drop out. In most systems, this is accomplished by moving the thermostat subbase switch to *off* and back to *cool* again.

HEATING RELAYS/TIME-DELAY RELAYS

In air conditioning installations, a *time-delay relay* is often installed in the control circuit to provide protection for the compressor and contactor. This device is activated or deactivated by the room thermostat. In operation, it causes a time delay between turning down the thermostat setting and the start of the compressor unit of approximately 20 to 45 seconds (depending on the manufacturer). On the shutoff cycle at the thermostat, the same

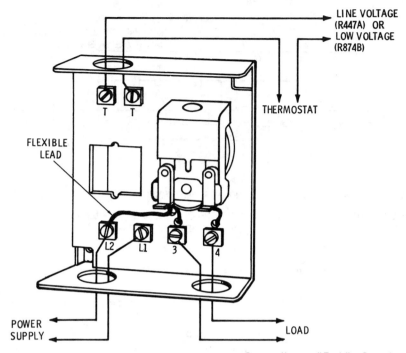

LINE VOLTAGE
(R447A) OR
LOW VOLTAGE
(R874B)

THERMOSTAT

FLEXIBLE
LEAD

L2   L1   3   4

POWER
SUPPLY

LOAD

Courtesy Honeywell Tradeline Controls

Fig. 6-35. Heavy-duty switching relay.

Courtesy Honeywell Tradeline Controls

Fig. 6-36. Impedance relay.

259

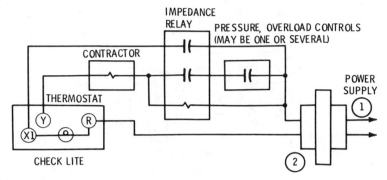

①PROVIDE DISCONNECT MEANS AND OVERLOAD PROTECTION AS REQUIRED.

②TRANSFORMER FOR LOW VOLTAGE WIRING OR R8231 RELAYS. R4231 ARE USED IN LINE VOLTAGE SYSTEMS (OMIT TRANSFORMER.

Courtesy Honeywell Tradeline Controls

Fig. 6-37. Impedance relay in a low-voltage circuit.

delay occurs. This control prevents rapid short cycles from occurring.

A time-delay relay is essentially a switching relay that contains a small heater wound around a bimetal element. The relay heater is energized through the cooling contacts of the thermostat and heats the bimetal element. After a time delay of approximately 20 to 45 seconds, the heated bimetal element bends to provide the switching force. In other words, it closes a set of snap-action contacts, thereby completing the circuit through the starter or contactor coil required to start the compressor. The relay may be wired to provide a delay in breaking the circuit after the thermostat is satisfied.

The time-delay relay shown in Fig. 6-38 is used with a two-wire low-voltage thermostat (and remote-mounted thermostat) to provide a time delay between stages for electric heaters in furnace ducts. Each relay can control up to 6000 'watts at 240 volts AC. One relay is required per stage or time increment. This particular model provides a delay of approximately 75 seconds between the *on* cycle of consecutive stages. The sequencing of heating loads is permitted by auxiliary contacts.

Because of the dual nature of their function, time-delay relays are also referred to as *time-delay switches, thermal switching*

Courtesy Honeywell Tradeline Controls

Fig. 6-38. Thermal switching relay used for control of electric furnaces or electric duct heaters.

relays, thermal time-delay relays, thermal relays, and *heating relays.* Other examples of time-delay/heating relays are shown in Figs. 6-39 and 6-40.

POTENTIAL RELAY

In air conditioning installations, the *potential (start) relay* serves as a switch to disconnect the starting capacitors when the compressor motor has overcome the initial starting torque.

An important fact to remember is that each potential relay is specifically designed for the compressor to which it is attached. Should this relay fail or become erratic in its operation, no attempt should be made to repair it. The relay must be replaced with an *identical* component.

PRESSURE SWITCHES

A *pressure switch* is a safety device designed to stop or start heating or air conditioning equipment in response to gas- or air-

261

Fig. 6-39. Electric heating relay used with a two-wire thermostat for control of electric boilers, duct heaters, fan coils, or electric furnaces.

pressure changes. These switches are used in either positive-pressure or differential-pressure systems.

Gas-pressure switches used in the control of gas-fired furnaces and boilers are described in the section "Pressure Switches" in Chapter 5 (Gas and Oil Controls). Pressure switches used as refrigerant controllers in a cooling system are described in this chapter (see below: "Low-Pressure Cutout Switches" and "High-Pressure Cutout Switches").

The National Electrical Code requires that a duct heater be interlocked with the system fan so that the heater cannot be energized unless the fan is also energized. This can be accomplished by using either a fan interlock relay or a built-in air-pressure switch.

An air-pressure switch designed to provide fan interlock control consists of an internal diaphragm that is actuated by positive air pressure. The switch sensor is mounted so that it extends into the air stream (Fig. 6-41). Movement of the diaphragm closes an electrical switch, which permits the duct heater to turn on. When there is no air flow or low air flow, the switch will open and turn off the duct heater.

SAIL SWITCHES

A *sail switch* consists of a steel or polyester film sail mounted on a switching device (Fig. 6-42). The combined unit is mounted so that the sail is located in an air duct. When the air velocity increases, the switch makes an electrical circuit. Fig. 6-43 illustrates the location of the sail switch in a gas control circuit.

Sail switches are used in forced warm-air heating systems, in air conditioning systems, and with gas-fired unit heaters. Some

Courtesy Singer Controls Co. of America

Fig. 6-40. Time-delay relay used in electric baseboard heating.

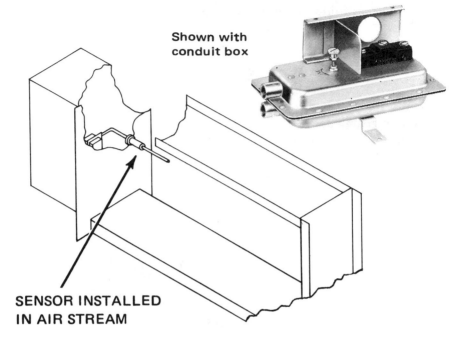

Shown with
conduit box

SENSOR INSTALLED
IN AIR STREAM

Courtesy Vulcan Radiator Co.

Fig. 6-41. Air-pressure switch.

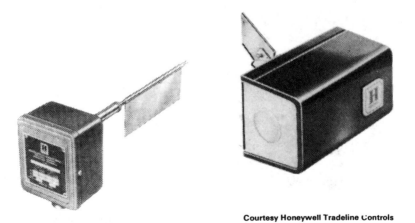

Courtesy Honeywell Tradeline Controls

Fig. 6-42. Sail switches.

264

sail switches are designed to provide on-off control of electronic air cleaners, odor control systems, humidifiers, and other equipment that is energized when the fan is operating. In these applications, the sail switch completes a power circuit to auxiliary equipment, which can be wired independently of the blower motor. Sail switches are also used in electric heating systems to provide minimum air flow.

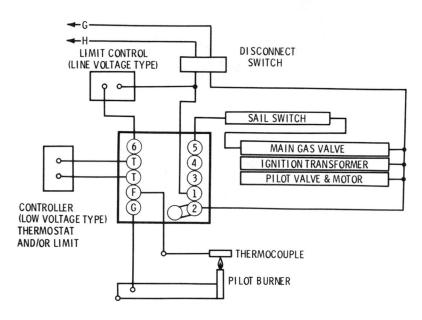

Fig. 6-43. Location of the sail switch in a gas control circuit.

OTHER SWITCHES AND RELAYS

Other switches and relays used in heating and/or cooling systems include the following:

1. Balancing relays.
2. Manual switches.
3. Auxiliary switches.

A *balancing relay* (Fig. 6-44) is used with an electric motor that does not have an integral balancing relay. The relay is mounted separately from the motor so that vibrations will not affect it.

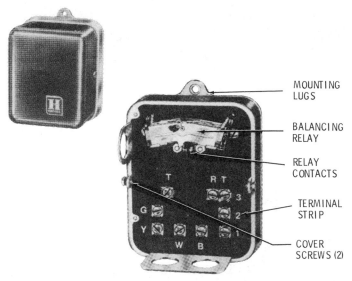

MOUNTING
LUGS

BALANCING
RELAY

RELAY
CONTACTS

TERMINAL
STRIP

COVER
SCREWS (2)

Courtesy Honeywell Tradeline Controls

Fig. 6-44. Balancing relay.

A *manual switch* is used to manually perform one or more operations in a heating and/or cooling installation. These switches are generally two-position types (on-off), although multiple-position switches are also available. An example of the latter would be the heat-off-cool switch on a heating and cooling thermostat. Examples of some special switching hookups in which manual switches are used are shown in Figs. 6-45, 6-46, and 6-47. In each of these examples, a relay could have been used instead of the manual switch.

An *auxiliary switch* (Fig. 6-48) is used in conjunction with an electric motor to provide control of auxiliary equipment. This control functions as a direct extension of motor operation. The auxiliary switch may be an integral part of the motor or an external unit fitted to the motor and adjusted to open or close at the

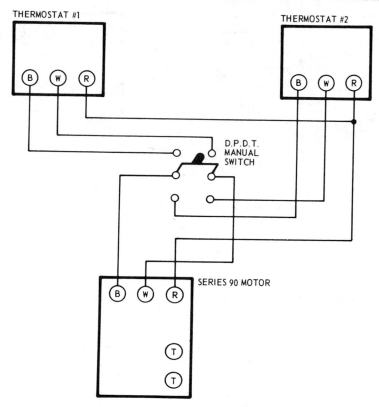

THERMOSTAT #1

THERMOSTAT #2

D.P.D.T.
MANUAL
SWITCH

SERIES 90 MOTOR

Fig. 6-45. Manual switch (DPDT) used to transfer control of an electric
motor from one thermostat to another.

desired point in the motor stroke (Fig. 6-49). As shown in Fig.
6-50, the internal auxiliary switches are in a single-pole double-
throw (SPDT) configuration.

SEQUENCE CONTROLLERS

A *sequence controller* (also referred to as a *sequencer* or *step
controller*) is a device used to operate two or more electric
switches in predetermined sequence. This function is accom-

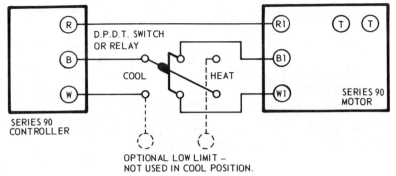

SERIES 90
CONTROLLER

D.P.D.T. SWITCH
OR RELAY

COOL   HEAT

OPTIONAL LOW LIMIT –
NOT USED IN COOL POSITION.

SERIES 90
MOTOR

Courtesy Honeywell Tradeline Controls

Fig. 6-46. Manual switch (DPDT) used for reversing control so that the same thermostat can be used for both heating and cooling.

plished by means of a proportional electric or pneumatic operator.

Sequence controllers are most commonly used to provide sequenced starting of a number of electric heating elements or compressor motors. This prevents the massive drawing on the current that simultaneous starting would cause.

Sequence controllers are manufactured in a number of different sizes and capacities. The Honeywell S984 step controller, shown in Fig. 6-51, provides up to ten adjustable switches. It can also be used to control at least two other step controllers when greater switching capacity is required.

Each switch in the Honeywell step controller is operated by a cam mounted on the main shaft (Fig. 6-52). Adjustments of the step controller can be made by setting each of the switches to make or break a circuit at the correct time or angle in the stroke. The procedure for setting the switches in the Honeywell step controller are:

1. Loosen the setscrew so that the cam assembly will turn freely on the motor shaft.
2. Run the step controller motor until the desired switch "make" point is reached. This will be determined by the time or degrees of cam shaft rotation.
3. Turn the cam until the switch just makes and tighten the setscrew.

4. Run the step controller *back* to the desired switch "break" point.
5. Loosen the lockscrew and turn the differential cam so the switch will "break" at this point.

The instructions for adjusting the Honeywell step controller recommend setting the "make" points of *all* the switches first and then setting all the "break" points. This procedure avoids unnecessary cycling of stages.

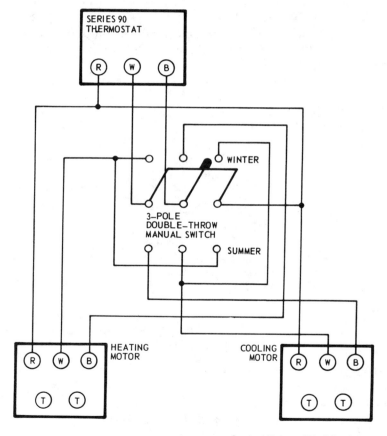

Courtesy Honeywell Tradeline Controls

Fig. 6-47. Manual switch (three-pole double-throw) used to transfer thermostat control to one or the other of two motors.

269

The sequence of controller, shown in Fig. 6-53, provides time-delay switching of up to eight electric heater banks and a fan or pump. When the thermostat calls for heat, it starts the low-voltage sequence motor, which operates the rotating cams. The first cam locks in the motor circuit so that if the thermostat should break immediately, the motor will continue running, rotating all cams back into the starting position.

The second cam starts the fan. The remaining cams switch on the heater banks with a specific time delay between the energizing of each bank. The motor will stop running after 180° rotation as long as the thermostat calls for heat. The fan and all heater banks will continue to be energized. When the thermostat is satisfied, the motor starts again and switches off the heater banks at the predetermined time interval. The fan is deenergized, and the motor stops when the cams are back in the starting position.

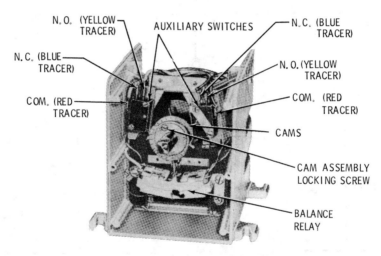

Courtesy Honeywell Tradeline Controls

Fig. 6-48. Electric motor with an integral auxiliary switch.

CONTACTORS

A *contactor* is essentially a switching relay device that functions as a primary control in a cooling system. Its operation is based on either magnetic or mercury-to-mercury contact action.

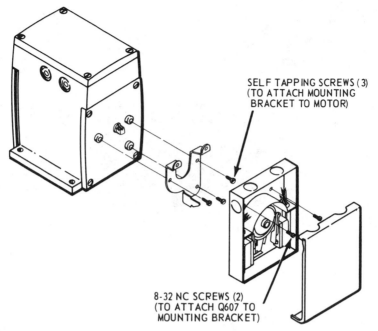

SELF TAPPING SCREWS (3)
(TO ATTACH MOUNTING
BRACKET TO MOTOR)

8-32 NC SCREWS (2)
(TO ATTACH Q607 TO
MOUNTING BRACKET)

Courtesy Honeywell Tradeline Controls

Fig. 6-49. Externally mounted auxiliary switch.

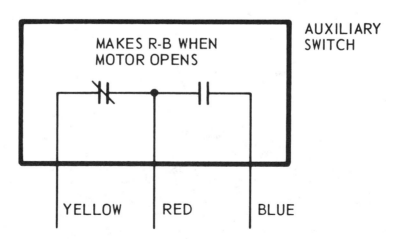

MAKES R-B WHEN
MOTOR OPENS

AUXILIARY
SWITCH

YELLOW   RED   BLUE

Courtesy Honeywell Tradeline Controls

**Fig. 6-50. Internal auxiliary switches in a single-pole double-throw (SPDT)
configuration.**

271

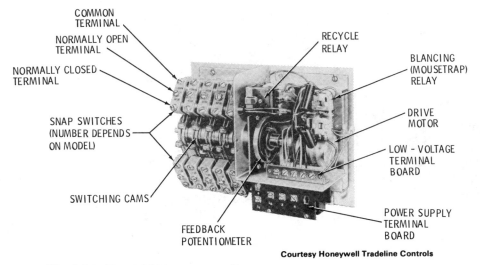

COMMON
TERMINAL

NORMALLY OPEN
TERMINAL

NORMALLY CLOSED
TERMINAL

SNAP SWITCHES
(NUMBER DEPENDS
ON MODEL)

SWITCHING CAMS

FEEDBACK
POTENTIOMETER

RECYCLE
RELAY

BLANCING
(MOUSETRAP)
RELAY

DRIVE
MOTOR

LOW - VOLTAGE
TERMINAL
BOARD

POWER SUPPLY
TERMINAL
BOARD

Courtesy Honeywell Tradeline Controls

Fig. 6-51. Model S984 step controller.

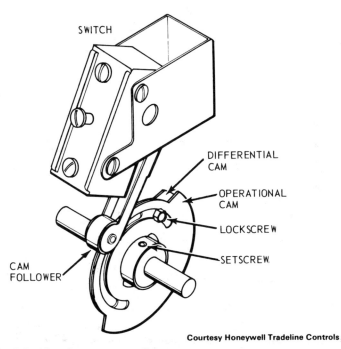

SWITCH

DIFFERENTIAL
CAM

OPERATIONAL
CAM

LOCKSCREW

SETSCREW

CAM
FOLLOWER

Courtesy Honeywell Tradeline Controls

Fig. 6-52. Adjusting the step controller.

Fig. 6-53. Model 38 sequence control.

A magnetic contactor (Fig. 6-54) is similar in design and operating principle to a relay, but is larger in size. In a mercury contactor (Fig. 6-55), the contacts are made and broken between two pools of mercury, separated by a ceramic insulator. Mercury-to-mercury contact action results in a quieter operation than magnetic contactors can provide, but a mercury contactor has the disadvantage of being position sensitive. It *must* be mounted in an upright position.

A contactor is used for applications requiring heavy current, high voltage, or a large number of poles; in other words, applications where the capacity of a relay would be inadequate.

In a cooling system, an electric-driven compressor motor may be cycled by a thermostat (either a low-voltage or line voltage type) or a low-pressure control. Because these controllers are usually unable to handle the high current drawn by the compressor motor, a contactor is installed between the thermostat or pressure control and the motor where it functions as a primary control. This contactor is the electrical contact switch to which the main electric power is supplied. When activated by the room thermostat, the contactor causes the compressor and condenser

273

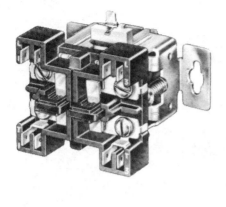

Courtesy Honeywell Tradeline Controls

Fig. 6-54. Magnetic contactors.

Courtesy Honeywell Tradeline Controls

Fig. 6-55. Mercury contactors.

blower motor in an air conditioning system to begin operating. Contactors are used to control all but the smallest compressor motors.

The size of the contactor selected for use in a cooling system will depend on such variables as the size and type of compressor motor and the auxiliary loads.

Both single-phase and three-phase motors are used in compressors. As shown in Table 6-1, the motor current (expressed in

Table 6-1. Compressor Size and Motor Current Ratings

Single-Phase Motor		Three-Phase Motor	
Compressor Size (Tons)	Motor Current (Amperes)	Compressor Size (Tons)	Motor Current (Amperes)
2	18	3	18
3	25–30	4	25–30
4	30–40	5	30–40
5	35–50	7½	35–50

Courtesy Honeywell Tradeline Controls

amperes) will vary according to the size of the compressor. Contactors are rated in amperes and should be selected to match the rating of the compressor motor. The number of pole contacts is also an important consideration in selecting a contactor. A one- or two-pole contactor is required for a single-phase compressor motor; a three-pole contactor for a three-phase motor. Auxiliary poles may be used for interlock switching, fan loads, or crankcase heaters (Fig. 6-56).

TROUBLESHOOTING CONTACTORS

Always check the fuse box (or circuit breakers) first to make certain the fuses are good and the switch is in the *on* position. If this is not the source of the problem, then set the thermostat and base to cooling position and check for continuity across contacts. If the circuit is open, repair or replace the thermostat.

If the system fails to start and the contactor is open and buzzing, check the voltage at the contactor coil (Fig. 6-57). Normal

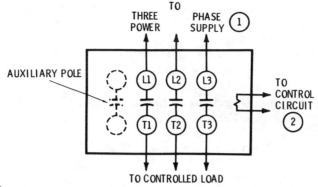

① ADD DISCONNECT MEANS AND OVERLOAD PROTECTION AS REQUIRED.

② INCLUDES LINE OR LOW VOLTAGE POWER SUPPLY AND THERMOSTAT.

Fig. 6-56. Typical hookup of a three-pole contactor.

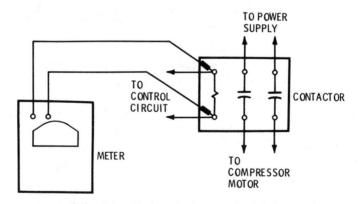

Fig. 6-57. Checking voltage to the contactor coil.

voltage will be within plus or minus 10 percent of the rated coil voltage. Subnormal voltage may be caused by an undersized transformer or low voltage at the supply side of the transformer. Thermostat wiring that is too long can also cause a subnormal voltage reading. A normal voltage reading under these circumstances (i.e., an open, buzzing contactor) is generally caused by a tight or fouled contactor armature. The armature should be cleaned or the contactor replaced.

If the system compressor will not start and the contactor is open but *not* buzzing, the contactor coil may not be powered. A voltage and continuity check must be run to locate the problem. This is more complicated procedure than the one described for an open buzzing contactor.

The voltage to the contactor coil should be checked first (Fig. 6-57). If the voltage is normal (within plus or minus 10 percent of the rated voltage), the contactor should be replaced. If a zero voltage reading is obtained, check the voltage at the transformer secondary (Fig. 6-58). If a zero voltage reading is obtained, check the line voltage side of the transformer (Fig. 6-59). A normal supply voltage reading here indicates the transformer is defective and should be replaced. A zero voltage reading, on the other hand, indicates a problem with the power supply. Check the fuses, circuit breakers, line disconnect switch, and the power at the service entrance.

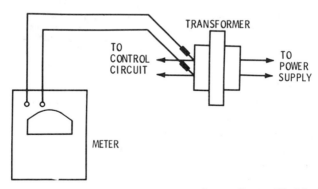

Courtesy Honeywell Tradeline Controls

Fig. 6-58. Checking voltage at the transformer secondary.

If the voltage checked at the transformer secondary (as shown in Fig. 6-58) is *normal,* the control circuit to the contactor coil is open. With a continuity checker, jump across the terminals of the following controls:

1. Low-pressure switch.
2. High-pressure switch.

277

3. Room thermostat.
4. Heating/cooling interlock switch.
5. Time-delay switch.
6. Lockout relay.

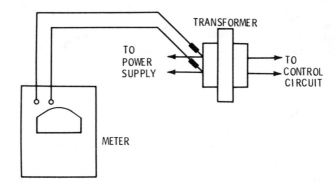

Fig. 6-59. Checking voltage at the line side of the transformer.

If the system starts with a particular control out of the circuit, the control is probably defective and needs to be replaced. If none of these controls is defective, check the circuits for broken wires or loose connections. An open internal thermostat or over-load relay switch in the compressor is another possible cause.

If the compressor hums but will not start, and the contactor remains closed, the problem may be in the motor-starting circuit. Check the motor circuit wiring for loose or broken wires. If there is no problem with the wiring, check the starting capacitors and motor starter. A defective motor is also a possibility. If the motor-starting circuit is not defective, check for abnormal system pressures. Another possible cause is a tight, stuck, or burned-out compressor.

If the contactor is closed and the compressor motor neither starts nor hums, check the continuity of the overload switch and the open compressor motor windings. Check also for broken or loose wiring. If none of these is the cause, the contacts in the contactor are probably burned. Replace the contacts.

CLEANING CONTACTORS

Sometimes a contactor will fail to operate because a layer of dust and lint has accumulated on the electrical contacts. This dust and lint can be removed by placing a file card between the contacts, closing the contacts against the card, and sliding the card back and forth. This will usually clean the contacts. Do not use abrasive material to clean the contacts because this will scratch and possibly ruin the surface.

REPLACING CONTACTORS

A contactor contains a stationary contact that operates in conjunction with a contact bar (Fig. 6-60). The contacts should be replaced if they show evidence of uneven wear.

Always compare the ratings of the replacement contactor with the old one. They should at least be equal in rating. To be on the safe side, it is better to overrate a contactor and replace the old one with a contactor of slightly higher rating.

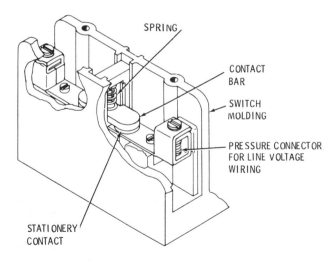

Fig. 6-60. Principal components of a contactor.

279

When replacing a contactor, make certain the terminal connections of the new one fit the installation. The mounting holes and dimensions should also be compatible.

Some manufacturers provide replacement coils for their contactors. *Never* replace a contactor coil until you have located and corrected the cause of the original coil failure. If you fail to do this, the replacement coil will probably burn out, too.

Always disconnect the power supply before attempting to remove a contactor. Be sure to tag the wiring connections to the contact terminals as soon as you have removed the contactor. This will minimize the possibility of confusion when the replacement is installed.

MOTOR STARTER

When an electric-driven compressor motor stalls or is overloaded, it draws current many times its full load rating. If the condition lasts any length of time, the motor windings overheat and a fire may start in the insulation. This will result in very expensive damage to the motor. One method of guarding against the occurrence of an overload condition is by installing a motor starter in the control circuit (Fig. 6-61).

A *motor starter* consists of a contactor plus one or more overload relays. Each overload relay consists of a bimetal contact in series with the motor contact coil and a heater in series with the compressor motor. The overload relay (or relays) disconnects the motor from the power supply when the motor temperature and/or the current drawn by the motor become excessive.

OVERLOAD RELAY HEATER

As shown in Fig. 6-61, an overload relay consists of a bimetal contact in series with the motor coil, and a heater in series with the compressor motor. An *overload relay heater* is a small electric heating device designed to work in conjunction with the bimetal contact. When the compressor motor becomes overloaded or stalled, the heavy continuous current through the heater causes

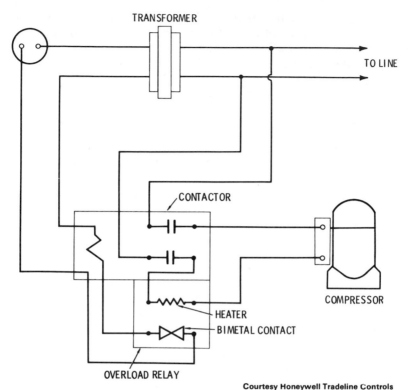

Courtesy Honeywell Tradeline Controls

Fig. 6-61. Motor starter in the control circuit.

the bimetal contact to bend until it opens the motor-starter coil circuit. This action stops the flow of current through the motor starter and results, in turn, in the opening of the load contacts to stop the flow of current to the compressor motor.

An overload relay heater must be accurately sized for the installation. Generally the manufacturer of the cooling equipment will provide instructions for sizing overload relay heaters.

INHERENT PROTECTOR

An *inherent protector* is another safety device used to protect an electric-driven compressor motor from overload damage. It

281

accomplishes this purpose by disconnecting the motor from the power supply when the motor temperature or current becomes excessive. Its function is similar to that of the overload relay.

An inherent protector is essentially a thermostat operated by the snap action of a bimetal disc. As shown in Fig. 6-62, it consists of a heater, thermostatic disc, and contacts.

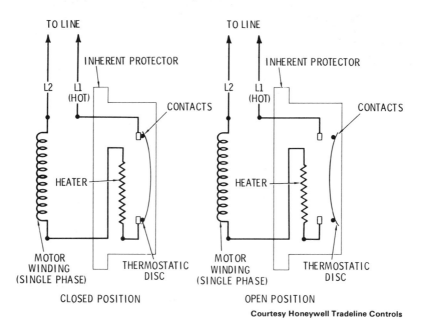

CLOSED POSITION   OPEN POSITION

Fig. 6-62. Inherent protector.

In operation, the motor current flows through both the motor winding and the heater. When an overload condition occurs, the critical temperature level is reached in the motor winding at exactly the same time as the protector reaches its tripping point. When the temperature rises to the rating of the bimetal disc, the disc snaps open, reversing its curvature, and cuts off the flow of current to the motor. When the temperature drops, the action is reversed.

Inherent protectors are available for all sizes of single-phase compressor motors. For three-phase motors, they are available for motor sizes up to 1½ hp.

PILOT DUTY MOTOR PROTECTOR

Some manufacturers will install a *pilot duty motor protector* (or *pilot duty thermostat*) on the compressor motor to protect the motor from overcurrent damage. This device is a small temperature-sensitive thermostat mounted inside the compressor on the motor windings. If the motor windings become overheated, the pilot duty thermostat breaks the circuit to the contactor relay and shuts off the compressor motor.

The pilot duty thermostat is usually wired in the 24-volt control circuit. This can be accomplished by interrupting the transformer secondary of the control circuit. A wiring diagram of a pilot duty thermostat connected in to 24-volt circuit is shown in Fig. 6-63. Pilot duty thermostat contacts can also interrupt 120-volt and 240-volt circuits.

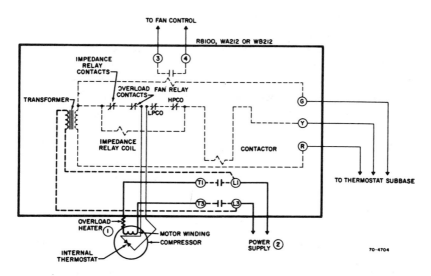

(1) Heinemann overload must be manual reset type if not used in reset circuit as shown.

(2) Provide disconnecting means and overload protection as required.

Courtesy Honeywell Tradeline Controls

Fig. 6-63. Pilot duty thermostat.

CAPACITORS

A *capacitor* is a device that provides the phase shift in the running and starting windings of an electric motor in order to increase the torque and efficiency of the motor-compressor assembly.

Capacitors are used in the following single-phase induction motors:

1. Capacitor-start motors.
2. Permanent-split capacitor motors.
3. Capacitor-start, capacitor-run motors.

A *capacitor-start motor* (Fig. 6-64) is used to power fans, blowers, and centrifugal pumps where constant speed drive is necessary. This motor develops high starting torque on the frictional horsepower ratings and moderate starting torque in the lower ratings.

In operation, an auxiliary winding is connected in series with a capacitor in the motor circuit. When the motor approaches running speed, a centrifugal switch cuts the capacitor and auxiliary winding out of the circuit.

A *capacitor-start, capacitor-run motor* (Fig. 6-65) also develops high starting torque. This is accomplished by employing a starting capacitor and a running capacitor. The starting capacitor gives good starting ability but is suited for short-time operation only. The starting capacitor is cut out of the circuit during the running period. The running capacitor provides high efficiency at running speed. Capacitor-start, capacitor-run motors are used to power compressors, reciprocating pumps, and similar types of equipment.

On some capacitor-start, capacitor-run motors, the starting capacitor may be cut out of the circuit with a centrifugal switch (Fig. 6-65). An alternate method of cutting out the starting capacitor is by using a back-emf relay (Fig. 6-66). Back-emf relays are usually used on hermetic (sealed) compressors where it is impractical to install a centrifugal switch. The coil of the back-emf relay is connected in parallel with the auxiliary winding. When the motor approaches running speed, the relay pulls in, opening its contacts and taking the capacitor out of the circuit.

284

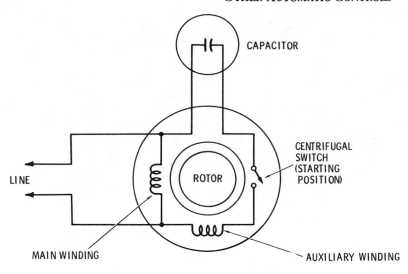

Courtesy Honeywell Tradeline Controls

Fig. 6-64. Capacitor-start motor.

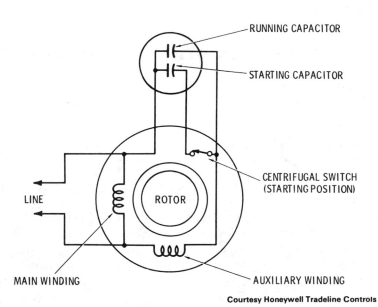

Courtesy Honeywell Tradeline Controls

Fig. 6-65. Capacitor-start, capacitor-run motor.

285

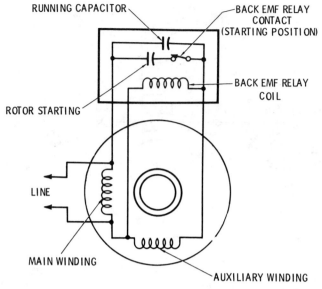

RUNNING CAPACITOR

BACK EMF RELAY
CONTACT
(STARTING POSITION)

BACK EMF RELAY
COIL

ROTOR STARTING

LINE

MAIN WINDING

AUXILIARY WINDING

Courtesy Honeywell Tradeline Controls

Fig. 6-66. FMF relay.

TROUBLESHOOTING CAPACITORS

A defective capacitor may be the cause of the following operating problems:

1. Condenser fan will not run.
2. Condenser fan will run, but compressor will not start.
3. Compressor hums but will not start.

A defective running (or run) capacitor is often the direct cause of the compressor motor cycling on overload.

A defective capacitor can be removed and replaced in the field, but first it should be tested to make certain that the capacitor is actually the source of the problem.

A ground in the run capacitor can be checked by connecting an ohmmeter or test neon lamp in series with the capacitor and the metal part of the case for each capacitor. A ground is indicated if a continuity of circuit exists. Both terminals should be tested to ground and must be done with the capacitor disconnected from the compressor.

When putting a running (or run) capacitor back into a unit, check the capacitor terminals for a marked or identified terminal. The marked terminal may be indicated by a dab of solder, a paint mark, or a stamping on the case. This terminal must always be connected with the wire leading directly back to the contactor. If this capacitor should become defective by a ground, a fuse in the power circuit will blow. If the capacitor is not properly connected, a defect by ground will cause a flow of power through the start and run windings before reaching a fuse and cause compressor damage.

HIGH-PRESSURE CUTOUT SWITCH

A *high-pressure cutout switch* is a pressure-actuated refrigerant controller connected to the pilot (low) voltage circuit of the main contactor in a refrigeration system (Fig. 6-67). These switches are

Courtesy Honeywell Tradeline Controls

Fig. 6-67. High-pressure cutout switch.

installed in all refrigerating units over 1 hp to provide protection against dangerously high head pressures. These excessive and unsafe pressures develop as a result of a number of different abnormal operating conditions, including: (1) air in the refrigerant lines, (2) excessive refrigerant charge, (3) dirty condenser, (4) faulty or inoperative condenser fan, and (5) insufficient water in a water-cooled condenser.

As shown in Fig. 6-68, the high-pressure cutout switch is connected to the so-called high side of the system. When the head

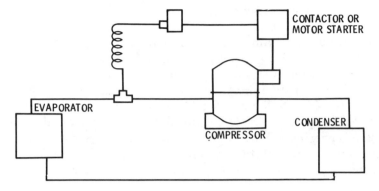

Fig. 6-68. Low-pressure cutout switch.

pressure exceeds the control setting, the high-pressure cutout switch opens the compressor power circuit and prevents excessive head pressures from building in the condenser by shutting off the compressor motor.

If automatic reset is used, the compressor motor will start again when the pressure returns to normal. More specifically, the contacts of the high-pressure cutout switch close automatically and reestablish the compressor power circuit when the head pressure drops the amount of control differential.

Automatic reset allows the equipment to cycle off the high-pressure control and continue to provide some degree of cooling while the high-pressure condition is being corrected. However, if the high-pressure condition is not corrected and the equipment continues to cycle over an extended period of time, damage may be caused to the motor or compressor. For obvious reasons, this cannot occur when a manual reset high-pressure cutout switch is used.

LOW-PRESSURE CUTOUT SWITCH

A *low-pressure cutout switch* is similar in design to a high-pressure switch (see above) except that it provides low-pressure cutout protection.

As shown in Fig. 6-69, this switch is connected to the so-called low side of the compressor and is designed to open the compressor power circuit if the low-side pressure drops below a desired level. These excessively low pressures are usually caused by dirty filters, evaporator fan failure, a stuck damper, damper motor failure, or some other interruption in the air supply.

The cutout action of the switch prevents the temperatures in the evaporator from falling below the temperature at which frost would form on the coils. It also prevents the feeding of an excessive amount of liquid to the compressor. Low-pressure cutout switches may be designed to provide either manual or automatic reset.

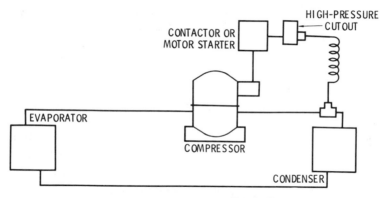

Courtesy Honeywell Tradeline Controls

Fig. 6-69. High-pressure cutout switch.

TRANSFORMERS

A transformer is an inductive stationary device designed to transfer electrical energy from one circuit to another. Each transformer contains a primary and secondary winding. A changing voltage applied to one of the windings induces a current to flow in the other winding. In this manner, electrical energy is transferred from one circuit to another. Usually the changing voltage is applied to the primary winding and a current is induced in the secondary. The electrical energy may be transferred at the same voltage (a *coupling* transformer), at a higher voltage (a *stepup* transformer), or at a lower voltage (a *stepdown* transformer).

The transformers used in heating and cooling systems are stepdown transformers. They are designed to reduce (step down) the higher line voltage power to the 24 to 30 volts required by low-voltage control circuits.

All wiring connections to transformers must be done in accordance with the recommendations of the National Electrical Code.

Interconnected transformer secondaries are not permitted by the National Electrical Code. One method of avoiding the need for interconnecting transformers is by using a single transformer rated to carry both the heating and cooling load. Using a thermostat and subbase combination with isolated heating and cooling circuits is also an acceptable method. Still another successful method utilizes an isolating relay to isolate the heating power supply from the cooling power supply.

SIZING TRANSFORMERS

Transformers are not 100 percent efficient. There will generally be some loss of energy between the primary and secondary coils. In any event, the secondary coil must have enough remaining energy to drive the load connected to it.

When the equipment in a heating and/or cooling system is not adequately powered, check the transformer primary and secondary voltages. If these readings are within plus or minus 10 percent of the rated voltage and there is no problem with the wiring, the transformer may not be large enough for the system. A transformer too small for the system can be a very serious matter because it will supply abnormally low voltage to the control circuit. As a result, contactors or motor starters will not operate properly and eventually the compressor will suffer possible damage.

When replacing a transformer, always select one the same size or larger than the one being replaced. For new installations, follow the equipment manufacturer s recommendations.

The capabilities of a transformer are described by its electrical rating. This information will include the primary voltage and frequency, the open-circuit secondary voltage, and the load rating in volt-amperes (VA).

The Class 2 transformers used in low-voltage control circuits have a maximum load rating of 100 VA and a maximum open-circuit secondary voltage of 30 volts. The secondary current must also be limited. This can be accomplished by using an energy-limiting transformer, or by adding a 3.2 amp (or less) fuse in the secondary. In the latter case, the maximum load rating of a typical 4-volt Class 2 transformer is 77 VA (24 volts $\times$ 3.2 amps = 77 VA).

CONTROL PANELS

A *control panel* combines many of the heating and cooling controls into a single, unified preassembled package. As a result, field wiring and control troubleshooting are greatly simplified.

A complete line of standard control panels is available for a variety of different functions. Depending upon the requirements of the installation, it is possible to obtain a variety of different combinations of the following control components in a panel:

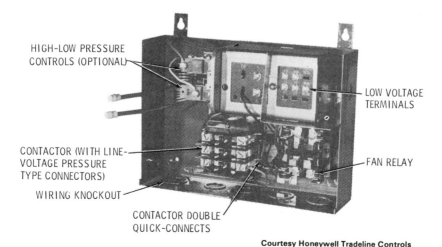

HIGH-LOW PRESSURE CONTROLS (OPTIONAL)

LOW VOLTAGE TERMINALS

CONTACTOR (WITH LINE-VOLTAGE PRESSURE TYPE CONNECTORS)

FAN RELAY

WIRING KNOCKOUT

CONTACTOR DOUBLE QUICK-CONNECTS

Courtesy Honeywell Tradeline Controls

Fig. 6-70. Heating-cooling control.

291

1. Transformer.
2. Line voltage and low-voltage wiring terminals.
3. High-low pressure cutout.
4. Two-stage cooling time delay circuit.
5. Compressor contactor or motor starter.
6. Auxiliary relays.

The internal view of the Honeywell heating-cooling panel illustrated in Fig. 6-70 shows the typical arrangement of these control components.

CHAPTER 7

Ducts and
Duct Systems

Air distribution systems based on the forced-air principle of delivery utilize a system of ducts to deliver the heated or cooled air to the various rooms and spaces within the structure. These air ducts are generally rectangular or round pipes made from a variety of different materials. When these ducts are accurately sized and the duct system correctly designed, the air will be delivered to the rooms and spaces with a minimum of resistance, the result being a more efficient operation with reduced operating costs. The purpose of this chapter is to suggest methods for sizing ducts and designing an efficient duct system.

Methods for sizing fans are described in Chapter 7 of Volume 3 (Fan Selection and Operation). Chapter 4 of Volume 1 (Heating Calculations) and Chapter 9 of Volume 3 (Air Conditioning Calculations) providing information and methods for sizing the heating and cooling units.

CODES AND STANDARDS

Always consult local codes and standards first before designing and installing a duct system. Any aspect of a duct system that does not comply with these codes and standards will have to be changed. These changes could be expansive.

Information about duct construction can be obtained from the *1972 ASHRAE Guide and Data Book (Equipment Volume)*. These materials will remain current until 1975, at which point they undergo revision. Information about duct sizing, air distribution, and air duct design methods is contained in the *1972 ASHRAE Handbook of Fundamentals* (current until 1976).

Other sources of information include the Commodity Standards Division of the U.S. Department of Commerce and the National Warm Air Heating and Air Conditioning Association. The former provides detailed information on ducts and duct fittings. Design and installation methods as well as information concerning codes and standards can be obtained from the latter organization.

TYPES OF DUCT SYSTEMS

The two duct systems most commonly used in forced warm-air heating are: (1) the perimeter duct system and (2) the extended plenum duct system. Each is available in several design modifications and is described in the sections that follow.

Details about the piping arrangements used with gravity warm-air furnaces are included in the section describing these furnaces in Chapter 10 of Volume 1 (Furnace Fundamentals).

An excellent source of information for designing a warm-air heating system (particularly designing gravity systems) is found in the *Code and Manual for the Design and Installation of Gravity Warm Air Heating Systems*, a publication of the National Warm Air Heating and Air Conditioning Association.

PERIMETER DUCT SYSTEMS

A *perimeter duct system* is one in which the supply outlets are located around the perimeter (i.e., outer edge) of the structure close to the floor of the outside wall, or on the floor itself. The return grilles are generally placed near the ceiling on the inside wall.

The two basic perimeter duct systems used in warm-air heating are: (1) the perimeter-loop duct system, and (2) the radial perimeter duct system.

The *perimeter-loop duct system* (Fig. 7-1) is characterized by feeder supply ducts that extend outward from the furnace plenum to a loop duct running around the perimeter. Warm-air supply outlets are located in the loop duct.

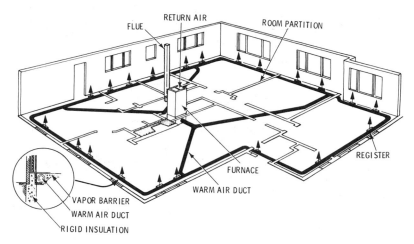

FLUE
RETURN AIR
ROOM PARTITION
VAPOR BARRIER
WARM AIR DUCT
RIGID INSULATION
FURNACE
WARM AIR DUCT
REGISTER

Courtesy U.S. Department of Agriculture
Fig. 7-1. Typical perimeter-loop system.

There is no loop duct in the *radial perimeter duct system* (Fig. 7-2). The feeder supply ducts extend from the furnace plenum to the warm-air supply outlets located on the outside walls or the floor next to the outside walls.

EXTENDED PLENUM SYSTEMS

In the *extended plenum system* (Fig. 7-3), a large rectangular duct extends straight out from the furnace plenum (hence the term "extended plenum") and generally in a straight line down the center of the basement, attic, or ceiling. Round or rectangular

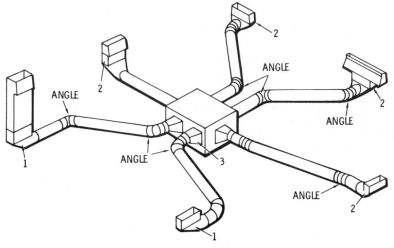

1. CENTER END REGISTER BOOT
2. REGISTER BOOT
3. ADJUSTABLE ROUND PIPE SIDE TAKE OFF

Fig. 7-2. Radial perimeter duct system.

supply ducts extend as branches from the plenum extension to the warm-air supply outlets. The large extension to the plenum permits a better air-flow rate with reduced resistance because of its large duct diameter. The branching ducts are usually located between joists and can be easily covered with a ceiling.

CRAWL-SPACE PLENUM SYSTEMS

It is possible to incorporate the entire crawl space into a heating system if the crawl-space walls are tight and well insulated. The heated air is forced down into the crawl space and enters the rooms through perimeter outlets, usually located beneath windows.

This type of duct arrangement may be referred to as a *crawl-space plenum system* and represents a modification of the extended plenum system. Because the entire crawl space is filled with warm air, this system provides relatively uniform temperatures throughout the structure.

296

DUCT MATERIALS

Ducts are manufactured from a variety of different materials. The material selected will depend upon the use for which it is intended. It is very important that the duct material be taken into consideration when designing a duct system because not every material is suitable for all conditions in which ducts are used.

It is possible to purchase ducts manufactured from the following materials:

1. Steel.
2. Galvanized sheet steel.
3. Aluminum.
4. Copper.
5. Glass fiber.

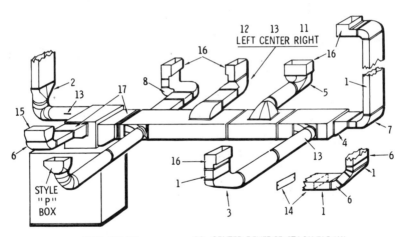

1. WALL STACK	10. CENTER REVERSE STACK ELBOW
2. STRAIGHT BOOT	11. RIGHT REVERSE STACK ELBOW
3. END BOOT	12. LEFT REVERSE STACK ELBOW
4. END	13. ROUND VOLUME DAMPER
5. ANGLE BOOT	14. STACK DAMPER
6. ANGLE	15. STACK HEAD
7. ELBOW	16. STACK HEAD
8. ELBOW	17. STARTING COLLAR
9. ELBOW	

Fig. 7-3. Extended plenum duct system.

297

6. Paper fiber.
7. Vitrified clay tile.
8. Asbestos-cement.

Plain steel and *galvanized sheet steel* ducts are available in thicknesses ranging from 0.0163 to 0.1419 in. Ducts manufactured from this material are preferred for use in warm-air gravity and forced circulation warm-air heating systems. Table 7-1 indicates the thicknesses, gauges, and weights in which plain steel and galvanized sheet steel ducts are available.

Aluminum ducts are available in thicknesses ranging from

Table 7-1. Thicknesses, Gauges, and Weights of Plain (Black) and Galvanized Sheet Metal

U.S. Std. Gauge		Approximate Thickness, In.		Weight Per Square Foot	
		Steel	Iron	Ounces	Pounds
Black Sheets	30	0.0123	0.0125	8	0.500
	28	0.0153	0.0156	10	0.625
	26	0.0184	0.0188	12	0.750
	24	0.0245	0.0250	16	1.000
	22	0.0306	0.0313	20	1.250
	20	0.0368	0.0375	24	1.500
	18	0.0490	0.0500	32	2.000
	16	0.0613	0.0625	40	2.500
	14	0.0766	0.0781	50	3.125
	12	0.1072	0.1094	70	4.375
	11	0.1225	0.1250	80	5.000
	10	0.1379	0.1406	90	5.625
Galvanized Sheets*	30	0.0163	0.0165	10.5	0.656
	28	0.0193	0.0196	12.5	0.781
	26	0.0224	0.0228	14.5	0.906
	24	0.0285	0.0290	18.5	1.156
	22	0.0346	0.0353	22.5	1.406
	20	0.0408	0.0415	26.5	1.656
	18	0.0530	0.0540	34.5	2.156
	16	0.0653	0.0665	42.5	2.656
	14	0.0806	0.0821	52.5	3.281
	12	0.1112	0.1134	72.5	4.531
	11	0.1265	0.1290	82.5	5.156
	10	0.1419	0.1446	92.5	5.781

*Galvanized sheets are gauged before galvanizing and are therefore approximately 0.004 in. thicker.

Courtesy *ASHRAE 1960 Guide*

0.012 to 0.064 in. (Table 7-2), and are used in the same types of heating systems as steel ducts. Although aluminum ducts are lighter than steel ones, they generally cost more. Aluminum ducts are frequently used in duct systems located on the outside of buildings.

Table 7-2. Thicknesses, Gauges, and Weights of 2S Aluminum (Density 0.098 lb./cu. in.)

B. & S. Gauge	Thickness, Inches		Weight per Square Foot	
	Decimal	Nearest Fraction	Pounds	Ounces
28	0.012	$\frac{1}{64}$	2.7	0.169
26	0.016	$\frac{1}{64}$	3.6	0.226
24	0.020	$\frac{1}{64}$	4.5	0.282
22	0.025	$\frac{1}{32}$	5.4	0.353
20	0.032	$\frac{1}{32}$	7.2	0.452
18	0.040	$\frac{3}{64}$	9.0	0.563
16	0.051	$\frac{3}{64}$	11.5	0.720
14	0.064	$\frac{1}{16}$	14.4	0.903

Courtesy *ASHRAE 1960 Guide*

Copper ducts are available in sizes and gauges matching the aluminum ones and are frequently used in outside ductwork.

Round *glass-fiber ducts* can be purchased in a number of different sizes ranging up to 14 in. in diameter with duct walls up to 1 in. in thickness. Square or rectangular glass-fiber ducts can also be made from flat glass-fiber board. Because of their composition, glass-fiber ducts dampen sound.

Paper-fiber ducts are laid in concrete and used in warm-air heating systems. They are accordingly not recommended for use in attics, basements, or other exposed areas.

Vitrified clay tile ducts represent another duct material suitable for installation under a concrete slab. These ducts range in outside diameter from 5 ⅛ to 42 ¼ in.

DUCT SYSTEM COMPONENTS

The components of a typical duct system are illustrated in Fig. 7-4. In a forced warm-air heating system, the warm air collects in an area at the top of the furnace called the *furnace hood* or

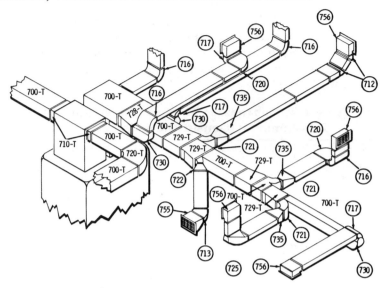

700-T DUCT SIZES 4 X 8 TO 36 X 8	721 SIDE TAKE OFFS 4 X 8, 5 X 8, 6 X 8.
710-T STARTING COLLAR, SIZED SAME/ AS/DUCT	722 SIDE TAKE OFFS 4 X 8, 5 X 8, 6 X8.
728-T & 729-T INCREASER-REDUCER SECTIONS, SIZES SAME AS DUCT (MAX. ING. 10" FOR 728-T AND 5" FOR 729-T)	224 - 725 - 726 REVERSE STACK ELBOWS 10 X 3 1/4, 14 X 3 1/4
	712 ANGLE 10 X 3 1/4, 12 X 3 1/4, 14 3 1/4
	713 ANGLE 10 X 3 1/4, 12 X 3 1/4, 14 X 3 1/4
712-T MAIN TRUNK ANGLES, SIZES SAME AS DUCT	716 ELBOW 10 X 3 1/4, 12 X 3 1/4, 14 X 3 1/4
713-T MAIN-TRUNK ANGLES SIZE SAME AS DUCT	717 ELBOW 10 X 3 1/4, 12 X 3 1/4, 14 X 3 1/4
717-T MAIN TRUNK ELLS, SIZES SAME AS DUCT	720 ELBOW 10 X 3 1/4, 12 X 3 1/4, 12 X 3 1/4, 14 X 3 1/4
720-T MAIN TRUNK ELLS, SIZES SAME AS DUCT	700 WALL STACK 10 X 3 1/4, 12 X 3 1/4, 14 X 3 1/4
	755 STACK HEAD 4, 5, 6, 8 X 10, 4, 5, 6, 8 X 12, 4, 5, 6, 8 X 14
	756 STACK HEAD 4, 5, 61 8 X10, 4, 5, 6,

Courtesy Clayton and Lamberι Mfg. Co.

Fig. 7-4. Principal components of a warm-air duct system.

plenum. An extended plenum duct system will have a large rectangular duct connected to the plenum by a *starting collar* and extending out along the ceiling in a straight line. *Round* (see Fig. 7-3) or *square supply ducts* are connected to the plenum (or plenum extension) usually by *adjustable side takeoffs* and extend to either a *register boot* or an *elbow*. Changes of direction in the round duct are accomplished with flexible *angle duct*. A *nonflexible elbow* provides the same function in rectangular ducts. A vertical duct or warm-air riser is sometimes referred to as a *stack*. A warm-air duct that carries the warm air horizontally in a straight line from the furnace plenum to the stack is often referred to as a *leader*. *Dampers* are located in the duct so that the quantity of warm air can be regulated manually or automatically by thermostatic control. Ducts that carry the warm air to the rooms are called *supply ducts*. All ducts that carry the return air back to the furnace are referred to as *return ducts*.

SUPPLY AIR REGISTERS, GRILLES, AND DIFFUSERS

The three basic types of supply air outlets used in an air distribution system are: (1) grilles, (2) registers, and (3) diffusers.

Grilles (Fig. 7-5) are used not only to admit the air flow but also to deflect it up or down, or to one side or the other, depending upon the direction in which the hand-operated bar moves. They are used primarily on high or low wall locations. Floor grilles are used extensively in gravity warm-air heating systems.

A *register* (Fig. 7-6) is similar in design and function to the grille but with the added feature of being able to regulate the *volume* of the air with a damper. They may be located on walls (high or low) or floors. Floor registers are often used when installing a new heating system in an old house. The major objection to floor registers is that they tend to collect dust and trash.

A *diffuser* (Fig. 7-7) is also used to deflect the air flow, but it differs fundamentally in design from the grille. Diffusers manufactured in the form of concentric cones or pyramids are usually mounted on ceilings or walls. Baseboard diffusers are used in perimeter forced warm-air heating systems. The major objection to ceiling diffusers is that they cause drafts when the air is discharged downward and dirt smudges when the air is discharged horizontally across the ceiling.

If the duct system is designed primarily for cooling, outlets are

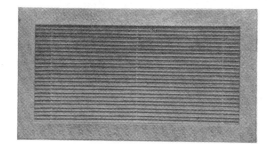

Courtesy A-J Mfg. Co.

Fig. 7-5. Examples of various grilles.

sometimes located on the ceiling or high on the wall; however, satisfactory heating and cooling can be achieved with baseboard outlets placed low on walls by increasing the air volume and velocity and by properly directing the air flow.

RETURN AIR AND EXHAUST AIR INLETS

Grilles and registers are the two principal types of air inlets used to exhaust the air from a space or to return the air to the centrally located heating or cooling unit. The grilles are generally

Fig. 7-6. Examples of various registers.

Courtesy United States Register Co.

303

Courtesy United States Register Co.

Fig. 7-7. Examples of diffusers.

fixed-angle types because there is no need to direct air circulation when return air is involved.

DUCT RUN FITTINGS

Round and rectangular duct run fittings are available in a variety of different shapes and sizes, depending upon the requirements of the air distribution system. Some duct run fittings are used only for cooling systems, others are designed for use in both heating and cooling systems. Duct run fittings can be purchased from manufacturers and local supply houses, or they can be made locally. Making your own fittings requires a knowledge of sheet-metal work. Components of a typical duct system are shown in Fig. 7-4. Based on their function, these duct run fittings can be divided into the following principal categories:

1. Supply air and return air bonnet or plenum (Fig. 7-8).
2. Plenum and extended plenum takeoffs (Fig. 7-9).
3. Trunk duct angles and elbows (Fig. 7-10).
4. Stack angles and elbows (Fig. 7-11).

5. Boot fittings (Figs. 7-12 and 7-13).
6. Wall sections (Fig. 7-14).

AIR SUPPLY AND VENTING

Any boiler or furnace fired with a combustible fuel (e.g., coal, oil, gas) must be equipped with a piping system to remove smoke

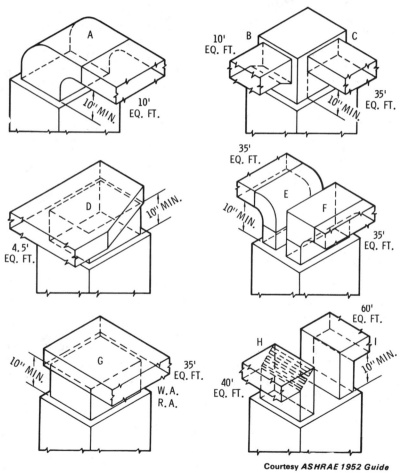

Courtesy *ASHRAE 1952 Guide*

Fig. 7-8. Warm-air and return-air bonnet or plenum.

305

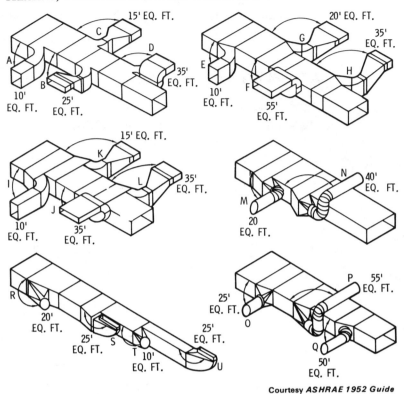

Courtesy *ASHRAE 1952 Guide*

Fig. 7-9. Trunk duct takeoff.

and other low-temperature flue gases and to provide sufficient air for combustion. These air supply and venting systems are composed of round pipes and fittings made from sheet metal and resemble ducts in design and construction.

The design and installation of air supply and venting systems are described in the several chapters dealing with furnaces and boilers. See, for example, the appropriate sections of Chapter 11 of Volume 1 (Gas-Fired Furnaces).

DUCT DAMPERS

A *duct damper* is a device used for controlling the direction or volume of air flowing through a duct.

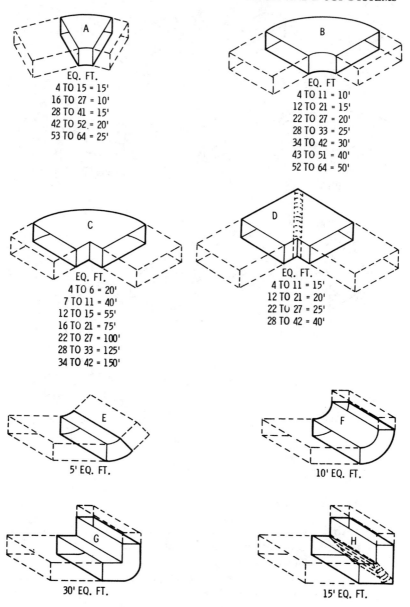

EQ. FT.
4 TO 15 = 15'
16 TO 27 = 10'
28 TO 41 = 15'
42 TO 52 = 20'
53 TO 64 = 25'

EQ. FT
4 TO 11 = 10'
12 TO 21 = 15'
22 TO 27 = 20'
28 TO 33 = 25'
34 TO 42 = 30'
43 TO 51 = 40'
52 TO 64 = 50'

EQ. FT.
4 TO 6 = 20'
7 TO 11 = 40'
12 TO 15 = 55'
16 TO 21 = 75'
22 TO 27 = 100'
28 TO 33 = 125'
34 TO 42 = 150'

EQ. FT.
4 TO 11 = 15'
12 TO 21 = 20'
22 TO 27 = 25'
28 TO 42 = 40'

5' EQ. FT.

10' EQ. FT.

30' EQ. FT.

15' EQ. FT.

Courtesy *ASHRAE 1952 Guide*

Fig. 7-10 Angles and elbows for trunk ducts.

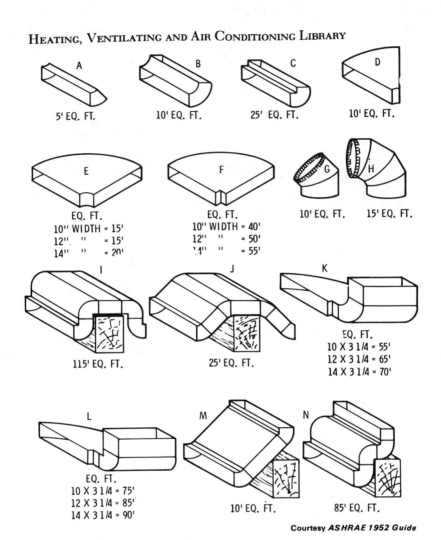

A
5' EQ. FT.

B
10' EQ. FT.

C
25' EQ. FT.

D
10' EQ. FT.

E
EQ. FT.
10" WIDTH = 15'
12" " = 15'
14" " = 20'

F
EQ. FT.
10" WIDTH = 40'
12" " = 50'
14" " = 55'

G
10' EQ. FT.

H
15' EQ. FT.

I
115' EQ. FT.

J
25' EQ. FT.

K
EQ. FT.
10 X 3 1/4 = 55'
12 X 3 1/4 = 65'
14 X 3 1/4 = 70'

L
EQ. FT.
10 X 3 1/4 = 75'
12 X 3 1/4 = 85'
14 X 3 1/4 = 90'

M
10' EQ. FT.

N
85' EQ. FT.

Courtesy *ASHRAE 1952 Guide*

Fig. 7-11. Stock angles and elbows.

The ASHRAE defines a damper as being "a device used to vary the volume of air passing through a confined cross-section by varying the cross-sectional area." In other words, a damper functions as an obstruction to air flow through the duct; however, it is a *movable* obstruction that can be adjusted to give various-size openings for the passage of air.

The air volume in an air distribution system may have to be

308

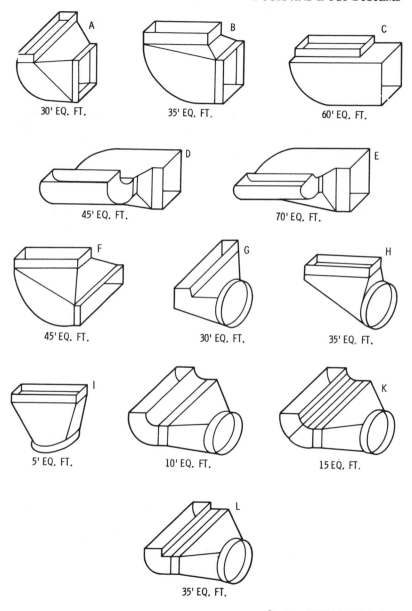

Fig. 7-12. Boot fittings from branch to stack.

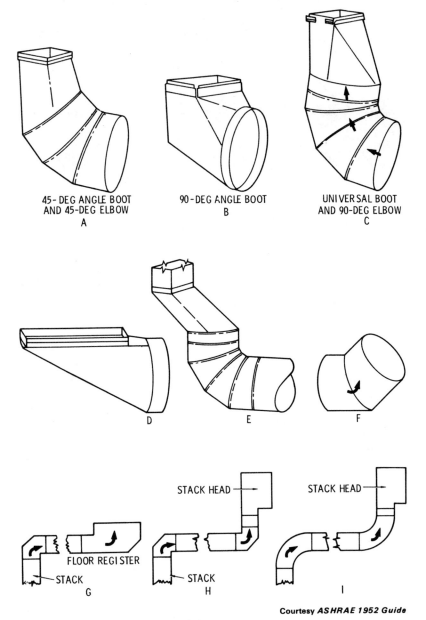

45-DEG ANGLE BOOT
AND 45-DEG ELBOW
A

90-DEG ANGLE BOOT
B

UNIVERSAL BOOT
AND 90-DEG ELBOW
C

D E F

STACK HEAD

STACK HEAD

FLOOR REGISTER

STACK
G

STACK
H

I

Courtesy *ASHRAE 1952 Guide*

Fig. 7-13. Combination warm-air boots.

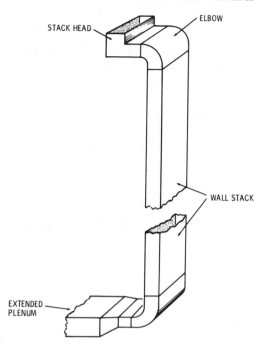

Fig. 7-14. Wall stack.

changed between the heating operation and the cooling operation because it generally requires *more* air at 50°F to cool a house to 75°F than warm air at 160°-170°F to heat a house to 75°F. Bearing this in mind, the ductwork and fan should be sized for whichever operation (i.e., heating or cooling) requires the greater air volume. *Volume control dampers* may then be installed in the duct system to reduce the air volume during the season that requires the smaller volume.

A damper is usually made in the form of a round or rectangular blade and can be either manually operated or motorized. The ASHRAE lists the following three basic types of dampers:

1. Volume dampers.
2. Splitter dampers.
3. Squeeze dampers.

A *volume damper* or *volume control damper* (Fig. 7-15) is installed in a duct to either completely cut off or regulate the

311

flow of air in the duct. Adjustments of the volume of air flow and resistance can be made with a *squeeze damper*. A *splitter damper* is a directional device consisting of a blade hinged at one end and used at locations where a branch run leaves the main duct.

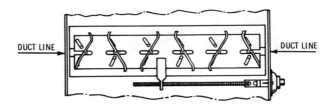

Fig. 7-15. Volume control damper used in branch ducts to control the volume of air.

Three examples of volume dampers are: (1) slide dampers, (2) hit-and-miss dampers, and (3) butterfly dampers. The *slide damper* is operated by sliding or pushing a metal plate across the duct. It operates either at full open or full closed positions, there being no practical intermediate setting. A *hit-and-miss damper* provides volume control with two slotted plates or discs placed adjacent to one another. At full open position, 50 percent of the duct cross section remains blocked. A *butterfly damper* consists of a single blade hinged in the middle. At full open position, almost the entire cross section of the duct is free of blockage.

The adjustment lever for a duct damper should be in an accessible location, and it should be labeled to indicate position settings, function, and area served. The damper should also be equipped with a positive locking device.

Dampers properly placed in the supply and return ducts of an air distribution system will control the supply of air to a room; however, care should be taken in the placement of these dampers. A damper in a supply duct placed too close to the supply air outlet may disturb the air flow. Excess noise is often the result of improperly placed dampers.

Splitters and *guide vanes* (Fig. 7-16) are nonmovable sheet-metal partitions placed in stack heads or elbows to reduce air turbulence and to guide the air flow. *Air directors* are similar devices, but are used to direct a portion of the air flow to a branch duct.

A *fire damper* is a damper designed to close off a duct in order to prevent the spread of flames and smoke. Fire dampers should be placed in branch duct connections which pass through fire walls, fire-rated partitions, or floors. *Never* install a fire damper in main exhaust ducts and risers. Installation of fire dampers should comply with local ordinances and National Fire Protection Association Standards.

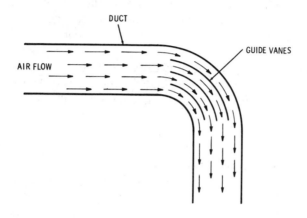

Fig. 7-16. Guide vanes in an air duct.

DAMPER MOTORS AND ACTUATORS

Damper motors or actuators are devices used in heating, ventilating, and air conditioning systems for the following applications:

1. Diverting air flow.
2. Decreasing air flow.
3. Controlling ventilation.
4. Zoning.

A damper motor or actuator can be used to position a diverting damper when a parallel air-flow pattern is required. The damper is used to direct air flow through either the heating or cooling

313

unit. The damper motor is connected to the heat-cool switch on a semiautomatic changeover thermostat or in parallel with the cooling contact of an automatic changeover thermostat.

A damper motor or actuator can also be used to operate a bypass damper for the following purposes:

1. To bypass a cooling coil for dehumidification.
2. To bypass the heat exchanger on cooling.
3. To bypass the heat exchanger on heating.

It is sometimes necessary to decrease the air flow during the heating cycle in some systems. This can be accomplished by using a damper motor or actuator to position a resistance damper.

Systems that require the introduction of outdoor air during the cooling season, but not during the heating season, can use damper motors or actuators to control the ventilation.

Damper motors or actuators are also frequently used in zoning a duct system by opening or closing dampers to the various zones.

The small, compact electric motors used to provide proportional control of dampers can also be used to drive valves or step controllers in heating, cooling, and ventilating systems. For this reason, additional information about these motors is included in Chapter 9 of Volume 2 (Valves and Valve Installation).

An example of a motor used to drive a damper is found in Fig. 7-17. This is a cutaway of a Honeywell modutrol motor, which functions as the drive unit in a *modulating control circuit*. The basic components of this motor are:

1. Reversible motor.
2. Balancing relay.
3. Feedback potentiometer.
4. Gear train.

The balancing relay controls the motor, which turns the motor drive shaft through the gear train. The motor is equipped with switches that limit its rotation to 90° or 160°. The gear train and other moving parts are immersed in oil to eliminate the need for periodic lubrication.

The motor is started, stopped, and reversed by the single-pole

Fig. 7-17. Cutaway view of a modutrol motor.

double-throw contacts of the balancing relay. The balancing relay consists of two solenoid coils with parallel axes, into which are inserted the legs on the U-shaped armature. The armature is pivoted at the center so that it can be tilted by the changing magnetic flux of the two coils to energize the relay. A contact arm is fastened to the armature so that it may touch either of the two stationary contacts as the armature moves back and forth on its pivot. When the relay is balanced, the contact arm floats between the two contacts, touching neither of them.

A feedback potentiometer consisting of a coil of wire and a sliding contact is included in the Honeywell modulating motor. The sliding contact is moved by the motor shaft so that it travels along the coil and establishes contact wherever it touches, according to the position of the motor.

Fig. 7-18 shows a typical wiring diagram for a Honeywell modutrol motor. Note that there are two separate circuits in the modulating motor powered from T1 and T2. The *motor circuit* consists of the reversible motor, the rotational limit switches, and the contacts of the balancing relay. The *control circuit* includes the feedback potentiometer, the coils of the balancing relay, and the controller potentiometer.

The control circuit offers two paths for current flow—one through each side of the balancing relay. Increasing resistance in the "B" leg of the motor control circuit, by changing the setting of the controller potentiometer, will run the motor toward the

315

FACTORY CALIBRATION POTENTIOMETER-
DO NOT ADJUST

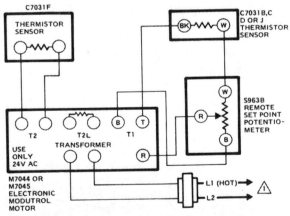

⚠ POWER SUPPLY. PROVIDE OVERLOAD PROTECTION AND DISCONNECT
MEANS AS REQUIRED.

Courtesy Honeywell Tradeline Controls

Fig. 7-18. Wiring diagram for an electronic modutrol motor.

closed position. Adding resistance to the "W" leg runs the motor open.

As the motor shaft turns, it moves a wiper over the feedback potentiometer. This makes the resistance in each side of the circuit the same. When the resistances are equal, the current flow through both sides of the balancing relay is equal. The balancing relay contacts open, stopping the motor. The circuit is said to be balanced.

The motor of a damper actuator should be completely sealed and immersed in oil. Such a motor can operate without maintenance or service for the life of the unit. An example of this type of motor is found on the ITT General Controls DHO series damper actuator shown in Fig. 7-19. These DHO series motors are availa-

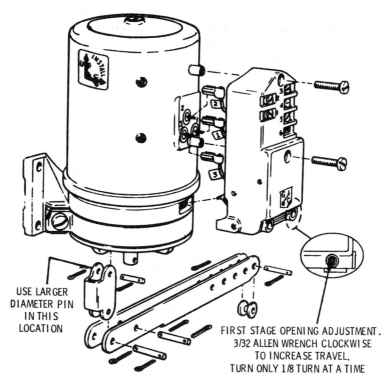

USE LARGER
DIAMETER PIN
IN THIS
LOCATION

FIRST STAGE OPENING ADJUSTMENT.
3/32 ALLEN WRENCH CLOCKWISE
TO INCREASE TRAVEL.
TURN ONLY 1/8 TURN AT A TIME

Courtesy ITT General Controls

Fig. 7-19. Damper actuator.

ble in two-position type (energized and deenergized) or three-position type (deenergized, first stage open, and second stage open). The first-stage intermediate position can be adjusted.

It is important the DHO series damper motors work within the manufacturer's specified load limit during all phases of operation in order to ensure delivery of rated forces under usual voltage variation. Maximum workload can be determined from the data given in Table 7-3. "Load" is the dead weight pull at the particu-

Table 7-3. Maximum Workload Data

Connection Position	1		2		3		4		5	
Length of Travel In.	3.25*		3.00*		2.75*		2.50*		2.25*	
Specifications	Max. Load Lbs.	Return Force Lbs.	Max. Load Lbs.	Return Force Lbs.	Max. Load Lbs.	Return Force Lbs.	Max. Load Lbs.	Return Force Lbs.	Max. Load Lbs.	Return Force Lbs.
25 VA	15	8½	17	9½	19	10	20	11	22	13
40 VA	30	8½	34	9½	37	10	42	11	45	13
120 VA	30	8½	34	9½	37	10	42	11	45	13
Length of Travel In.	2.625†		2.375†		2.125†		1.875†		1.625†	
25 VA	18	10	21	11	24	13	27	15	30	17
40 VA	36	10	42	11	48	13	54	15	60	17
120 VA	36	10	42	11	48	13	54	15	60	17

*When using dual damper arm combined load must not exceed load maximums shown.
†Table for reversed damper arm position.

Courtesy ITT General Controls

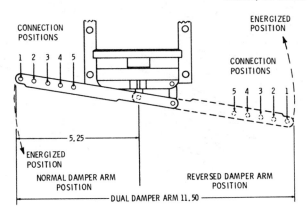

lar stroke position. The imposed "load" can be measured with a small spring scale.

The Honeywell M833A damper actuator shown in Fig. 7-20 is used to regulate duct damper condition according to zone thermostat requirements. It attaches directly to a damper shaft ½ in. in diameter or ⅜-in. shaft with adapter provided. It mounts in any position directly on a duct, or inside a standard wiring junction box where Class 1 wiring is required.

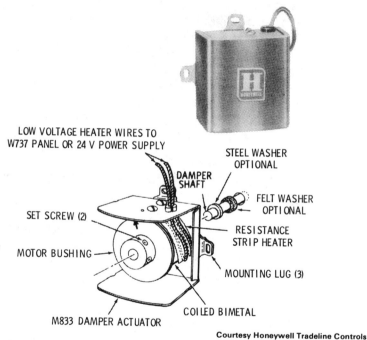

Courtesy Honeywell Tradeline Controls

Fig. 7-20. Damper actuator.

INSTALLING DAMPER MOTORS

The location of a damper motor will be determined by the design of the ductwork. If possible, the motor should be easily accessible for servicing and maintenance. The following suggestions should be considered when seeking a location for a duct motor:

1. The ambient temperature must be within the ratings of the motor being installed.
2. A motor should not be installed where corrosive vapors or sprays, or other contaminants, may present a problem to motor operation. It is especially necessary to prevent dust or dirt from getting on the feedback potentiometer.
3. The linkage between the motor and the damper must be free to operate without binding on pipes or other obstructions.
4. The motor must be installed so that it is free to run from one end of its stroke to the other.
5. Honeywell modutrol motors must be mounted with the shaft in a horizontal position.
6. A *remote* balancing relay mounted on a firm support must be used if motor location is subject to severe vibration.
7. Weatherproof motors should be used when the unit is located in a position exposed to the elements.

TROUBLESHOOTING DAMPER MOTORS

Although damper motors are completely sealed, they do malfunction from time to time. The principal malfunctions, their possible causes, and suggested remedies are listed in Table 7-4.

Some manufacturers produce special compact control units for mounting on damper motors. These control units (e.g., the

Table 7-4. Troubleshooting Damper Motor

	Possible Cause	Remedy
Motor not operating	Power off	Check switches and fuse
	Loose wiring	Check connections
	Motor damaged	Replace actuator
Motor operates but output shaft does not move	Linkage binding	Check for free movement
	Actuator damaged	Replace actuator
Motor stalls at maximum travel position	Limit switch damage	Inspect and replace if necessary

Courtesy ITT General Controls

Honeywell W859 Economizer shown in Fig. 7-21) give damper motors the capacity to regulate the outside and recirculated air dampers for the following applications:

1. To bring in outside air (when it is colder than the inside air) for cooling or ventilating.
2. To close outdoor air dampers to minimum position during the heating cycle.
3. To prevent short-cycling of the cooling compressor to avoid coil icing.

Courtesy Honeywell Tradeline Controls

Fig. 7-21. Economizer control.

BLOWERS (OR FANS) FOR DUCT SYSTEMS

The ductwork and blower (or fan) are sized for either heating or cooling (whichever requires the greater air volume); however, their correct sizing depends upon an accurate determination of the following facts about a structure or space:

1. The total heat loss.
2. The total cooling load.
3. Air delivery required (CFM).
4. External static pressure for ductwork.

Once these facts are known, it is simply a matter of referring to the performance data provided by the blower or fan manufacturer and selecting the most suitable equipment for the system.

Detailed methods for calculating heat loss, the cooling load, air volume (CFM), and external static pressure for the ductwork are described in Chapter 4 of Volume 1 (Heating Calculations); Chapter 7 of Volume 3 (Fan Selection and Operation); and Chapter 9 of Volume 3 (Air Conditioning Calculations).

DESIGNING A DUCT SYSTEM

The purpose of a duct system is to convey air from the blower or fan to the air supply outlets located in the various rooms and spaces of the structure, and then to return it to its point of origin.

Accuracy in estimating the resistance to the flow of air through the duct system is important in the selection of a suitable blower or fan. Resistance should be kept as low as possible in the interest of economy; however, underestimating the resistance will result in the failure of the blower or fan to deliver the required volume of air.

Every precaution should be taken in the design of a duct system to ensure a smooth and efficient flow of air. Careful study should be made of the building drawings with consideration being given to the construction of duct locations and clearances.

The following recommendations may be of use to you in designing a duct system:

1. Keep all duct runs as short as possible, bearing in mind that the air flow should be conducted as directly as possible between its source and delivery points, with the fewest possible changes in direction.

2. Select location for duct outlets that will ensure proper air distribution.
3. Provide ducts with cross-sectional areas that will permit air to flow at suitable velocities.
4. Use moderate velocities in all ventilating work to avoid waste of power and to reduce noise.
5. Use lower velocities in schools, churches, theaters, etc., than in factories and other places where noise due to air flow is not objectionable.

DUCT SYSTEM CALCULATIONS

An *ideal* duct system would be one in which total pressure remained constant. In other words, there would be no pressure losses anywhere in the system. Under ideal operating conditions, any change in velocity pressure would be compensated for by an equal change in static pressure, thereby maintaining a constant total pressure (i.e., the sum of the velocity and static pressures). A drop in velocity pressure would trigger a corresponding rise in static pressure, and vice versa. Unfortunately, this is not the way things work in actual practice because pressure losses very definitely do occur.

In the ductwork, pressure losses result from the resistance of the ducts to the passage of the air. This resistance occurs as a result of two effects: (1) friction loss and (2) dynamic loss. The former is caused primarily by the friction of the moving air against the surface of the duct. Dynamic loss results from sudden changes of direction (e.g., in sharp elbows) in the air stream.

An important aspect of duct system calculations is determining the *total* external static pressure drop (i.e., total resistance) of the duct system. In large part, a blower or fan is selected on the basis of its capacity to operate against the total resistance of the duct work. This resistance of the ductwork to the flow of air is referred to as the *external* static pressure (or external static pressure *drop*), because it represents the pressure drop occurring *outside* the heating or cooling unit. In addition to external static pressure, a blower or fan must also overcome resistance due to *internal* static pressure drop caused by the passage of air through

323

heaters, coils, filters, and washers. This data can be obtained from rating tables provided by manufacturers. Duct resistance (i.e., external static pressure drop) must be calculated for each duct system.

The equal friction method is frequently used to calculate the external static pressure of a duct system (see "Equal Friction Method" in this chapter). Two other methods used for making these calculations are: (1) the velocity-reduction method, and (2) the static-regain method. Detailed descriptions of both of these methods can be found in the *ASHRAE Guide.*

DUCT HEAT LOSS AND GAIN

Duct heat loss or gain is another important factor to consider when designing a duct system. This aspect of heat transmission will depend upon some or all of the following factors:

1. Temperature of the air in the duct.
2. Ambient temperature.
3. Air velocity in the duct.
4. Duct insulation.

If the temperature of the air inside the duct is different from the temperature surrounding it (i.e., the ambient temperature), then either a loss or gain of heat will occur.

If the ducts are used to convey heat, excessive heat loss from the ducts will reduce the efficiency of the heating system. This will result in a total loss of heating effect if the heat loss occurs in ductwork passing through an unheated area, but it can be considerably reduced with proper duct insulation (see "Duct Insulation"). Such heat loss can also occur in heated spaces, but here it is a problem of poor air distribution.

The same conditions exist when the ducts carry cool air. When air passes through spaces subject to the cooling effect of air conditioning, insulating the ducts will effectively reduce the amount of heat gain. Proper air distribution will also minimize heat gain in spaces that are partially cooled.

Air velocity will also influence the amount of heat gain in the ducts. High air velocities are recommended when the ducts are carrying cool air because their effect is to reduce the amount of heat gain pickup in the ducts; however, they must be maintained consistent with the acoustic requirements of the installation.

AIR LEAKAGE

Air leakage through duct seams and holes will result in the loss of a portion of the air flowing through the duct and a proportionate reduction in the heating or cooling effectiveness of the system. Depending on the seriousness of the problem, the ductwork can lose up to ⅓ of the air supply in this manner. It is usually a matter of poor workmanship and can be corrected by sealing the seam cracks and holes by calking, soldering, or covering with asbestos tape.

DUCT INSULATION

Ducts are insulated to prevent excessive heat loss or gain. If the ducts are used to convey heat, excessive heat loss from the ducts will reduce the efficiency of the heating system. The reverse is true of ducts used in cooling systems. If the ducts are not insulated, they will absorb heat from the air around them, and system performance will be impaired.

To maintain the proper level of performance in a heating or cooling system, the following ducts should be insulated:

1. Supply ducts running through spaces that are neither heated nor cooled (e.g., attics, basements, garages, crawl spaces).
2. Long supply ducts (particularly those over 45 ft. in length).
3. All ducts located on the outside of buildings.
4. Cool-air return ducts passing through hot areas (e.g., furnace and boiler rooms, kitchens).

Round ducts are insulated with flexible fiberglass insulation. Both flexible and slab (board) insulation are used for rectangular ducts. The latter is made of spun fibrous glass wool. These lightweight, semirigid panels are available with a variety of facings (e.g., 0.0025 embossed aluminum foil) for appearance and functional use.

325

Flexible and slab insulation are also produced as duct liners for absorbing duct system noise. One or both sides are coated to reduce air friction loss and bind surface fibers. They also insulate thermally.

Flexible insulation is secured in place with light-gauge wire ties. Slab insulation is secured to the duct surface with adhesive or mechanical clips, and the cracks (seams) are sealed with asbestos tape.

It is recommended that a vapor barrier be placed between the insulating material and the duct surface to prevent the formation of condensation. The vapor barrier should be used when the temperature of the air inside the duct is lower than the dew point temperature of the air surrounding the duct.

EQUAL FRICTION METHOD

The *equal friction* method of sizing ducts is recommended because it does not require a great deal of experience in the selection of proper velocities in the various sections of the duct system. It is necessary to select the main duct velocity consistent with good practice from a standpoint of noise for a particular building or application. In this duct-sizing method, the duct design is based primarily on a consistent pressure loss for each foot of duct.

Proportioning for equal friction is more advantageous than reducing the velocity in a haphazard manner because the friction calculation is greatly simplified. In calculating the friction, it is only necessary to know the length of the longest run, the number and size of elbows, and the diameter and velocity of the largest duct. The friction loss is exactly the same as though the entire amount of air were carried the whole distance through the largest duct.

The air velocities listed in Table 7-5 have been found to accomplish satisfactory results in engineering practice. Where quiet operation is essential, the blower or fan should be selected on the basis of a low outlet velocity. This will also result in lower operating costs.

Table 7-5. Recommended and Maximum Air Velocities

Designation	Recommended Velocities, FPM		
	Residences	Schools, Theaters, Public Buildings	Industrial Buildings
Outdoor Air Intakes*	500	500	500
Filters*	250	300	350
Heating Coils*	450	500	600
Air Washers	500	500	500
Fan Outlets	1000-1600	1300-2000	1600-2400
Main Ducts	700-900	1000-1300	1200-1800
Branch Ducts	600	600-900	800-1000
Branch Risers	500	600-700	800
	Maximum Velocities, FPM		
Outdoor Air Intakes*	800	900	1200
Filters*	300	350	350
Heating Coils*	500	600	700
Air Washers	500	500	500
Fan Outlets	1700	1500-2200	1700-2800
Main Ducts	800-1200	1100-1600	1300-2200
Branch Ducts	700-1000	800-1300	1000-1800
Branch Risers	650-800	800-1200	1000-1600

*These velocities are for total face area, not the net free area; other velocities in table are for net free area.

Courtesy *ASHRAE 1960 Guide*

The following steps are involved in sizing ducts by the equal friction method:

1. Compute the total volume (in cubic feet) of the structure.
2. Compute the cubic foot volume of each room in the structure to be supplied with heated or cooled air. The volume of each room should be expressed as a percentage of the total volume of the structure. For example, a room having a volume of 6000 cu. ft. would represent 10 percent of a structure having a total volume of 60,000 cu. ft.
3. Compute the total amount of air to be handled by the blower or fan. This will be the total CFM for the entire structure and can be computed by the air change method:

$$\text{CFM} = \frac{\text{Building Volume in Cubic Feet}}{\text{Minutes Air Change}}$$

327

4. Determine the portion of the total amount of air to be delivered to each room. This is computed by multiplying the total CFM for the structure by the room volume percentage (see Step 2). For example, if the blower or of fan handles a required 15,000 CFM, then the room described in Step 2 would receive 1500 CFM (15,000 CFM × 10% = 1500 CFM).

5. Determine the design and location of the duct runs (supply, branch, and return runs) (Fig. 7-22) and then locate the supply air outlets and return air openings for each room to give the most uniform distribution of air. Each outlet should be selected for suitable air velocity and throw (the manufacturer's catalog will provide the necessary data).

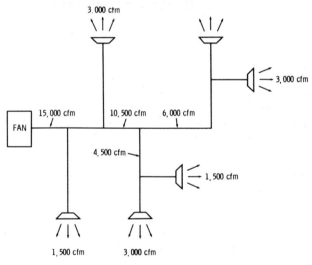

Fig. 7-22. Air velocities of a duct system.

6. Determine the total CFM for the main supply duct *before* any branch ducts are reached. This will be equal to the total CFM for the entire structure.

7. Determine the allowable air velocity in the main supply duct (Table 7-5). For a commercial building such as the one represented by this example (i.e., 60,000 cu. ft.), the air velocity in the main duct will be 1300 FPM.

8. Determine the static pressure drop from Fig. 7-23. Since the supply duct must carry 15,000 CFM, locate 15,000 on the vertical scale on the left side of the friction chart in Fig. 7-23. This will be found a little over half way between 10,000 and 20,000 on the scale. Draw a line horizontally across the chart

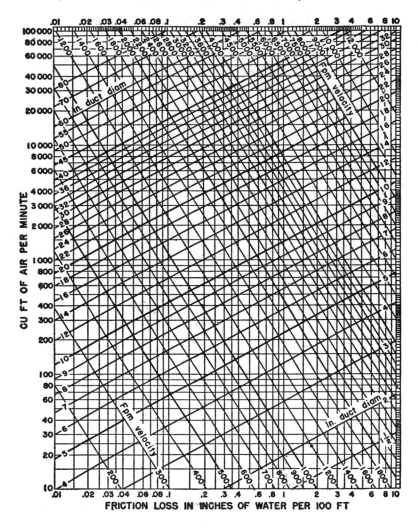

Fig. 7-23. Friction of air in ducts.

to where it intersects at a point halfway between the diagonals representing 1200 and 1400 FPM (i.e., 1300 FPM). This gives a static pressure drop for the main duct of approximately .04 in. of water per 100 ft. of duct. Use the static pressure drop of .04 in. as a constant for the entire duct system.

9. Use Fig. 7-23 to compute the round duct diameter for each branch duct. For example, branch No. 10 in Fig. 7-22 will have a diameter of approximately 19 in. This is determined by finding the 1500 CFM on the left-hand column of the chart in Fig. 7-23 (halfway between 1000 and 2000 CFM) and moving horizontally in a straight line to the right until it intersects with the vertical line representing a static pressure drop of .04 in. The two lines intersect at a point representing an approximate round duct diameter of 19 in.

10. Using the same data in Step 9, determine the air velocity for each branch duct. The air velocity of duct branch No. 10, for example, will be approximately 750 FPM.

11. Size the return air duct system by first determining the amount of return air required. In Step 4, the amount of supply air for the structure was stated as being 15,000 CFM. The amount of return air (CFM) required can be determined by subtracting the amount of fresh air intake from the supply air (CFM). The amount of fresh air intake (CFM) can be determined from the following formula:

$$CFM = \frac{\text{Volume of Structure}}{60 \text{ Min./Hr.}} \times \begin{array}{c}\text{Number of}\\\text{Air Changes}\\\text{per Hour}\end{array}$$

12. After the duct sizes have been determined, it is necessary to compute the external static pressure of the system so that a suitable fan can be selected which will handle the required volume of air (in this case 15,000 CFM) against the total static pressure of the system. In the equal friction method, the total external static pressure drop of the system is obtained by calculating the external static pressure for the duct run having the highest total resistance. For the duct system shown in Fig. 7-22, this can be obtained

by adding the total length of the main supply duct from the point where the air enters the system (i.e., duct sections numbered 1–6) *and* the equivalent straight pipe length of all elbows and transitions. This total length figure is then multiplied by the .04 in. static pressure constant (see Step 9) to obtain the *total* resistance (i.e., the total *external* static pressure drop) for the duct system. If, for example, the total length of duct was found to be 406 ft., the total resistance would be 0.1624 in. (406 ft. ÷ 100 ft. = 4.06 ft. × 0.04 = 0.1624 in.).

13. The total resistance for the return duct run is determined in the same manner as described for the supply air run. Because it represents less air volume than the supply air (return air = supply air − fresh air intake), it will always be a smaller figure than the static pressure determined for the supply duct run. As a rule-of-thumb, the return run static pressure will be about 25 percent of the supply run static pressure.

14. The total static pressure against which the blower or fan must operate includes:

a. Total resistance of the supply air duct system (i.e., the total *external* static pressure drop).

b. Total resistance of the return air duct system (i.e., the total external static pressure drop).

c. Total *internal* static pressure losses (i.e., resistance through filters, cooling coils, and other forms of equipment).

BALANCING AN AIR DISTRIBUTION SYSTEM

After the air distribution system has been installed, it should be tested to determine if the air delivery and distribution correspond with the system design. Engineers and experienced field workers use a number of different instruments to make the various measurements required for balancing the system. Because

331

these measuring devices are not generally available to the average person, the following simplified balancing procedure is suggested:

1. Open all duct and outlet dampers.
2. Check drafts, noise, and temperature differences from room to room while the fan or blower is operating.
3. Adjust the dampers to provide the greatest uniformity in operating characteristics.

This procedure is an abbreviated form of the one recommended for balancing a warm-air heating system. Read the appropriate sections of Chapter 6 of Volume 1 (Warm-Air Heating Systems) for more information.

DUCT MAINTENANCE

The ducts used in heating, ventilating, and air conditioning systems require very little maintenance other than periodic cleaning. Evidence that dust and dirt will accumulate in the ducts is often indicated by dirt streaks from ceiling and wall air supply outlets. Accumulations of dust and dirt in the ducts can be dangerous because they represent potential fire hazards. These are combustible materials that can be ignited when conditions are right. Ducts should be cleaned periodically to prevent these accumulations from forming. Doors should be placed in the ducts to provide access for cleaning.

Corrosion is very seldom a problem unless there is an accumulation of condensation over a long period of time. This can be prevented by keeping the ducts dry. A vapor barrier will accomplish this purpose (see "Duct Insulation").

ROOF PLENUM UNITS

Air distribution systems for commercial and industrial buildings sometimes require variable amounts of outside air for ventilation purposes to be mixed with return air from conditioned spaces within the structure. A roof-mounted insulated plenum to which distribution supply and return air ducts may be connected through a roof opening is required for such an installation. Outside air dampers are designed to adjust from closed to full open, and may be manually controlled by a multiposition remote potentiometer. When full open, the dampers will permit the introduction of a maximum of 80 percent outside air for ventilation purposes.

The Janitrol roof plenum unit shown in Fig. 7-24 is located next to the control box to which it is connected by means of a flexible conduit. The construction of the roof base and curb is illustrated in Fig. 7-25. Roof plenum units that do not provide for the addition of outside air to the system are also available.

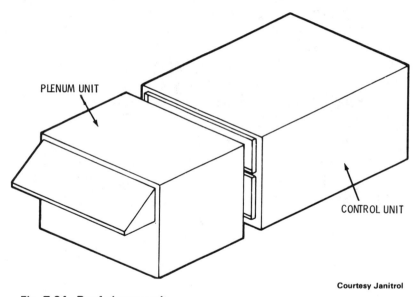

Courtesy Janitrol

Fig. 7-24. Roof plenum unit.

MOBILE HOME DUCT SYSTEMS

Duct kits for adding central air conditioning to forced warm-air heating systems are available from some manufacturers for installation in mobile homes, and particularly for installation in the "double-wide" combination mobile homes.

333

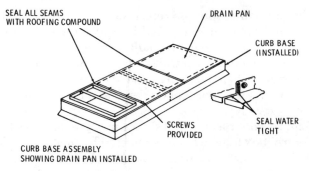

SEAL ALL SEAMS
WITH ROOFING COMPOUND

DRAIN PAN

CURB BASE
(INSTALLED)

SCREWS
PROVIDED

SEAL WATER
TIGHT

CURB BASE ASSEMBLY
SHOWING DRAIN PAN INSTALLED

Courtesy Janitrol

Fig. 7-25. Roof base and curb assembly.

A typical air conditioning duct kit installation is illustrated in Fig. 7-26. The air conditioner is located outdoors and the cool air is supplied to the mobile home by a round (12-in. diameter) flexible duct that leads from the air conditioning unit to the long

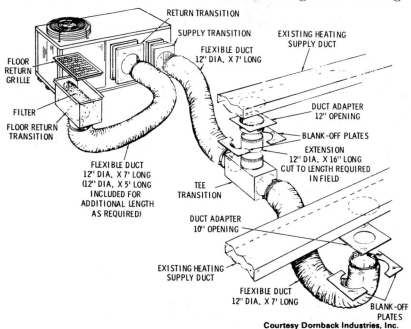

RETURN TRANSITION

SUPPLY TRANSITION

EXISTING HEATING
SUPPLY DUCT

FLEXIBLE DUCT
12" DIA. X 7' LONG

FLOOR
RETURN
GRILLE

FILTER

FLOOR RETURN
TRANSITION

DUCT ADAPTER
12" OPENING

BLANK-OFF PLATES

EXTENSION
12" DIA. X 16" LONG
CUT TO LENGTH REQUIRED
IN FIELD

FLEXIBLE DUCT
12" DIA. X 7' LONG
(12" DIA. X 5' LONG
INCLUDED FOR
ADDITIONAL LENGTH
AS REQUIRED)

TEE
TRANSITION

DUCT ADAPTER
10" OPENING

EXISTING HEATING
SUPPLY DUCT

FLEXIBLE DUCT
12" DIA. X 7' LONG

BLANK-OFF
PLATES

Courtesy Dornback Industries, Inc.

Fig. 7-26. Duct kit for installation in the double wide combination mobile home.

metal heating ducts located in the center of the structure. A transition tee with a short length of round, nonflexible metal duct connects the flexible duct leading from the air conditioner to the metal heating duct running the length of the mobile home through a *supply air opening* (not to be confused with the warm-air *registers* on the top of the heat supply duct and which are a part of the existing heating system).

A return air grille is located in the floor near the outside wall closest to the air conditioner. As shown in Fig. 7-27, the return

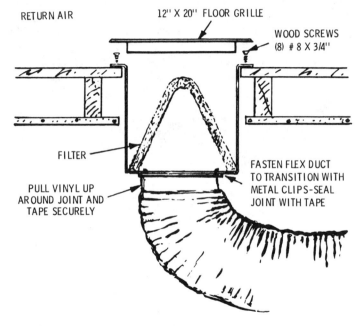

RETURN AIR   12" X 20" FLOOR GRILLE

WOOD SCREWS (8) # 8 X 3/4"

FILTER

PULL VINYL UP AROUND JOINT AND TAPE SECURELY

FASTEN FLEX DUCT TO TRANSITION WITH METAL CLIPS-SEAL JOINT WITH TAPE

CUT 12-1/4" X 20-1/4" OPENING IN BOTH MAIN AND LOWER FLOORS.
LOWER RETURN AIR DUCT ENCLOSURE THROUGH OPENING FROM INSIDE OF COACH.
SCREW DUCT TO MAIN FLOOR WITH #8 X 3/4 WOOD SCREWS.

Courtesy Dornback Industries, Inc.

Fig. 7-27. Construction details of the return air grille and filter.

grille is located over an enclosure which contains a filter for cleaning the air returning to the air conditioner.

The flexible, cool air supply duct is connected by cutting a 16-in. by 16-in. square opening in the flooring directly beneath the main heat supply duct at the location decided upon. A smaller 12½-in.-wide hole is then cut through the bottom of the heating duct. An adapter plate with a round opening is placed over the opening in the heating duct and secured in place with screws. Tape is then run around the outer edge of the adapter plate to seal it against air leaks (Fig. 7-28). Supply air outlets closest to the air conditioner use an adapter plate with a 12-in. opening. Those farthest away use an adapter plate with a 10-in. opening.

SUPPLY AIR (FURTHEST FROM UNIT)

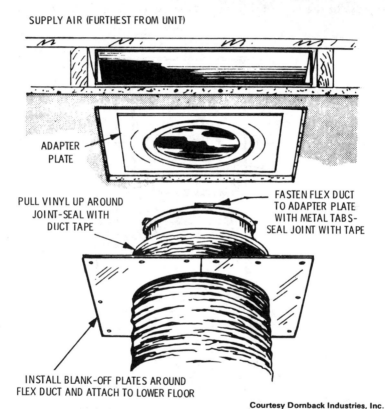

ADAPTER PLATE

PULL VINYL UP AROUND JOINT-SEAL WITH DUCT TAPE

FASTEN FLEX DUCT TO ADAPTER PLATE WITH METAL TABS-SEAL JOINT WITH TAPE

INSTALL BLANK-OFF PLATES AROUND FLEX DUCT AND ATTACH TO LOWER FLOOR

Courtesy Dornback Industries, Inc.

Fig. 7-28. Construction details of a supply air grille.

Before locating the best position for the air conditioner and routing the flexible duct, the positions of the supply air outlets and the return air grille should be determined. When doing this, the following recommendations are offered:

1. *Never* locate the supply air opening closer than 4 ft. from the furnace.
2. *Never* locate the supply air opening directly under a supply register.
3. Connect the *flexible*, cool air supply duct to the main heating duct as close to the center of the mobile home as possible.
4. Locate the return air grille in a room or area open to the rest of the mobile home so that proper air circulation may be maintained.
5. Provide adequate accessibility to the return air enclosure for cleaning or replacing the filter.

PROPRIETARY AIR DISTRIBUTION SYSTEMS

Proprietary air distribution systems are available from several manufacturers to provide supplementary heating or cooling in existing installations. They are also available as a complete heating and/or cooling system for installation in a structure where none previously existed or in which the existing system is clearly inadequate. They can also be used in new construction.

The Dunham-Bush Space-Pak Air Distribution System illustrated in Fig. 7-29 is an example of a proprietary system that provides total comfort conditioning, including heating, cooling, air cleaning, and humidifying.

The Dunham-Bush air distribution system is generally an attic or overhead installation, but it may also be installed in basements or any other suitable area with no impairment of its operating performance. Attic or overhead installation requires the construction of a mounting platform for the blower unit (Fig. 7-30). It is recommended that isolation pads or strips be placed between the blower-coil unit and the mounting platform to prevent vibration transmission through walls or floors (Fig. 7-31). If the blower-coil unit is suspended from a ceiling, then both the blower-coil and the electric heater are mounted on separate platforms.

The supply air is moved through a 7-in. insulated plenum duct with 2-in. insulated flexible tubing runs to air outlets. All compo-

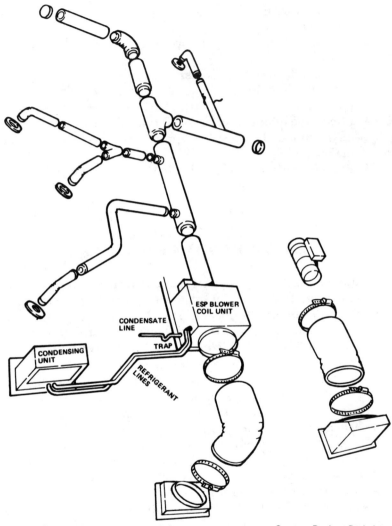

Fig. 7-29. Space-Pak air distribution system.

nents in the Dunham-Bush air distribution system snap or twist together. As in all duct systems, the number of tees and elbows should be limited to as few as possible in order to keep system pressure drop on larger layouts to a minimum.

338

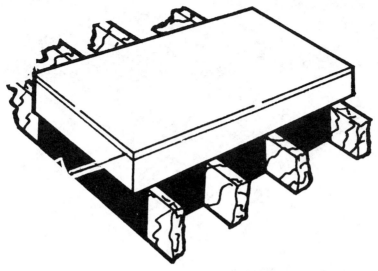

Fig. 7-30. Mounting platform for blower coil in attic or overhead installation.

A blower-coil unit and an attached electric heater (Fig. 7-32) provide either heat or cool air depending upon the control setting.

A cross section of a blower-coil unit is shown in Fig. 7-33. The float switch is used to interrupt compressor operation when the condensation level exceeds a normal operating level. A further safeguard would be to provide a secondary drain pan.

DUCT FURNACES

A *duct furnace* is essentially a unit heater designed for installation in a duct system. It is usually designed to operate on oil, natural gas, or propane gas, although electric duct heaters are also available and growing in popularity.

An example of a gas-fired (natural gas) furnace is the Janitrol 72 Series Duct Furnace illustrated in Fig. 7-34, which is available in 13 sizes depending upon the requirements of the installation. These are factory-assembled units inspected and tested before

339

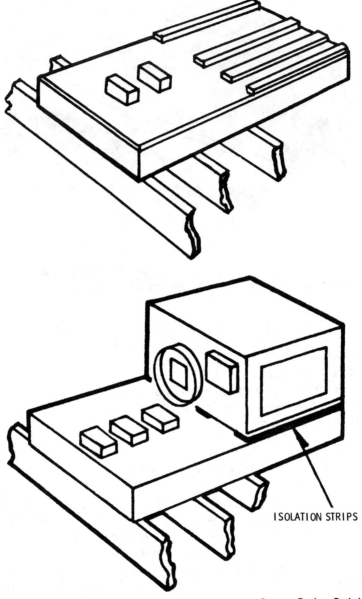

ISOLATION STRIPS

Fig. 7-31. Using isolation strips for vibration control.

340

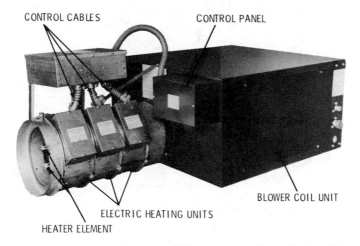

CONTROL CABLES

CONTROL PANEL

BLOWER COIL UNIT

ELECTRIC HEATING UNITS

HEATER ELEMENT

Courtesy Dunham-Bush, Inc.

Fig. 7-32. Blower-coil unit and electric heater.

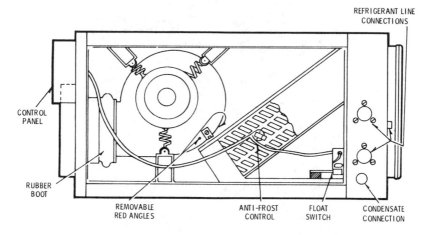

REFRIGERANT LINE
CONNECTIONS

CONTROL
PANEL

RUBBER
BOOT

REMOVABLE
RED ANGLES

ANTI-FROST
CONTROL

FLOAT
SWITCH

CONDENSATE
CONNECTION

Courtesy Dunham-Bush, Inc.

Fig. 7-33. Cross section of blower coil unit.

they are shipped. Once they reach their destination they should be unpacked and the contents carefully checked against the packing list. Missing or damaged parts should be reported to the supplier immediately. This is a procedure you should follow habitually whenever you receive a shipment of equipment.

341

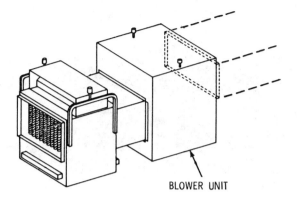

BLOWER UNIT

Courtesy Janitrol

Fig. 7-34. Series 72 gas-fired duct furnace.

342

The design of these Janitrol duct furnaces is AGA certified for use in a duct system with static pressure up to 2 in. of water and with temperature rises as shown in column 4 of Table 7-6.

Duct furnaces used in conjunction with cooling equipment should be installed in parallel with or on the *upstream* side of the cooling coils. If a parallel flow arrangement is used, the dampers (or other means of controlling the air flow) should be made tight enough to prevent the circulation of cooled air through the unit (see ANSI Z21.30).

When equipped with a suitable condensation pan, a duct furnace may be AGA certified for installation *downstream* from cooling coils, air washers, and evaporative coolers when operating as air-cooling systems. In the Janitrol 72 Series Duct Furnace illustrated in Fig. 7-34, the condensation pan is added in the following manner:

1. Remove the bottom panel of the duct furnace.
2. Remove the circular knock-out section in the panel.
3. Place the condensation drain connection of the condensation pan in the opening.
4. Reinstall the bottom panel.
5. Connect the drain line.
6. Provide disconnect adjacent to bottom panel.

If space limitations prevent bottom-panel removal, access is possible through the front panel and burners (both of which must be removed *before* inserting the condensation panel).

Duct furnaces may be installed downstream of evaporative coolers or air washers considered as air cooling system (i.e., operating with chilled water, which delivers air below the ambient air at the duct furnace).

The minimum clearances between the duct furnace (and its draft hood) and the nearest adjacent walls, ceilings, and floors of combustionable construction should be *at least* 6 in. (see ANSI Z21.30 and NEPA No. 31, *Installation of Oil Burning Equipment 1972*). Under certain circumstances the minimum clearance between the bottom of the duct furnace and the floor can be as close as 2 in., provided the other clearances (i.e., the 6-in. minimums) are maintained for the rest of the unit.

343

Table 7-6. Engineering Data for Series 72 Gas-fired Duct Furnaces

Unit Size	*AGA, Rating—Btu/hr.		Temp. Rise	EDR‡ steam	72-XXX-3		72-XXX-4	
	Input†	Output			Net Wt. Lbs.	Ship Wt. Lbs. Approx.	Net Wt. Lbs.	Approx. Ship Wt. Lbs.
72-100	100,000	80,000	25-100	333	148	160	140	152
72-125	125,000	100,000	25-100	418	168	180	154	166
72-150	150,000	120,000	25-100	500	168	180	158	170
72-175	175,000	140,000	25-90	586	156	178	148	170
72-200	200,000	160,000	25-100	667	210	232	188	210
72-225	225,000	180,000	25-100	750	200	222	188	210
72-250	250,000	200,000	25-100	834	212	300	260	282
72-300	300,000	240,000	25-100	1000	270	298	256	284
72-350	350,000	280,000	25-90	1167	262	290	256	284
72-400	400,000	320,000	25-100	1333	399	440	383	424
72-500	500,000	400,000	25-100	1667	469	518	455	504
72-600	600,000	480,000	25-100	2000	565	622	525	582
72-700	700,000	560,000	25-90	2333	567	624	545	602

*Tabled rating for elevations up to 2000 feet above sea level. From 2000 to 7000 feet, input must be reduced 4% per 1000 feet above sea level by manually adjusting manifold pressure. For elevation above 7000 feet, when ordering from factory, order must state elevation at which unit is to operate.

†Input of unit (Btu/hr.) ÷ heat value of gas (Btu/cu. ft.) = gas consumption (cu. ft./hr.)

‡Output ÷ 240 Btu.

When planning duct furnace clearances, consideration must be given to accessibility for the following:

1. Cleaning the heat exchanger.
2. Removal of burners.
3. Servicing the controls.

The minimum clearance should be 18 in. for front removal of the burners and 10 in. for bottom removal.

The inlet and outlet ducts should be attached to the duct connection flange. An access panel (or removable duct section) should be provided at both the inlet and outlet ducts to provide for servicing the limit and fan control elements. Moreover, such access should be so constructed as to provide for visual inspection of the heat exchangers.

In gas-fired duct furnaces, *all* gas piping *must* be run in accordance with requirements outlined in the American Gas Association's publication *Installation of Gas Appliances and Gas Piping*.

It is recommended that a pipe joint compound certified for use with LPG be used on all pipelines. If possible, run a *new* gas supply line directly to the duct furnace from the meter. Support the gas line with hangers positioned close enough together to prevent strain on the unit. A trap consisting of a tee with a capped nipple should be provided in the gas line when the unit is installed. After installing the gas line, test it for leaks with a soap solution (*never* with a flame).

Oil- and gas-fired units *must* be properly vented. The draft-head is built into these units, and the flue pipe must be the same size as the outlet of the flue collector on the duct furnace. *Never* reduce the size of the flue pipe or install a damper in it. Install the flue pipe to provide minimum clearances of 18 in. between it and combustible material. *Always* examine the chimney for proper construction and repair *before* connecting the flue pipe. Other details concerning venting practices are found in ANSI Z21.30 and NEPA No. 31, *Installation of Oil Burning Equipment 1972*. It is not necessary to vent electric-fired duct furnaces.

All electrical wiring for duct furnaces must be done in accordance with the National Electric Code, ANSI CI-1971, and local code requirements. The unit must be grounded in accordance with these codes.

If a pilot flame is used for ignition, the flame should extend 1 in. beyond the pilot burner. Flame adjustment is made either with the pilot flame adjustment device located on the gas valve (on units without a pilot gas valve), or with the built-in adjusting screw (on units supplied with a pilot gas valve). Pilot gas pressure regulators are used on natural gas-fired units in areas where gas-pressure variations are great. If a pilot regulator is supplied with the unit, pilot flame length is adjusted by adjusting the gas pressure regulator. These pressure regulators are *not* supplied with propane gas-fired units. In any event, the length of the pilot flame on propane gas-fired units is not adjustable. A normal flame on a natural gas-fired unit will be blue in color with an inner cone approximately 1 in. high. On a propane gas-fired unit, a normal flame will be green in color with a distinct inner cone approximately ⅛ to ¼ in. high. Fig. 7-35 shows a typical pilot assembly for a Janitrol gas-fired duct furnace.

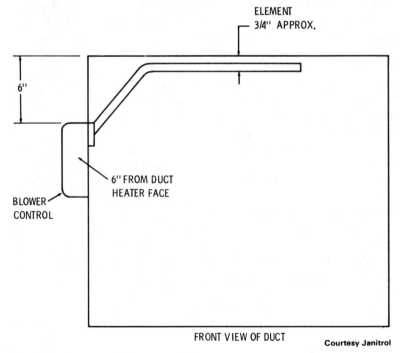

ELEMENT
3/4" APPROX.

6"

6" FROM DUCT HEATER FACE

BLOWER CONTROL

FRONT VIEW OF DUCT

Courtesy Janitrol

Fig. 7-35. Location of fan control element.

The thermocouple pilot and thermopilot relay on Janitrol gas-fired duct furnaces can be checked for proper functioning as follows:

1. Read the steps outlined on the *operating instruction plate*.
2. Start the unit and allow the pilot to heat for at least 3 minutes (do *not* turn on the main burners at this time).
3. Close the pilot valve, and wait for the pilot relay switch to open and cause the electric gas valve to close.
4. Under normal circumstances, the length of time between the closing of the pilot valve and the opening of the pilot relay switch should be *less* than 2 minutes.
5. If the length of time is *greater* than 2 minutes, then the pilot relay must be replaced.

The gas input for a gas-fired duct furnace must not be greater than specified on the rating plate of the unit. Duct furnaces are shipped with spuds containing orifices sized for the particular gas (natural gas or propane) with which they are fired. Table 7-7 lists the spud sizes for natural gas and propane available for use in Janitrol duct furnaces.

Table 7-7. Orifice Sizes

Unit Size	Natural	Propane
100	40	54-38
125	43	.0492-48
150	40	54-38
175	37	53-30
200	40	54-38
225	37	53-30
250	43	.0492-48
300	40	54-38
350	37	53-30
400	40	54-38
500	40	54-38
600	40	54-38
700	37	53-30

Courtesy Janitrol

In natural gas-fired duct furnaces, the main burner gas-pressure regulator must be adjusted for the correct gas input. This may be done either by timing the test dial on the meter or by checking the manifold pressure (Table 7-8). The following steps are recommended for adjusting the gas input:

1. Remove the cap on the top of the regulator.
2. Turn the adjusting screw in (or clockwise) to increase the gas input.
3. Back out the adjusting screw (turn it counterclockwise) to reduce the gas input.

Table 7-8. Natural Gas Manifold Pressures

Btu Per Cu. Ft.	Sp. Gr.	Man. Press. (In. Water)	Btu Per Cu. Ft.	Sp. Gr.	Man. Press. (In. Water)
900	0.50	3.4	1000	0.55	3.0
	.55	3.7		.60	3.3
	.60	4.1		.65	3.6
	.65	4.4		.70	3.9
925	0.50	3.2	1025	0.55	2.9
	.55	3.5		.60	3.1
	.60	3.9		.65	3.4
	.65	4.2		.70	3.7
950	0.50	3.1	1050	0.55	2.7
	.55	3.4		.60	3.0
	.60	3.7		.65	3.2
	.65	4.0		.70	3.5
	.70	4.3			
975	0.50	3.2	1075	0.55	2.6
	.60	3.5		.60	2.9
	.65	3.8		.65	3.2
	.70	4.1		.70	3.3
			1100	0.55	2.5
				.60	2.7
				.65	2.9
				.70	3.2

NOTE: Manifold pressures on this table are based on orifice sizes as shown in orifice table in installation manual. Pressures given in this table apply to sizes of units. This table does not apply to units used in high altitude areas. See supplement for high altitude manifold pressure table.

Courtesy Janitrol

The gas pressure at the inlet to the regulator on natural gas-fired duct furnaces should *not* be allowed to exceed 12 in. of water. Propane gas-fired units *must* maintain a manifold pressure of 11 in. of water for proper operation. These units are not supplied with appliance gas-pressure regulators.

Some duct furnace installations require the use of a fan control. When this is the case, the element of the fan control should be located as shown in Fig. 7-35. Adjust the fan control for an "off" temperature as low as possible without causing the occupants to experience a feeling of cold air.

A proper maintenance schedule for duct furnaces will increase the life of the equipment and result in more efficient and economical operating characteristics. Frequently check burners, pilot, and the interior and exterior of the heat exchanger for a buildup of residue. Clean the exterior of the heat exchanger brushes, compressed air, or a heavy-duty vacuum. The following steps are recommended for the more complicated procedure of cleaning the interior of the heat exchanger:

1. Disconnect the flue pipe from the unit.
2. Remove the pilot and burners.
3. Remove the flue collector and heat exchanger tube baffles.
4. Brush the interior of the heat exchanger with a flexible 1 ½-in. diameter bristle brush.
5. Remove debris with a vacuum.
6. Replace parts in reverse order to which they were removed.

On a gas-fired duct furnace, leave the pilot on during the summer (*except* when the unit is part of an air conditioning system). Inspect the flue pipe for deterioration at the beginning of the heating season. Make a similar inspection of the heat exchanger (and related components) for carbon deposit, rust, or corrosion. Clean or replace when required.

Never light the pilot on gas-fired duct furnaces until you have first read the lighting instructions on the unit plate. Select and maintain a thermostat setting that provides adequate comfort. Do *not* keep changing the thermostat setting. It will only result in higher heating costs.

ELECTRIC DUCT HEATERS

Electric duct heaters are designed and manufactured to function in the same manner as oil- or gas-fired duct furnaces. These

349

are prewired factory assembled units available in a wide range of sizes and heating capacities for a variety of different installations. Typical duct heater construction is illustrated by the Vulcan unit in Fig. 7-36. The frame for the electric resistors is designed for insertion into the duct as shown in Figs. 7-37 through 7-44.

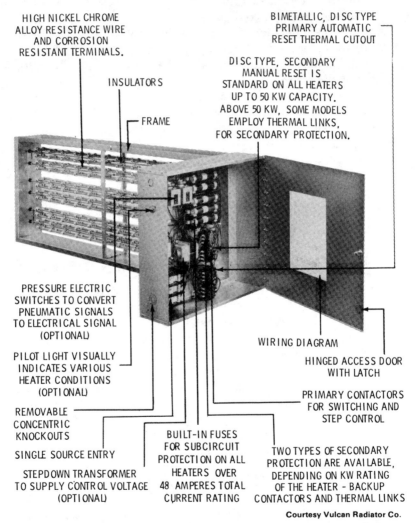

HIGH NICKEL CHROME ALLOY RESISTANCE WIRE AND CORROSION RESISTANT TERMINALS.

INSULATORS

FRAME

BIMETALLIC, DISC TYPE PRIMARY AUTOMATIC RESET THERMAL CUTOUT

DISC TYPE, SECONDARY MANUAL RESET IS STANDARD ON ALL HEATERS UP TO 50 KW CAPACITY. ABOVE 50 KW, SOME MODELS EMPLOY THERMAL LINKS, FOR SECONDARY PROTECTION.

PRESSURE ELECTRIC SWITCHES TO CONVERT PNEUMATIC SIGNALS TO ELECTRICAL SIGNAL (OPTIONAL)

PILOT LIGHT VISUALLY INDICATES VARIOUS HEATER CONDITIONS (OPTIONAL)

REMOVABLE CONCENTRIC KNOCKOUTS

SINGLE SOURCE ENTRY

STEPDOWN TRANSFORMER TO SUPPLY CONTROL VOLTAGE (OPTIONAL)

WIRING DIAGRAM

HINGED ACCESS DOOR WITH LATCH

PRIMARY CONTACTORS FOR SWITCHING AND STEP CONTROL

BUILT-IN FUSES FOR SUBCIRCUIT PROTECTION ON ALL HEATERS OVER 48 AMPERES TOTAL CURRENT RATING

TWO TYPES OF SECONDARY PROTECTION ARE AVAILABLE, DEPENDING ON KW RATING OF THE HEATER - BACKUP CONTACTORS AND THERMAL LINKS

Courtesy Vulcan Radiator Co.

Fig. 7-36. Construction of a Vulcan electric duct heater.

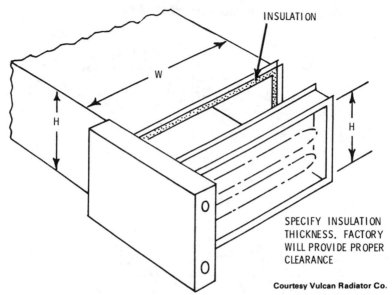

INSULATION

W

H

H

SPECIFY INSULATION
THICKNESS. FACTORY
WILL PROVIDE PROPER
CLEARANCE

Courtesy Vulcan Radiator Co.

Fig. 7-37. Flanged heater with internally insulated duct.

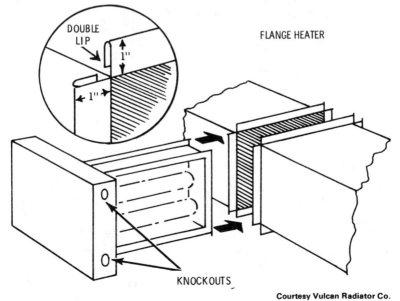

DOUBLE
LIP

1"

1"

FLANGE HEATER

KNOCKOUTS

Courtesy Vulcan Radiator Co.

Fig. 7-38. Flange heater.

351

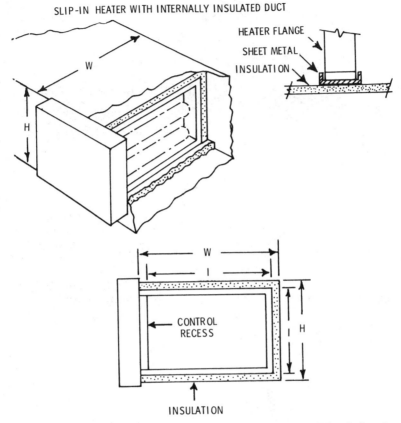

SLIP-IN HEATER WITH INTERNALLY INSULATED DUCT

HEATER FLANGE

SHEET METAL

INSULATION

W

H

W

CONTROL
RECESS

H

INSULATION

Courtesy Vulcan Radiator Co.

Fig. 7-39. Slip-in heater with internally insulated duct.

These Vulcan duct heaters are available in five standard supply voltage ratings (120, 277 single phase; 208, 240, and 480 single or three phase) or a specific voltage depending on the requirements of the installation.

The control housing is located on the outside of the duct and should be located so that the control panel is accessible for inspection and service. Various standard and optional built-in control components are available, including: (1) staging contactors, (2) control transformers, (3) sequencers, (4) fuses, and (5) primary and secondary protection.

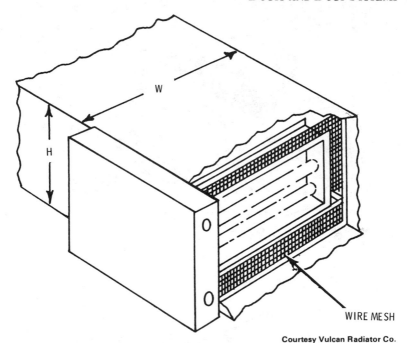

WIRE MESH

Courtesy Vulcan Radiator Co.

Fig. 7-40. When a heater is smaller than the duct area, the opening between the heater frame and duct must be filled with wire mesh or expanded metal.

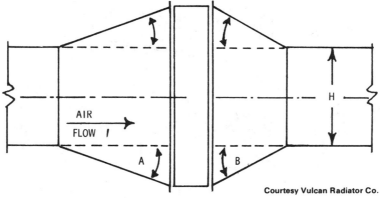

AIR FLOW

A

B

H

Courtesy Vulcan Radiator Co.

Fig. 7-41. When a heater is larger than the duct area, the duct cross section may be increased by a sheet-metal transition as shown in this drawing.

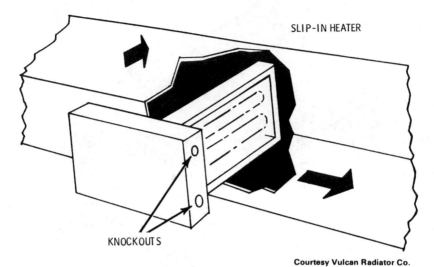

SLIP-IN HEATER

KNOCKOUTS

Courtesy Vulcan Radiator Co.

Fig. 7-42. The installation of a slip-in heater.

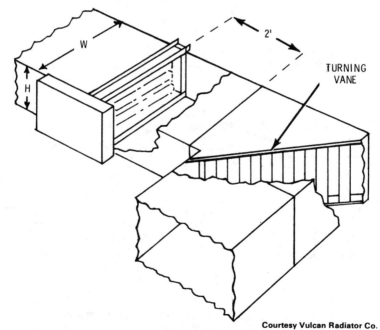

W

H

2'

TURNING
VANE

Courtesy Vulcan Radiator Co.

Fig. 7-43. The installation of a turning vane.

354

There are three types of UL listed thermal cutouts used as safety limit controls in Vulcan duct heaters. These are: (1) primary automatic reset, (2) secondary manual reset, or (3) secondary heat limiters (thermal links). The function of each of these limit controls is to shut off the duct heater when there is no air or when the air flow is too low for efficient operation.

The *primary automatic reset*, a high-limit control, is a snap-acting device that is sensitive to both radiant and convected heat. It is designed to deenergize the duct heater at a preselected, higher than normal temperature and to automatically reenergize the unit when it cools.

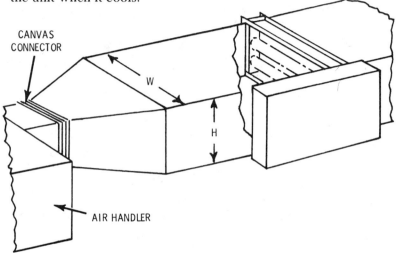

CANVAS
CONNECTOR

W

H

AIR HANDLER

Fig. 7-44. The installation of a canvas connector to minimize vibration.

The *secondary manual reset* control is a warning device set to open at a temperature higher than the primary reset. It cannot be reset until the heater has cooled below the set point, and it must be reset manually.

The *heat limiters (thermal links)* belong to the secondary protection system. These are fusible, *one-time* protective devices that must be replaced in the event they cut out.

Duct heater circuits may be subdivided into equal or unequal heating increments or stages in order to more closely match heater output with temperature variations.

CHAPTER 8

Pipes, Pipe Fittings, and Piping Details

Pipes are available in a variety of materials, sizes, and weights. The type selected for a particular installation will depend on a number of factors, including government codes and standards, specifications, system requirements, availability of materials, and cost. Some of these factors, such as codes and standards, will have priorities over others, but all should be given equal consideration before deciding which pipe to use. The purpose of this chapter is to offer some guidance for making these decisions.

TYPES OF PIPE MATERIALS

The most common metals used in the manufacture of piping for heating, ventilating, and air conditioning systems are: (1) iron, (2) steel, and (3) copper and brass.

Steel piping leads in popularity and is available either in the form of wrought steel or galvanized steel. Wrought-iron and cast-iron pipe are also used, but less frequently because of their higher cost. Copper and brass are employed in the manufacture of both pipes and tubes, and find their greatest application in air conditioning and refrigeration systems. Other metals (e.g., aluminum, bronze, and alloy metals) have also been used in the manufacture of this pipe, but with limited application due to cost, lack of availability, and other factors.

This chapter concentrates on a description of iron, steel, copper, and brass pipes; the pipe fittings used with them; and the methods employed in installing them.

Wrought-Iron Pipe

The pipe ordinarily used on heating jobs and for many other applications is what is known as *wrought pipe*. The material commonly used is made of wrought iron or wrought steel.

Wrought-iron pipe differs very little in appearance from wrought-steel pipe. Indeed, were it not for special markings, the casual observer would not be able to distinguish between the two. Some wrought-iron pipe is stamped "genuine wrought iron" on each length. More frequently, the distinguishing mark is a spiral line marked into each length of pipe. The spiral indentification line may be painted on to the surface in some bright color (usually red) or knurled into the metal.

The pipe used in heating installations is manufactured for different pressures and is available in various rated sizes (also referred to as the *nominal inside diameter*). The three *grades* of wrought-iron pipe are:

1. Standard.
2. Extra strong (or heavy).
3. Double extra strong (or very heavy).

The pipe sections shown in Fig. 8-1 are approximately half their actual size. Fig. 8-2 illustrates their actual sizes, showing proportions of the three grades of wrought pipes.

The diameters given for pipes are far from the *actual* diameters, especially in the small sizes. Thus, a pipe known as ¼ in.

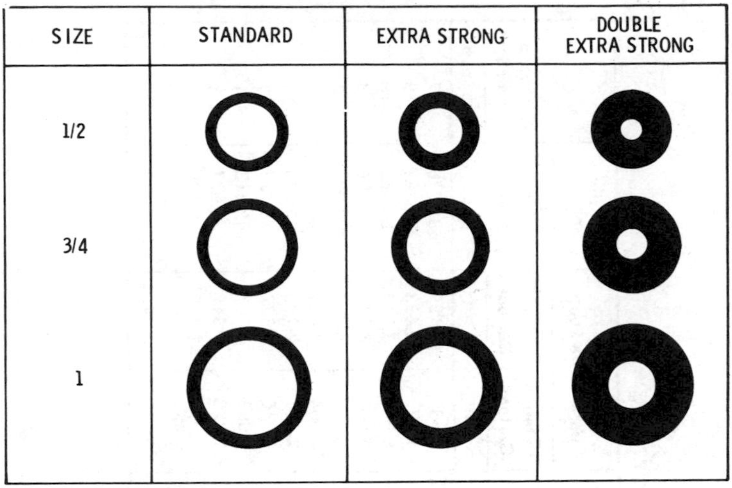

SIZE	STANDARD	EXTRA STRONG	DOUBLE EXTRA STRONG
1/2			
3/4			
1			

Fig. 8-1. Three sizes of standard, extra strong, and double extra strong wrought pipe.

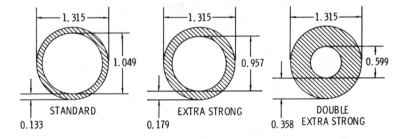

Fig. 8-2. The three weights of wrought pipe shown actual size.

(rated size) has an outside diameter of 0.54 in. and an inside diameter of 0.364 in. Dimensions for standard, extra strong, and double extra strong pipe are given in Tables 8-1, 8-2, and 8-3.

Wrought-iron pipes are adapted to higher pressures by making the walls thicker, but without changing the outside diameters. It is the inside diameter that is reduced.

Wrought-Steel Pipe

Wrought-steel pipe is cheaper than wrought-iron pipe, and consequently is used more widely in heating, ventilating, and air

359

Table 8-1. Dimensions of Standard Pipe

Nominal Size Inches	Diameter		Nominal Thickness Inches	Nominal Weight per Foot Threaded and Coupled Pounds	Transverse Area		Length of Pipe per Square Foot		Length of Pipe Containing 1 Cubic Foot Feet	Number of Threads Per Inch
	External Inches	Internal Inches			External Sq. In.	Internal Sq. In.	External Surface Feet	Internal Surface Feet		
1/8	0.405	0.269	0.068	0.245	0.129	0.057	9.431	14.199	2533.775	27
1/4	0.540	0.364	0.088	0.425	0.229	0.104	7.073	10.493	1383.789	18
3/8	0.675	0.493	0.091	0.568	0.358	0.191	5.658	7.748	754.360	18
1/2	0.840	0.622	0.109	0.852	0.554	0.304	4.547	6.141	473.906	14
3/4	1.050	0.824	0.113	1.134	0.866	0.533	3.637	4.635	270.034	14
1	1.315	1.049	0.133	1.684	1.358	0.864	2.904	3.641	166.618	11 1/2
1 1/4	1.660	1.380	0.140	2.281	2.164	1.495	2.301	2.768	96.275	11 1/2
1 1/2	1.900	1.610	0.145	2.731	2.835	2.036	2.010	2.372	70.733	11 1/2
2	2.375	2.067	0.154	3.678	4.430	3.355	1.608	1.847	42.913	11 1/2
2 1/2	2.875	2.469	0.203	5.819	6.492	4.788	1.328	1.547	30.077	8
3	3.500	3.068	0.216	7.616	9.621	7.393	1.091	1.245	19.479	8
3 1/2	4.000	3.548	0.226	9.202	12.566	9.886	0.954	1.076	14.565	8
4	4.500	4.026	0.237	10.889	15.904	12.730	0.848	0.948	11.312	8

Table 8-1. Dimensions of Standard Pipe (Cont'd)

Nominal Size	Diameter		Nominal Thickness	Nominal Weight per Foot Threaded and Coupled	Transverse Area		Length of Pipe per Square Foot		Length of Pipe Containing 1 Cubic Foot	Number of Threads
	External	Internal			External	Internal	External Surface	Internal Surface		
Inches	Inches	Inches	Inches	Pounds	Sq. In.	Sq. In.	Feet	Feet	Feet	Per Inch
5	5.563	5.047	0.258	14.810	24.306	20.006	0.686	0.756	7.198	8
6	6.625	6.065	0.280	19.185	34.472	28.891	0.576	0.629	4.984	8
8	8.625	8.071	0.277	25.000	58.426	51.161	0.442	0.473	2.815	8
8	8.625	7.981	0.322	28.809	58.426	50.027	0.442	0.478	2.878	8
10	10.750	10.192	0.279	32.000	90.763	81.585	0.355	0.374	1.765	8
10	10.750	10.136	0.307	35.000	90.763	80.691	0.355	0.376	1.785	8
10	10.750	10.020	0.365	41.132	90.763	78.855	0.355	0.381	1.826	8
12	12.750	12.090	0.330	45.000	127.676	114.800	0.299	0.315	1.254	8
12	12.750	12.000	0.375	50.706	127.676	113.097	0.299	0.318	1.273	8
14	15.000	14.250	0.375	60.375	176.715	159.485	0.254	0.268	0.903	8
17	17.000	16.214	0.393	72.602	226.980	206.476	0.224	0.235	0.697	8
18	18.000	17.182	0.409	80.482	254.469	231.866	0.212	0.222	0.621	8
20	20.000	19.182	0.409	89.617	314.159	288.986	0.190	0.199	0.498	8

Table 8-2. Dimensions of Extra Strong Pipe

Nominal Size	Diameter		Nominal Thickness	Weight Nominal per Foot Plain Ends	Transverse Area		Length of Pipe per Square Foot		Length of Pipe Containing 1 Cubic Foot
	External	Internal			External	Internal	External Surface	Internal Surface	
Inches	Inches	Inches	Inches	Pounds	Sq. In.	Sq. In.	Feet	Feet	Feet
1/8	0.405	0.215	0.095	0.314	0.129	0.036	9.431	17.766	3966.393
1/4	0.540	0.320	0.119	0.535	0.229	0.072	7.073	12.648	2010.290
3/8	0.675	0.423	0.126	0.738	0.358	0.141	5.658	9.030	1024.689
1/2	0.840	0.546	0.147	1.087	0.554	0.234	4.547	6.995	615.017
3/4	1.050	0.742	0.154	1.473	0.866	0.433	3.637	5.147	333.016
1	1.315	0.957	0.179	2.171	1.358	0.719	2.904	3.991	200.198
1 1/4	1.660	1.278	0.191	2.996	2.164	1.283	2.301	2.988	112.256
1 1/2	1.900	1.500	0.200	3.631	2.835	1.767	2.010	2.546	81.487
2	2.375	1.939	0.218	5.022	4.430	2.953	1.608	1.969	48.766
2 1/2	2.875	2.323	0.276	7.661	6.492	4.238	1.328	1.644	33.976
3	3.500	2.900	0.300	10.252	9.621	6.605	1.091	1.317	21.801
3 1/2	4.000	3.364	0.318	12.505	12.566	8.888	0.954	1.135	16.202
4	4.500	3.826	0.337	14.983	15.904	11.497	0.848	0.998	12.525
5	5.563	4.813	0.375	20.778	24.306	18.194	0.686	0.793	7.915
6	6.625	5.761	0.432	28.573	34.472	26.067	0.576	0.663	5.524
8	8.625	7.625	0.500	43.388	58.426	45.663	0.442	0.500	3.154
10	10.750	9.750	0.500	54.735	90.063	74.662	0.355	0.391	1.929
12	12.750	11.750	0.500	65.415	127.676	108.434	0.299	0.325	1.328

Table 8-3. Dimensions of Double Extra Strong Pipe

Nominal Size	Diameter		Nominal Thickness	Nominal Weight per Foot Plain Ends	Transverse Area		Length of Pipe per Square Foot		Length of Pipe Containing 1 Cubic Foot
	External	Internal			External	Internal	External Surface	Internal Surface	
Inches	Inches	Inches	Inches	Pounds	Sq. In.	Sq. In.	Feet	Feet	Feet
½	0.840	0.252	0.294	1.714	0.554	0.050	4.547	15.157	2887.165
¾	1.050	0.434	0.308	2.440	0.866	0.148	3.637	8.801	973.404
1	1.315	0.599	0.358	3.659	1.358	0.282	2.904	6.376	510.998
1¼	1.660	0.896	0.382	5.214	2.164	0.630	2.301	4.263	228.379
1½	1.900	1.100	0.400	6.408	2.835	0.950	2.010	3.472	151.526
2	2.375	1.503	0.436	9.029	4.430	1.774	1.608	2.541	81.162
2½	2.875	1.771	0.552	13.695	6.492	2.464	1.328	2.156	58.457
3	3.500	2.300	0.600	18.583	9.621	4.155	1.091	1.660	34.659
4	4.500	3.152	0.674	27.541	15.904	7.803	0.848	1.211	18.454
5	5.563	4.063	0.750	38.552	24.306	12.966	0.686	0.940	11.107
6	6.625	4.897	0.864	53.160	34.472	18.835	0.576	0.780	7.646
8	8.625	6.875	0.875	72.424	58.426	37.122	0.443	0.555	3.879

conditioning than the latter. Depending on the method of manufacture, wrought-steel pipe is available either as *welded pipe* or *seamless pipe*. Seamless wrought-steel pipe finds frequent application in high-pressure work.

The wall thicknesses and weights of wrought-steel pipe are approximately the same as those for wrought-iron pipe. As with wrought-iron pipe, the two most commonly used weights are *standard* and *extra strong*. Theoretical bursting and working pressures for wrought-steel pipe are listed in Table 8-4.

Galvanized Pipe

Galvanized steel or iron pipe is covered with a protective coating to resist corrosion. This type of pipe is often used underground or in other areas subject to corrosion. The coating is not permanent, and care should be used when handling it to avoid nicks and scratches. If the surface coating is broken, corrosion will begin that much sooner. Galvanized pipe is cheaper than copper pipe but more expensive than either wrought-iron or wrought-steel pipe.

Copper and Brass Pipes

Copper and brass are used in the manufacture of both pipes and tubes and find their greatest application in air conditioning and refrigeration applications. One major advantage of using copper or brass is that both metals offer resistance to corrosion.

Copper tubing is available in the form of hard grained copper tubes or soft copper (or annealed) tubes. The former is subject to freezing, which can cause the tubes to twist in almost the same way as steel pipes. On the other hand, the stiffness of hard grained copper tubes enables them to hold their shape better than the softer ones. They are, therefore, often used for exposed lines such as mains hung from the ceiling. Because of the relative ease with which soft copper tubes can be bent and shaped, they are especially well adapted for making connections around furnaces, boilers, oil-burning equipment, and other obstructions.

Sometimes the terms "copper tube" and "copper pipe" are used interchangeably as if they were synonymous. Another semantic idiosyncrasy is to refer to the same product as "tube" or

Table 8-4. Wrought-Steel Pipe

Theoretical bursting and working pressures pounds per square inch.

Size Inches	Size Millimeters	Standard Bursting Pressure Barlow's Formula	Standard Working Pressure Factor 8	Extra Strong Bursting Pressure Barlow's Formula	Extra Strong Working Pressure Factor 8	Double Extra Strong Bursting Pressure Barlow's Formula	Double Extra Strong Working Pressure Factor 8	Large O.D 3/8-Inch Thick Bursting Pressure Barlow's Formula	Large O.D 3/8-Inch Thick Working Pressure Factor 8	Large O.D 1/2-Inch Thick Bursting Pressure Barlow's Formula	Large O.D 1/2-Inch Thick Working Pressure Factor 8
1/8	3	13,432	1679	18,760	2345						
1/4	6	13,032	1629	17,624	2204						
3/8	10	10,784	1348	14,928	1866						
1/2	13	10,384	1298	14,000	1750						
3/4	19	8,608	1076	11,728	1466						
1	25	8,088	1011	10,888	1361	28,000	3500				
1 1/4	32	6,744	843	9,200	1150	23,464	2933				
1 1/2	38	6,104	763	8,416	1052	21,776	2722				
2	50	5,184	648	7,336	917	18,408	2301				
2 1/2	64	5,648	706	7,680	960	16,840	2105				
3	76	4,936	617	6,856	857	14,680	1835				
3 1/2	90	5,610	701	7,950	994	15,360	1920				
4	100	5,266	658	7,480	935	13,714	1714				
4 1/2	113	4,940	618	7,100	887	15,900	1987				
5	125	4,630	579	6,740	842	14,970	1871				
6	150	4,220	528	6,520	815	14,200	1775				
7	175	3,940	493	6,550	819	13,480	1685				
8	200	3,730	466	5,780	722	13,040	1630				
9	225	3,550	444	5,190	649	11,470	1434				
10	300	3,390	424	4,650	581	10,140	1267				
12	300	2,940	368	3,920	490						
14	350							2,680	335	3,570	446
15	375							2,500	313	3,333	417
16	400							2,340	293	3,120	390
18	450							2,080	260	2,770	346
20	500							1,870	234	2,500	313
22	550							1,700	213	2,270	284
24	600							1,560	195	2,080	260

In the above table, butt welded pipe was figured on sizes 3 in. and smaller and lap-welded pipe sizes 3 1/2 in. and larger.

"tubing" when still on the inventory at the supply house, but as "pipe" or "piping" when installed. Both usages can be confusing because there are differences between the two. For example, copper pipe is often made thicker than tubing because it *can* be used with threaded fittings if soldering is not desired for making the joint connection. Another point to remember is that copper pipes have the same outside diameters as standard steel pipes. Copper tubing, on the other hand, is standardized on the basis of use into three standard wall-thickness schedules (Table 8-5): (1) Type K, (2) Type L, and (3) Type M. Type L and Type M are used in heating, ventilating, and air conditioning systems.

Brass tube and brass pipe should also be distinguished from one another. Brass tubing is generally manufactured from yellow brass (about 65% copper, 35% zinc). Brass piping is more frequently a red brass (about 85% copper, 15% zinc), and is much stronger than the tubing. As a result, brass pipe is sometimes used in heating, ventilating, and air conditioning systems, particularly when it is necessary to use a pipe material that strongly resists corrosion.

PIPE FITTINGS

Pipe cannot be obtained in unlimited lengths. In most pipe installations, it is frequently necessary to join together two or more shorter lengths of pipe in order to create the longer one required by the blueprints. Furthermore, in practically all pipe installations there are numerous changes in directions and branches that require joining the pipes together in special arrangements. Pipe fittings have been devised for the necessary connections.

A *pipe fitting* may therefore be defined as any piece attached to pipes in order to lengthen a pipe, to alter its direction, to connect a branch to a main, to connect two pipes of different sizes, or to close an end.

Classification of Pipe Fittings

The pipe fittings used in heating, ventilating, and air conditioning installations are most commonly either screwed or flanged

PIPES, PIPE FITTINGS, AND PIPING DETAILS

Table 8-5. Sizes and Dimensions of Copper Water Tubes

Sizes		For Underground Services and General Plumbing Purposes, Used with Solder or Flared Fittings		For General Plumbing and House Heating Purposes, Used with Solder or Flared Fittings		For General Plumbing and House Heating Purposes, with Normal Water Conditions. Used with Solder Fittings Only	
		Type K Hard or Soft		Type L Hard or Soft		Type M Hard	
Nominal Size Inches	Outside Diameter Inches	Wall Thickness	Pounds Per Ft.	Wall Thickness	Pounds Per Ft.	Wall Thickness	Pounds Per Ft.
1/8	.250	.032	.085	.025	.068	.025	.068
1/4	.375	.032	.133	.030	.126	.025	.106
3/8	.500	.049	.269	.035	.198	.025	.144
1/2	.625	.049	.344	.040	.285	.028	.203
5/8	.750	.049	.418	.042	.362	.030	.263
3/4	.875	.065	.641	.045	.455	.032	.328
1	1.125	.065	.839	.050	.655	.035	.464
1 1/4	1.375	.065	1.04	.055	.884	.042	.681
1 1/2	1.625	.072	1.36	.060	1.14	.049	.94
2	2.125	.083	2.06	.070	1.75	.058	1.46
2 1/2	2.625	.095	2.92	.080	2.48	.065	2.03
3	3.125	.109	4.00	.090	3.33	.072	2.68
3 1/2	3.625	.109	5.12	.100	4.29	.083	3.58
4	4.125	.134	6.51	.110	5.38	.095	4.66
5	5.125	.160	9.67	.125	7.61	.109	6.66
6	6.125	.192	13.87	.140	10.20	.122	8.91
8	8.125	.271	25.90	.200	19.29	.170	16.46
10	10.125	.338	40.26	.250	30.04	.212	25.57
12	12.125	.405	57.76	.280	40.36	.254	36.69

types. Screwed pipe fittings use a male and female thread combination, which tightens together to form the joint. Screwed pipe fittings are designated as either male or female, depending upon the location of the thread. A female thread is an internal thread, and a male thread is an external one.

367

A flanged pipe fitting has a lip or extension projecting at a right angle to its surface. This lip is bolted to the facing lip of the adjacent fitting or additional strength. As a result, flanged fittings are generally recommended for 4-in. pipe and above.

Screwed and flanged pipe fittings are used to make "temporary" joints. If soldering, brazing, or welding is used in joining the separate lengths of pipe, the joint is considered a "permanent" one. The advantage of a so-called temporary joint is that it can be easily disassembled for repair.

The great multiplicity of pipe fittings can be divided on the basis of their functions into the following six general classes:

1. Extension or joining fittings.
2. Reducing or enlarging fittings.
3. Directional fittings.
4. Branching fittings.
5. Union or make-up fittings.
6. Shutoff or closing fittings.

Extension or Joining Fittings

Nipples, locknuts, couplings, offsets, joints, and unions are all examples of *extension* or *joining fittings*. With the possible exception of an offset, these fittings are designed to join and extend (but not change the direction) of a length of pipe. An offset is used to reposition a length of piping so that it is parallel but not in alignment with another section of its length. The offset *itself* constitutes a change of direction; however, its function is to create a piping run parallel but not in alignment with the rest of its length.

Nipples

Nipples are known as *close, short,* and *long* (Fig. 8-3). Standard lengths of nipples are listed in Table 8-6. The length of close nipples varies with the pipe size, it being determined by the length of thread necessary to make a satisfactory joint. There is one length of short nipple for each size pipe. Long nipples are made in many lengths up to 12 in. Anything over 12 in. is known as *cut pipe*.

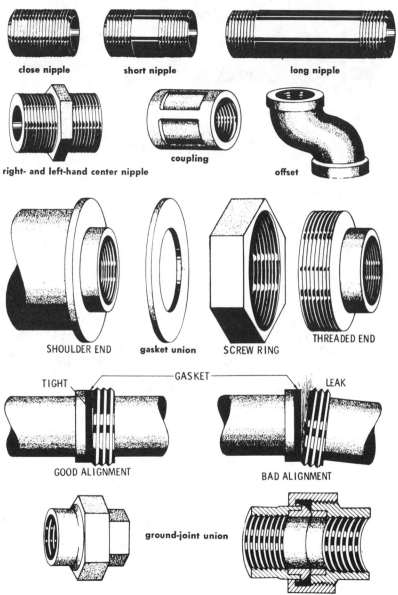

Fig. 8-3. Various pipe fittings used in heating and air conditioning installations.

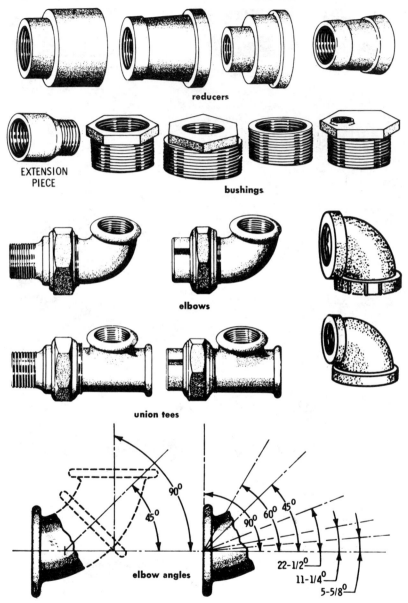

Fig. 8-3. Various pipe fittings used in heating and air conditioning installations (Cont'd).

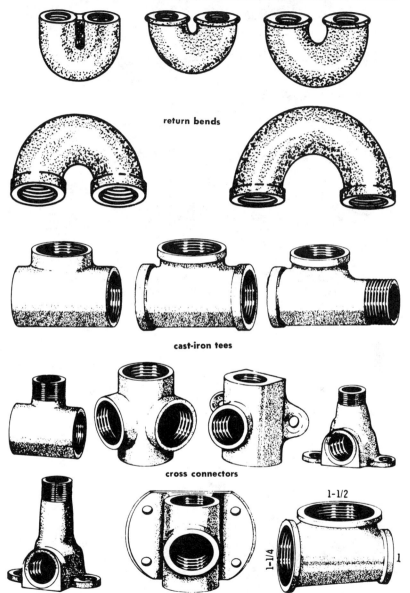

return bends

cast-iron tees

cross connectors

1-1/2

1-1/4

1

Fig. 8-3. Various pipe fittings used in heating and air conditioning installations (Cont'd).

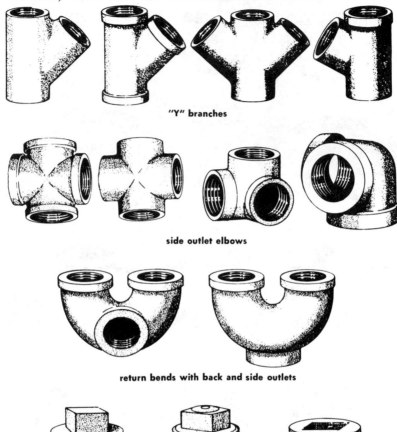

"Y" branches

side outlet elbows

return bends with back and side outlets

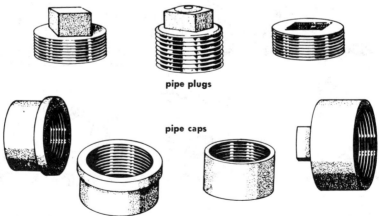

pipe plugs

pipe caps

Fig. 8-3. Various pipe fittings used in heating and air conditioning installations (Cont'd).

372

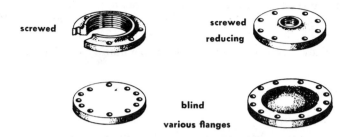

Fig. 8-3. Various pipe fittings used in heating and air conditioning installations (Cont'd).

Locknuts

Locknuts are commonly used on long screw nipples having couplings (Fig. 8-3). A recessed or grooved end on the locknut fitting is used to hold packing when a particularly tight joint is required, If at all possible, a union is preferred to a locknut because a much tighter joint is obtained.

Table 8-6. Standard Lengths of Nipples Carried in Stock

Standard Black, Right Hand

Size Inches	Kind of Nipples			
$\frac{1}{8}$ to $\frac{1}{2}$	Close, Short, then by	$\frac{1}{2}$-inch lengths from 2	inches long to 6 inches long	
	then by	1-inch lengths from 6	inches long to 12 inches long	
$\frac{3}{4}$ and 1	Close, Short, then by	$\frac{1}{2}$-inch lengths from 2 $\frac{1}{2}$	inches long to 6 inches long	
	then by	1-inch lengths from 6	inches long to 12 inches long	
1 $\frac{1}{4}$ to 2	Close, Short, then by	$\frac{1}{2}$-inch lengths from 3	inches long to 6 inches long	
	then by	1-inch lengths from 6	inches long to 12 inches long	
2 $\frac{1}{2}$ and 3	Close, Short, then by	$\frac{1}{2}$-inch lengths from 3 $\frac{1}{2}$	inches long to 6 inches long	
	then by	1-inch lengths from 6	inches long to 12 inches long	
3 $\frac{1}{2}$ and 4	Close, Short, then by	$\frac{1}{2}$-inch lengths from 4 $\frac{1}{2}$	inches long to 6 inches long	
	then by	1-inch lengths from 6	inches long to 12 inches long	
5 and 6	Close, Short, then by	$\frac{1}{2}$-inch lengths from 5	inches long to 6 inches long	
	then by	1-inch lengths from 6	inches long to 12 inches long	
8	Close, Short, then by	$\frac{1}{2}$-inch lengths from 5 $\frac{1}{2}$	inches long to 6 inches long	
	then by	1-inch lengths from 6	inches long to 12 inches long	
10 and 12	Close, Short, then by	1-inch lengths from 8	inches long to 12 inches long	

Standard Galvanized, Right Hand
Up to and including 8-inch size, same lengths as black, right hand.
Standard Black, Right and Left Hand
Up to and including 4-inch size, same lengths as black, right hand.
Extra Strong Black, Right Hand
Up to and including 2-inch size, same lengths as standard, black, right hand.

The standard lengths for all sizes of locknut (or tank nipples) is 6 in. They are made from standard weight pipe threaded (for locknut) 4 in. long on one end and with regular pipe thread on the other. Tank nipples longer than 6 in. are made to order only.

Couplings

A *coupling* is a pipe fitting used to couple or connect two lengths of pipe. They are available in numerous sizes and types. The *standard coupling* (Fig. 8-3) is threaded with right-hand threads. Others are available with both right-hand and left-hand threads. The *extension piece coupling* illustrated in Fig. 8-3 has a male thread at one end. Couplings used as reducers (Fig. 8-7) are also common.

Offsets

An *offset fitting* (Fig. 8-3) is used when it is necessary to pass the pipeline around an obstruction which blocks its path. The new path will be parallel to the old one, but not aligned with it.

Joints

A *joint* (also referred to as an *expansion joint* or *bend)* is a pipe fitting designed to accommodate the linear expansion and contraction of the pipe metal caused by the temperature differences between the water or steam inside the pipe and the air on the outside of it. The amount of expansion or contraction at different temperatures for a variety of metals used in steam pipes is listed in Table 8-7.

Unions

A union is another form of extension fitting used to join two pipes. The two most common types of unions are: (1) the ground-joint union and (2) the plain or gasket union.

A *ground-joint union* (Fig. 8-3) consists of a composition ring

Table 8-7. Expansion of Pipe
(Increase in Inches per 100 Feet)

The linear expansion and contraction of pipe, due to difference of temperature of the fluid carried and the surrounding air, must be cared for by suitable expansion joints or bends.

In order to determine the amount of expansion or contraction in a pipe line, the table below shows the increase in length of a pipe 100 feet long at various temperatures.

The expansion for any length of pipe may be found by taking the difference in increased length at the minimum and maximum temperatures, dividing by 100 and multiplying by the length in feet of the line under consideration.

Temperature, Degrees Fahrenheit	Cast Iron	Wrought Iron	Steel	Brass and Copper	Temperature, Degrees Fahrenheit	Cast Iron	Wrought Iron	Steel	Brass and Copper
0	0.00	0.00	0.00	0.00	450	3.89	4.28	4.08	6.18
50	0.36	0.40	0.38	0.57	475	4.20	4.62	4.41	6.68
100	0.72	0.79	0.76	1.14	500	4.45	4.90	4.67	7.06
125	0.88	0.97	0.92	1.40	525	4.75	5.22	4.99	7.55
150	1.10	1.21	1.15	1.75	550	5.05	5.55	5.30	8.03
175	1.28	1.41	1.34	2.04	575	5.36	5.90	5.63	8.52
200	1.50	1.65	1.57	2.38	600	5.70	6.26	5.98	9.06
225	1.70	1.87	1.78	2.70	625	6.05	6.65	6.35	9.62
250	1.90	2.09	1.99	3.02	650	6.40	7.05	6.71	10.18
275	2.15	2.36	2.26	3.42	675	6.78	7.46	7.12	10.78
300	2.35	2.58	2.47	3.74	700	7.15	7.86	7.50	11.37
325	2.60	2.86	2.73	4.13	725	7.58	8.33	7.96	12.06
350	2.80	3.08	2.94	4.45	750	7.96	8.75	8.36	12.66
375	3.15	3.46	3.31	5.01	775	8.42	9.26	8.84	13.38
400	3.30	3.63	3.46	5.24	800	8.87	9.76	9.31	14.10
425	3.68	4.05	3.86	5.85					

Note: Expansion given is approximate, but correct to the best known information.

375

pressing against iron or both contact surfaces of composition. A joint using a ground-joint union is characterized by spherical contact. Because no gasket is used, perfect alignment of the two pipes is not as important in making up the joint as it is when a plain or gasket union is used.

A disassembled *plain* or *gasket union* fitting is shown in Fig. 8-3. It consists of three basic parts and a gasket. In assembling, the gasket *(A)* is placed over the projection on the shoulder so that it is in contact with its surface *(B)*. The ring is slipped over the shoulder end and the threaded end placed in position so that the flat surface *(C)* of the threaded end presses against the gasket. The ring is then screwed firmly into the threaded end. Since the shoulder on the shoulder end cannot back off the ring, the two ends are pressed firmly together against the gasket by the ring, thus securing a tight joint.

The limitations of the plain or gasket union are also illustrated in Fig. 8-3. The alignment *must* be good to secure a tight joint. In the illustration, the ring section is omitted for clearness. If both ends are in line and firmly pressed together against the gasket by the ring, the gasket will bear evenly over the entire contact surface and the joint will be tight. If the two ends are out of alignment when the ring is screwed tight, it will bring great pressure on the gasket at point *A*, and the surfaces will not come together at the opposite point *B*, thus causing a leak.

Reducing or Enlarging Fittings

Both bushings and reducers are examples of reducing or enlarging fittings. Their function in pipe installations is to connect pipes of different sizes. In other words, they may be regarded as either reducing or enlarging fittings.

Bushings and reducers may be distinguished by their construction. A reducer is a coupling device with female threads at *both* ends (Fig. 8-3). A bushing has both male and female threads.

A bushing (Fig. 8-3) is a pipe fitting designed in the form of a hollow plug, and is used to connect the *male thread* of a pipe end to a fitting of larger size. Bushings are sold according to the pipe size of the male thread. Thus a ¼-in. bushing (or, more specifically, a ¼-in. × ⅛-in. bushing) is one used to connect a ¼-in. fitting to a ⅛-in. pipe. This may be less confusing if you remember that a bushing has one male and one female thread and that the *female* thread of the bushing must be tightened into the *male* thread of the pipe end.

Ordinary bushings are sometimes used when a reducing fitting of proper size is unavailable. If the reduction is considerable, it may be necessary to use two bushings. A common application of eccentric bushings is to avoid water pockets on horizontal pipelines. This application is illustrated in Fig. 8-51.

Directional Fittings

Directional fittings such as offsets, elbows, and return bends are used to change the direction of a pipe. Because of an overlap in function, offsets also may be considered to be a type of extension or joining fitting.

Elbows

An *elbow* (Fig. 8-3) is a pipe fitting used to change the direction of a gas, water, or steam pipeline. Elbows are available in standard angles of 45° and 90°, or special angles of 22½° and 60° (Fig. 8-3).

Return Bends

A *return bend* (Fig. 8-3) is a U-shaped pipe fitting commonly used for making up pipe coils for both water and steam heating boilers. They are commonly available in three patterns (close, medium, and open) with female threads at both ends. It is important to know the dimensions between centers when making up heating coils in order to avoid possible interference (Fig. 8-47).

Branching Fittings

As the name implies, a branching fitting is used to join a branch pipe to the main pipeline. The principal branching fittings used for this purpose are:

1. Tees.
2. Crosses.
3. Y branches.
4. Elbows with side outlets.
5. Return bends with back or side outlets.

Tees

Tees (Fig. 8-3) are made in a variety of sizes and patterns and represent the most widely used branch fitting. As the name suggests, a tee fitting is used for starting a branch pipe at a 90° angle to the main pipe.

A tee fitting is specified by first giving the run and then the branch. The *run* of a tee refers to its body with the outlets opposite each other (i.e., at 180°).

Tees are available with all three outlets the same size, with the branch outlet a size different from the two outlets of the run, or with all three outlets different sizes. Whatever the size configuration of outlets, the run is always specified first. Tee specifications generally take the following form:

1. 1″ (1-in. outlets on run and tee).
2. 1″ × ½″ (1-in. outlets on run; ½-in. tee outlet).
3. 1″ × ½″ × ½″ (outlets of 1 in. and ½ in. on the run; ½-in. tee outlet).

Crosses

A *cross fitting* (Fig. 8-3) is simply a tee with two branch outlets instead of one. The branch outlets are located directly opposite one another on the main pipe so that the fitting forms the shape of a cross (hence its name). Both branch outlets are *always* the same size, regardless of the size of the outlets.

Y Branches

A *Y branch* (Fig. 8-3) is a pipe fitting with side outlets located at 45° or 60° angles to the main pipe. Y branches are available in a number of pipe sizes. They may be straight or reducing, and single or double branching.

Elbows with Side Outlets

An elbow with three outlets is classified as a branching rather than a directional fitting because the third outlet serves as the connection for a branch pipe line (Fig. 8-3). The branching outlet should be at a 90° angle to the plane of the elbow run.

Return Bends with Back or Side Outlets

An ordinary return bend is a U-shaped fitting with two outlets (Fig. 8-3). Some return bends are designed with three outlets, the third outlet being located on either the back or the side of the fitting and used for connecting to a branch pipeline. This type of return bend is more properly classified as a branching fitting.

Shutoff or Closing Fittings

Sometimes it is necessary to close the end of a fitting or pipe. This is accomplished with a shutoff or closing fitting, and the two types used for this purpose are:

(1) Plugs.
(2) Caps.

Plugs

A *plug* (Fig. 8-3) is used to close the end of a pipe or fitting when it has a female thread. In other words, it is designed to be inserted *into* the end of the pipe or fitting. Plugs are made in a variety of sizes (⅛ in. to 12 in.), designs (hexagon, square head, or countersunk heads), and materials (iron, brass, etc.).

Caps

A *cap* (Fig. 8-3) performs the same function as a plug except that it is used to close the end of a pipe or fitting having a male thread. They are also available in a variety of sizes, designs, and materials.

Union or Make-up Fittings

Union or *make-up fittings* (Fig. 8-3) are represented by union elbows and union tees. This type of pipe fitting combines both a union and an elbow or tee in a single unit. They are available with female or male and female threads.

Flanges

Flanges (Fig. 8-3) are pipe fittings used to close flanged pipelines or fittings. They are manufactured in the form of cast-iron discs and are available in many sizes, thicknesses, and types.

Pipe Expansion

The linear expansion and contraction of pipe due to the surrounding air must be provided for (especially in the case of long lines) by suitable expansion joints, bends, or equivalent provisions.

In order to determine the amount of expansion or contraction in a pipeline, Table 8-8 shows the increase in length of a pipe 100 ft. long at various temperatures.

The expansion for any length of pipe may be found by taking the difference in increased length at the minimum and maximum temperatures, dividing by 100, and multiplying by the length in feet of the line under consideration. See Table 8-8.

VALVES

Some authorities regard a valve as simply another type of pipe fitting and distinguish it from others by its capacity to control the flow of steam or hot water through the pipe. Be that as it may, the subject of valves is so extensive it warrants a chapter of its own (see Chapter 9 of Volume 2, Valves and Valve Installation).

PIPE THREADS

The threads used on pipes are referred to as *pipe threads*. The distinguishing characteristic of pipe threads is that they are tapered. This results in a greater number of turns in screwing the

Table 8-8. Expansion of Steam Pipes (Inches Increase per 100 Feet)

Temperature (Degrees F)	Steel	Wrought Iron	Cast Iron	Brass and Copper
0	0	0	0	0
20	.15	.15	.10	.25
40	.30	.30	.30	.45
60	.45	.45	.40	.65
80	.60	.60	.55	.90
100	.75	.80	.75	1.15
120	.90	.95	.85	1.40
140	1.10	1.15	1.00	1.65
160	1.25	1.35	1.15	1.90
180	1.45	1.50	1.30	2.15
200	1.60	1.65	1.50	2.40
220	1.80	1.85	1.65	2.65
240	2.00	2.05	1.80	2.90
260	2.15	2.20	1.95	3.15
280	2.35	2.40	2.15	3.45
300	2.50	2.60	2.35	3.75
320	2.70	2.80	2.50	4.05
340	2.90	3.05	2.70	4.35
360	3.05	3.25	2.90	4.65
380	3.25	3.45	3.10	4.95
400	3.45	3.65	3.30	5.25
420	3.70	3.90	3.50	5.60
440	3.95	4.20	3.75	5.95
460	4.20	4.45	4.00	6.30
480	4.45	4.70	4.25	6.65
500	4.70	4.90	4.45	7.05
520	4.95	5.15	4.70	7.45
540	5.20	5.40	4.95	7.85
560	5.45	5.70	5.20	8.25
580	5.70	6.00	5.45	8.65
600	6.00	6.25	5.70	9.05
620	6.30	6.55	5.95	9.50
640	6.55	6.85	6.25	9.95
660	6.90	7.20	6.55	10.40
680	7.20	7.50	6.85	10.95
700	7.50	7.85	7.15	11.40
720	7.80	8.20	7.45	11.90
740	8.20	8.55	7.80	12.40
760	8.55	8.90	8.15	12.95
780	8.95	9.30	8.50	13.50
800	9.30	9.75	8.90	14.10

pipe to another length of pipe or pipe fitting. When properly done, this will result in a tight, leak-free joint. Care must be taken, however, not to exceed the elastic limit or the joint will leak.

The total taper used on pipe threads is ¾ in. per foot. The total number of threads per inch will vary from 27 threads for ⅛-in. pipe to 8 threads for 2½-in. pipe and larger sizes. Standard pipe threads are listed in Table 8-9.

381

Table 8-9. Standard Pipe Threads

Nominal Size of Pipe Inches	A Pitch Dia. at End of Pipe Inches	B Pitch Dia. at Gauging Notch Inches	E Length of Effective Thread Inches	F Normal Engagement by Hand Between Male and Female Thread Inches	G Outside Dia. of Pipe Inches	H Actual Inside Dia. of Pipe Inches	Number of Threads Per Inch	P Pitch of Thread Inch	Depth of Thread Inch
$\frac{1}{8}$	.36351	.37476	.2638	.180	.405	.269	27	.0370	.02963
$\frac{1}{4}$	.47739	.48989	.4018	.200	.540	.364	18	.0556	.04444
$\frac{3}{8}$	.61201	.62701	.4078	.240	.675	.493	18	.0556	.04444
$\frac{1}{2}$	.75843	.77843	.5337	.320	.840	.622	14	.0714	.05714
$\frac{3}{4}$	.96768	.98886	.5457	.339	1.050	.824	14	.0714	.05714
1	1.21363	1.23863	.6828	.400	1.315	1.049	11½	.0870	.06954
1¼	1.55713	1.58338	.7068	.420	1.660	1.380	11½	.0870	.06954
1½	1.79609	1.82334	.7235	.420	1.900	1.610	11½	.0870	.06956
2	2.26902	2.29627	.7565	.436	2.375	2.067	11½	.0870	.06956
2½	2.71953	2.76216	1.1375	.681	2.875	2.469	8	.1250	.10000
3	3.34063	3.38850	1.2000	.766	3.500	3.068	8	.1250	.10000
3½	3.83750	3.88881	1.2500	.821	4.000	3.548	8	.1250	.10000
4	4.33438	4.38713	1.3000	.844	4.500	4.026	8	.1250	.10000
4½	4.83125	4.88594	1.3500	.875	5.000	4.506	8	.1250	.10000
5	5.39073	5.44929	1.4063	.937	5.563	5.047	8	.1250	.10000
6	6.44609	6.50597	1.5125	.958	6.625	6.055	8	.1250	.10000
7	7.43984	7.50234	1.6125	1.000	7.625	7.023	8	.1250	.10000
8	8.43359	8.50003	1.7125	1.063	8.625	7.981	8	.1250	.10000
9	9.42734	9.49797	1.8125	1.130	9.625	8.941	8	.1250	.10000
10	10.54531	10.62094	1.9250	1.210	10.750	10.020	8	.1250	.10000
12	12.53281	12.61781	2.1250	1.360	12.750	12.000	8	.1250	.10000
14 O.D.	13.77500	13.87262	2.250	1.562	14.000	—	8	.1250	.10000
15 O.D.	14.76875	14.87419	2.350	1.687	15.000	—	8	.1250	.10000
16 O.D.	15.76250	15.87575	2.450	1.812	16.000	—	8	.1250	.10000
18 O.D.	17.75000	17.87500	2.650	2.000	18.000	—	8	.1250	.10000
20 O.D.	19.73750	19.87031	2.850	2.125	20.000	—	8	.1250	.10000
22 O.D.	21.72500	21.86562	3.050	2.250	22.000	—	8	.1250	.10000
24 O.D.	23.71250	23.86094	3.250	2.375	24.000	—	8	.1250	.10000

Data abstracted from the American Standard for Pipe Threads A.S.A.-B2—1919.

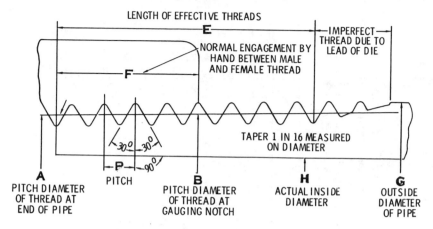

LENGTH OF EFFECTIVE THREADS

NORMAL ENGAGEMENT BY HAND BETWEEN MALE AND FEMALE THREAD

IMPERFECT THREAD DUE TO LEAD OF DIE

TAPER 1 IN 16 MEASURED ON DIAMETER

A
PITCH DIAMETER OF THREAD AT END OF PIPE

PITCH

B
PITCH DIAMETER OF THREAD AT GAUGING NOTCH

H
ACTUAL INSIDE DIAMETER

G
OUTSIDE DIAMETER OF PIPE

$A = G - (0.05 G + 1.1)P$
$B = A + 0.0625 F$
$E = P (0.8G + 6.8)$
$DEPTH\ OF\ THREAD = 0.8P$
TOTAL TAPER 3/4" - IN. PER FOOT

Illustration for Table 8-9.

PIPE SIZING

Pipe sizing refers to the procedure of determining the projected capacities of a piping installation and selecting the pipe sizes most capable of meeting these capacities. Most methods used for determining pipe sizes are only approximate calculations, and they should be considered as such when you are using them.

Both the American Society of Heating, Refrigerating and Air Conditioning Engineers (ASHRAE) and the Institute of Boiler and Radiator Manufacturers (IBR) issue publications that contain considerable data for designing the piping arrangements of various steam and hot-water space heating system. Manufacturers of proprietary heating systems, pipes, pipe fittings, and valves also provide data for sizing pipes and valves. Examples of these data are illustrated in Tables 8-10 and 8-11. Methods for pipe sizing are explained in detail with accompanying examples. These pub-

383

Table 8-10. Relative Discharging Capacities of Standard Pipe

The figure opposite the intersection of any two sizes is the number of smaller size pipes required to equal one of the larger.
Example: How many 2-inch standard pipes will it take to equal the discharge of one 8-inch standard pipe?
Solution: The figure in the first table below, opposite the intersection of these two sizes gives 29—2-inch pipes.

Pipe Size (In.)	D (In.)	$D^{5/2}$	1/8	1/4	3/8	1/2	3/4	1	1¼	1½	2	2½	3	3½	4	5	6	8	10	12
1/8	.269	.037530	1.0	—	—	—	—	—	—	—	—	—	—	—	—	—	—	—	—	—
1/4	.364	.079938	2.1	1.0	—	—	—	—	—	—	—	—	—	—	—	—	—	—	—	—
3/8	.493	.17065	4.5	2.1	1.0	—	—	—	—	—	—	—	—	—	—	—	—	—	—	—
1/2	.622	.30512	8.1	3.8	1.8	1.0	—	—	—	—	—	—	—	—	—	—	—	—	—	—
3/4	.824	.61634	16	7.7	3.6	2.0	1.0	—	—	—	—	—	—	—	—	—	—	—	—	—
1	1.049	1.1270	30	14	6.6	3.7	1.8	1.0	—	—	—	—	—	—	—	—	—	—	—	—
1¼	1.380	2.2372	60	28	13	7.3	3.6	2.0	1.0	—	—	—	—	—	—	—	—	—	—	—
1½	1.610	3.2890	88	41	19	11	5.3	2.9	1.5	1.0	—	—	—	—	—	—	—	—	—	—
2	2.067	6.1426	164	77	36	20	10	5.5	2.7	1.9	1.0	—	—	—	—	—	—	—	—	—
2½	2.469	9.5786	255	120	56	31	16	8.5	4.3	2.9	1.6	1.0	—	—	—	—	—	—	—	—
3	3.068	16.487	439	206	97	54	27	15	7.4	5.0	2.7	1.7	1.0	—	—	—	—	—	—	—
3½	3.548	23.711	632	297	139	78	38	21	11	7.2	3.9	2.5	1.4	1.0	—	—	—	—	—	—
4	4.026	32.523	867	407	191	107	53	29	15	9.9	5.3	3.4	2.0	1.4	1.0	—	—	—	—	—
5	5.047	57.225	1526	716	335	188	93	51	26	17	9.3	6.0	3.5	2.4	1.8	1.0	—	—	—	—
6	6.065	90.589	2414	1163	531	297	147	80	40	28	15	9.5	5.5	3.8	2.8	1.6	1.0	—	—	—
8	7.981	179.95	4795	2251	1054	590	292	160	80	55	29	19	11	7.6	5.5	3.1	2.0	1.0	—	—
10	10.020	317.81	8468	3976	1862	1042	516	282	142	97	52	33	19	13	9.8	5.6	3.5	1.8	1.0	—
12	12.000	498.83	13292	6240	2923	1635	809	443	223	152	81	52	30	21	15	8.7	5.5	2.8	1.6	1.0

Pipe Size in Inches

Table 8-11. Diagram Showing Resistance of Valves and Fittings of the Flow of Liquids

Example: The dotted line shows that the resistance of a 6-inch Standard Elbow is equivalent to approximatly 16 feet of 6-inch Standard Pipe.

Note. For sudden enlargements or sudden contractions, use the smaller diameter, d, on the pipe size scale.

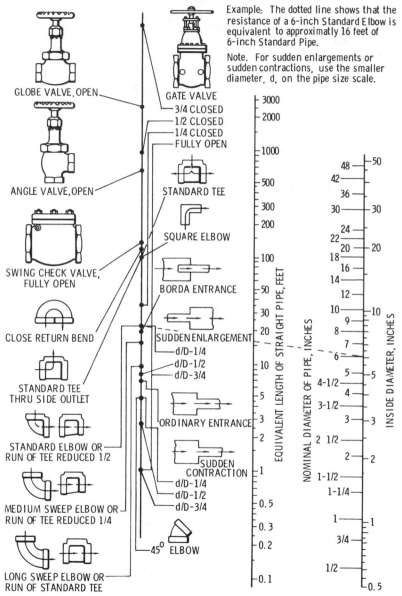

lications can be obtained by writing to these organizations. Their addresses are given elsewhere in this book. Sometimes copies are also available at a local library.

Because pipe sizing is specific to the piping arrangement and other variables within a system, no attempt is made in this chapter to cover the subject with the detail it requires. The basic principles of the methods used for sizing steam and hot-water heating pipes, along with recommendations for their application, are described in the sections that follow.

SIZING STEAM PIPES

Many manufacturers of proprietary or patented steam heating systems provide their own pipe-sizing schedules. For nonproprietary systems, the projected capacities of the piping installation must be determined by a number of sizing calculations.

The principal factors used in determining pipe sizes for a given load of steam in a heating system are:

1. Initial pressure.
2. Total pressure drop allowed between the boiler and the end of the return line.
3. Equivalent length of run from the boiler to the farthest radiator or convector.
4. Pressure drop per 100 ft. of equivalent length.

The total pressure drop should not exceed the initial gauge pressure of the system. As a general rule, it should not exceed 50 percent of the initial gauge pressure.

The equivalent length of run equals the actual measured length of pipe plus the equivalent straight pipe length of the fittings and valves. Table 8-12 lists equivalent lengths of the more common fittings and valves, and is an example of the data provided by the ASHRAE for sizing pipes.

The pressure drop in pounds per square inch per 100 ft. is determined by dividing 50 percent of the initial pressure by the equivalent length of the longest piping circuit.

For the sake of illustration, assume that you must calculate the pressure drop and determine the pipe size for a steam heating system in which the initial pressure is 2 psig. In order to do this, the following steps are necessary:

1. Measure the length of the longest run of pipe. For our problem, use the figure 500 ft.
2. Determine the *assumed* equivalent length This will not exceed twice the measured length (i.e., 1000 ft.).
3. Determine the pressure drop for the equivalent length of pipe (50 percent of the initial pressure, or 1 psig).
4. Determine the pressure drop per 100 ft. of equivalent length of pipe.

Table 8-12. Length of Feet of Pipe for Fittings to be Added to Actual Length of Run in Order to Obtain Equivalent Length

Size of Pipe	Length in Feet to Be Added to Run				
Inches	Standard Elbow	Side Outlet Tee	Gate Valve[a]	Globe Valve[a]	Valve[a]
½	1.3	3	0.3	14	7
¾	1.8	4	0.4	18	10
1	2.2	5	0.5	23	12
1¼	3.0	6	0.6	29	15
1½	3.5	7	0.8	34	18
2	4.3	8	1.0	46	22
2½	5.0	11	1.1	54	27
3	6.5	13	1.4	66	34
3½	8	15	1.6	80	40
4	9	18	1.9	92	45
5	11	22	2.2	112	56
6	13	27	2.8	136	67
8	17	35	3.7	180	92
10	21	45	4.6	230	112
12	27	53	5.5	270	132
14	30	63	6.4	310	152

[a]Valve in full open position.

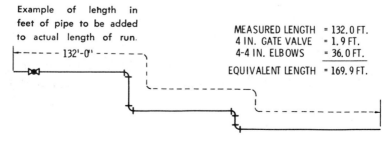

Example of length in feet of pipe to be added to actual length of run.

MEASURED LENGTH = 132.0 FT.
4 IN. GATE VALVE = 1.9 FT.
4-4 IN. ELBOWS = 36.0 FT.

EQUIVALENT LENGTH = 169.9 FT.

Pressure drop
per 100 ft.
of equivalent
length
$= \dfrac{1 \text{ psig}}{1000 \text{ ft.}} = 0.1$ psig per 100 ft.

5. Size pipes for a desired capacity based on a pressure drop of 0.1 psig per 100 ft. These can be determined from ASHRAE charts and tables.
6. Check the pressure drop by calculating the equivalent length of run of the longest circuit from the pipe sizes determined. The pipe size determined in Step 5 will be correct if the calculated pressure drop is less than the assumed pressure drop.

A steam supply main should not pitch less than ¼ in. per 10 ft. of run, and its diameter should not be smaller than 2 in. In gravity one-pipe systems, the diameter of the supply main at the farthest point should not be smaller than 50 percent of its largest diameter.

A rule-of-thumb method for determining the size of steam mains is to take the total amount of direct radiation and add to it 25 percent of the total for the piping allowance. Next, find the square root of this total and divide by 10. The result will be the size steam main to use for a one-pipe system. For a two-pipe system, one size smaller is generally sufficient for the supply main, and the return can be one or two sizes smaller than the supply main. A steam main should not decrease in size according to the area of its branches but very much more gradually.

The above method for sizing steam mains can be illustrated by using a structure with an assumed direct radiation of 500 sq. ft. Adding 25 percent for piping allowance gives 625. The square root of 625 is 25, which divided by 10 gives 2½, or the size of the steam main (2½ in.). For reference and practical use, refer to Table 8-13 when making calculations.

The pitch of runouts to risers and radiators should be *at least* ½ in. per foot toward the main. Runouts over 8 ft. in length, but with *less* than ½-in. pitch per foot, should be one size larger than specified in the pipe-sizing tables.

Table 8-13. Size of Steam Mains

Radiation	One-pipe work	Two-pipe work
125 square feet	1 ½ inch	1 ¼ × 1 inch
250 " "	2 ½ "	1 ½ × 1 ¼ "
400 " "	3 "	2 × 1 ½ "
650 " "	3 ½ "	2 ½ × 2 "
900 " "	4 "	3 × 2 ½ "
1,250 " "	4 ½ "	3 ½ × 3 "
1,600 " "	5 "	4 × 3 ½ "
2,050 " "	5 ½ "	4 ½ × 4 "
2,500 " "	6 "	5 × 4 ½ "
3,600 " "	7 "	6 × 5 "
5,000 " "	8 "	7 × 6 "
6,500 " "	9 "	8 × 6 "
8,100 " "	10 "	9 × 6 "

SIZING HOT-WATER PIPES

Simplified pipe sizing tables for hot-water heating systems are also available from organizations such as the ASHRAE.

Pipe sizing hot-water lines is similar in some respects to the calculations used in duct sizing because the pipe sizes are selected on the basis of the quantity and rate of water flow (expressed in gallons per minute, or gpm) and the constant friction loss. This friction loss (or drop) is expressed in milinches (thousandths of an inch) per foot of pipe length.

The velocity of water in the smaller residential pipes should not exceed 4 fps (feet per second), or there will be a noise problem.

In *forced* hot-water heating systems, the problem of friction drop in the pipes can be overcome by the pump or circulator. In this respect, a forced hot-water heating system is much easier to size than a steam heating system.

The rule-of-thumb method used to size steam mains (see above) can also be used to determine the approximate sizes of hot-water mains; however, certain important differences should be noted.

When sizing hot-water mains, the mains may be reduced in size

389

in proportion to the branches taken off. They should, however, have as large area as the sum of all branches beyond this point. It is advisable that the horizontal branches be one size larger than the risers. Returns should be the same size as the supply mains. Table 8-14 lists sizes of hot-water mains and the equivalent radiation ranges in square feet. Sizes for mains and branches are given in Table 8-15.

Table 8-14. Sizes of Hot-Water Mains

Radiation	Pipe
75 to 125 square feet	1 1/4 inch
125 to 175 " "	1 1/2 "
175 to 300 " "	2 "
300 to 475 " "	2 1/2 "
475 to 700 " "	3 "
700 to 950 " "	3 1/2 "
950 to 1,200 " "	4 "
1,200 to 1,575 " "	4 1/2 "
1,575 to 1,975 " "	5 "
1,975 to 2,375 " "	5 1/2 "
2,375 to 2,850 " "	6 "

PIPE FITTING MEASUREMENTS

Pipe fitting may be done either by making close visual judgments or entirely by measurements scaled on a drawing. The first method is a hit-or-miss process and requires an experienced fitter to do a good job; the second method is one of precision and is the better way. In actual practice, a combination of the two methods will, in some cases, save time and give satisfactory results.

Working from a drawing with all the necessary dimensions has certain advantages. Especially in the case of a big job, all the pipe may be cut and threaded in the shop so that at the place of installation the only work to be done is assembling.

In making a drawing, the measurements are based on the distances between the centers of fittings. The data necessary to locate these centers are given in a more general dimension drawing with an accompanying table such as the one illustrated in

390

Table 8-15. Table of Mains and Branches

Main	Branch
1 in. will supply 2	.... ¾ in.
1¼ in. " " 2	....1 in.
1½ in. " " 2	....1¼ in.
2 in. " " 2	....1½ in.
2½ in. " " .... 2, 1½ in.	and 1, 1¼ in., or 1, 2 in. and 1, 1¼ in.
3 in. " " .... 1, 2½ in.	and 1, 2 in., or 2, 2 in. and 1, 1½ in.
3½ in. " " .... 2, 2½ in.	or 1, 3 in., and 1, 2 in. or 3, 2 in.
4 in. " " .... 1, 3½ in.	and 1, 2½ in., or 2, 3 in. and 4, 2 in.
4½ in. " " .... 1, 3½ in.	and 1, 3 in., or 1, 4 in. and 1, 2½ in.
5 in. " " .... 1, 4 in.	and 1, 3 in., or 1, 4½ in. and 1, 2½ in.
6 in. " " .... 2, 4 in.	and 1, 3 in., or 4, 3 in. or 10, 2 in.
7 in. " " .... 1, 6 in.	and 1, 4 in., or 3, 4 in. and 1, 2 in.
8 in. " " .... 2, 6 in.	and 1, 5 in., or 5, 4 in. and 2, 2 in.

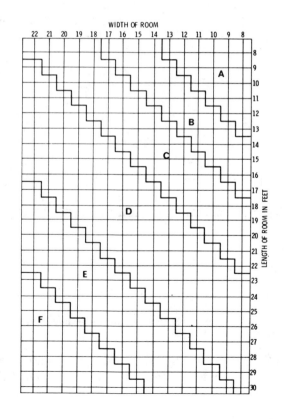

Table 8-16. These dimension drawings and tables should be the ones corresponding to the make of fittings used; otherwise, there might be the possibility of slight variation. In general, however, the different makes are pretty well standardized.

Fig. 8-4 illustrates how the *actual* length of pipe connecting the two fittings is obtained. The actual length of pipe is equal to the distance between centers (of fittings) *minus* twice the distance from the center of the face of the fittings *plus* twice the allowance for threads. This is expressed by the following equation:

$$D = A - 2B + 2C$$

The allowance for the length of thread that is screwed into the fitting (dimension C in Fig. 8-4) is obtained from a table furnished by the manufacturer. This allowance (called A in Fig. 8-5) corresponds to the values given in the accompanying table (Table 8-17). Note that the dimension A in Fig. 8-5 is the same as dimension C in Fig. 8-4.

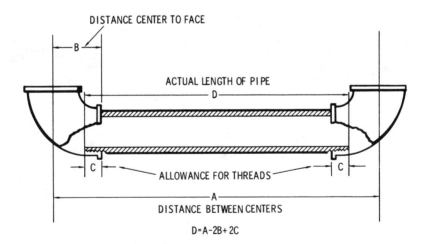

Fig. 8-4. Diagram showing how to obtain the actual length of a pipe connecting two fittings.

A working drawing showing center lines and distances between centers of an installation is shown in Fig. 8-6. This gives all the information *except* the actual length of the pipes.

392

Table 8-16. General Dimensions of Straight Sizes

General dimensions of Crane standard malleable iron screwed fittings. The reference letters refer to the accompanying table.

Fitting types (with reference letters): 90° ELBOW, 45° ELBOW, 90° STREET ELBOW, 45° STREET ELBOW, TEE, SERVICE TEE, CROSS, Y BRANCH, REDUCING COUPLING, PLAIN COUPLING, FLAT BAND COUPLING, CAP, CLOSE RETURN BEND, OPEN RETURN BEND, THREAD ENGAGEMENT, FOR NORMAL THREAD ENGAGEMENT

Dimensions, Inches

Size Inches	A	C	W	W	P	D_1	D_3	T
$\frac{1}{8}$	$\frac{11}{16}$	$\frac{3}{4}$	—	—	$\frac{7}{8}$	$1\frac{1}{4}$	—	$\frac{31}{32}$
$\frac{1}{4}$	$\frac{13}{16}$	$\frac{3}{4}$	—	—	$\frac{15}{16}$	$1\frac{1}{4}$	—	$1\frac{1}{16}$
$\frac{3}{8}$	$\frac{15}{16}$	$\frac{13}{16}$	$1\frac{3}{4}$	$1\frac{7}{8}$	$\frac{1}{32}$	$1\frac{3}{8}$	—	$1\frac{5}{32}$
$\frac{1}{2}$	$1\frac{1}{8}$	$\frac{7}{8}$	$2\frac{1}{32}$	$2\frac{5}{32}$	$1\frac{5}{32}$	$1\frac{7}{16}$	$1\frac{1}{6}$	$1\frac{11}{32}$
$\frac{3}{4}$	$1\frac{5}{16}$	1	$2\frac{5}{16}$	$2\frac{25}{32}$	$1\frac{15}{16}$	$1\frac{11}{16}$	$1\frac{3}{8}$	$1\frac{1}{2}$
1	$1\frac{1}{2}$	$1\frac{1}{16}$	$2\frac{5}{32}$	$3\frac{5}{16}$	$1\frac{15}{16}$	$2\frac{1}{16}$	$1\frac{3}{4}$	$1\frac{11}{16}$
$1\frac{1}{4}$	$1\frac{3}{4}$	$1\frac{5}{8}$	$2\frac{13}{16}$	$3\frac{11}{16}$	$2\frac{1}{8}$	$2\frac{5}{16}$	$2\frac{1}{6}$	$1\frac{15}{16}$
$1\frac{1}{2}$	$1\frac{15}{16}$	$1\frac{7}{16}$	$3\frac{7}{16}$	$4\frac{1}{4}$	$2\frac{11}{16}$	$2\frac{5}{8}$	$2\frac{1}{2}$	$2\frac{5}{32}$
2	$2\frac{1}{4}$	$1\frac{11}{16}$	$3\frac{7}{8}$	$5\frac{5}{32}$	$3\frac{1}{4}$	$3\frac{1}{16}$	3	$2\frac{17}{32}$
$2\frac{1}{2}$	$2\frac{9}{16}$	$1\frac{15}{16}$	—	$5\frac{13}{16}$	4	$3\frac{9}{16}$	$3\frac{1}{2}$	$2\frac{7}{8}$
3	$3\frac{1}{16}$	$2\frac{3}{16}$	—	$6\frac{5}{16}$	$4\frac{11}{16}$	4	4	$3\frac{1}{2}$
$3\frac{1}{2}$	$3\frac{27}{32}$	$2\frac{19}{32}$	—	$6\frac{15}{16}$	$5\frac{11}{16}$	$4\frac{3}{8}$	$4\frac{1}{2}$	—
4	$3\frac{13}{16}$	$2\frac{5}{8}$	—	—	—	$5\frac{1}{8}$	5	$4\frac{5}{16}$
5	$5\frac{1}{2}$	$3\frac{1}{8}$	—	—	—	$5\frac{5}{8}$	6	—
6	$5\frac{1}{8}$	$3\frac{15}{32}$	—	—	—	$5\frac{7}{8}$	—	—

Table 8-17. Length of Thread on Pipe

Size inches	Dimension A inches	Size inches	Dimension A inches	Size inches	Dimension A inches	Size inches	Dimension A inches
$\frac{1}{8}$	$\frac{1}{4}$	1	$\frac{9}{16}$	3	1	6	$1\frac{1}{4}$
$\frac{1}{4}$	$\frac{3}{8}$	$1\frac{1}{4}$	$\frac{5}{8}$	$3\frac{1}{2}$	$1\frac{1}{16}$	7	$1\frac{1}{4}$
$\frac{3}{8}$	$\frac{3}{8}$	$1\frac{1}{2}$	$\frac{5}{8}$	4	$1\frac{1}{16}$	8	$1\frac{5}{16}$
—	—	—	—	—	—	9	$1\frac{3}{8}$
$\frac{1}{2}$	$\frac{1}{2}$	2	$\frac{11}{16}$	$4\frac{1}{2}$	$1\frac{1}{8}$	10	$1\frac{1}{2}$
$\frac{3}{4}$	$\frac{1}{2}$	$2\frac{1}{2}$	$\frac{15}{16}$	5	$1\frac{3}{16}$	12	$1\frac{5}{8}$

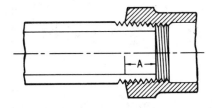

Fig. 8-5. Joint made up showing length of thread on pipe (A) screwed into fitting.

Problem: Find the length of the pipe connecting the 2-in. elbows in Fig. 8-7. Using Fig. 8-4 as a guide, the following dimensions are provided: A = 12 ft., B = $2\frac{1}{4}$ in., and C = $\frac{11}{16}$ in. The problem is solved as follows:

1. $D = A$ $- 2B$ $+ 2C$
2. $D = 12$ ft. $- 2 \times 2\frac{1}{4}$ in. $+ 2 \times \frac{11}{16}$ in.
3. $D = 12$ ft. $- 4\frac{1}{2}$ in. $+ \frac{22}{32}$ (or $1\frac{3}{8}$) in.
4. $D = 11$ ft. $7\frac{1}{2}$ in. $+ 1\frac{3}{8}$ in.
5. $D = 11$ ft. $8\frac{7}{8}$ in.

CALCULATING OFFSETS

In pipe fitting, an *offset* is a change of direction (other than 90°) in a pipe bringing one part out of (but parallel with) the line of another.

An example of an offset is illustrated in Fig. 8-8. As shown here, the problem is an obstruction *(E)*, such as a wall, blocking the path of a pipeline *(L)*. It is necessary to change the position of

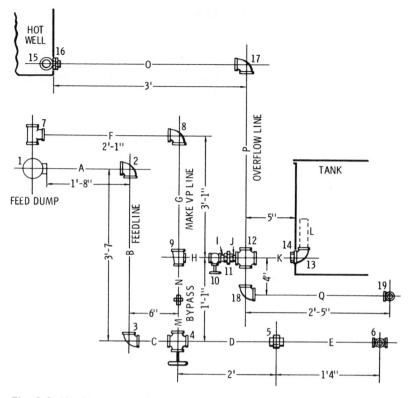

Fig. 8-6. Working drawing showing center line and distances between centers of an installation.

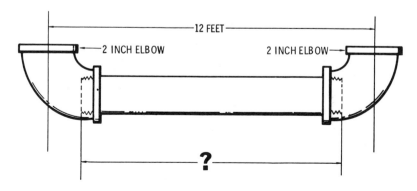

Fig. 8-7. Pipe connecting two elbows.

395

pipeline *L* at point *A* to some parallel position such as line *F* in order to move around the obstruction. When two lines such as *L* and *F* are to be piped with elbows other than 90° elbows, the pipe fitter is confronted with the following two problems: (1) finding the length of pipe *H* and (2) determining the distance *BC*. By determining the distance *BC*, the pipe fitter will be able to fix point *A* so that the two elbows *A* and *C* will be in alignment.

Of course, in the triangle *ABC*, the length of pipe *AC* and either offset *(AB or BC)* that may be required are quickly calculated by solving the triangle *ABC* for the desired member, but this involves taking the square root, which is not always easily understood by the average worker. Alternative methods are suggested below.

First Method

If in Fig. 8-8, the distance between pipelines *L* and *F* is 20 in. (offset *AB*), what length of pipe *H* is required to connect with the 45° elbows *A* and *C*?

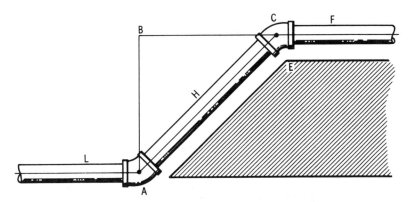

Fig. 8-8. Pipeline connected with two 45° elbows illustrating offsets and method of finding length of connecting pipe *H*.

Based on the triangle *ABC*, the following equation is offered for solving this problem:

1. $\overline{AC}^2 = \overline{AB}^2 + \overline{BC}^2$ from which:

2. $AC = \sqrt{\overline{AB}^2 = \overline{BC}^2}$ or substituting:

3. $AC = \sqrt{20^2 + 20^2} = \sqrt{800} = 28.28$ in.

It should be remembered that when 45° elbows are used, *both* offsets are equal. Therefore, if offset *AB* is 20 in. long, offset *BC* also must be the same length.

Note that the value (i.e., 28.28 in.) for the length of pipe *H* obtained by the above equation is the *calculated length* and does not allow for the projections of the elbows. In other words, pipe *H* (as calculated by this equation) is too long and must be shortened so that the elbows will fit.

Fig. 8-8 illustrates the difference between the calculated length (i.e. the measurement from point *A* to point *C*) and the actual length of connecting pipe *H* when used with elbows other than 90°. *Actual length* is obtained by deducting the *allowance for projection of the elbows* from the *calculated length*.

Second Method

Another method of calculating the connecting pipe length (i.e. the length of pipe *H* in Fig. 8-9), is by multiplying the offset by $^{53}/_{128}$ in. and adding the product to the original offset figure. Thus, if offset *AB* is 20 in., the following calculations are possible:

1. $20 \times \dfrac{53}{128} = \dfrac{1060}{128} = 8\dfrac{9}{32}$

2. $20 + 8\dfrac{9}{32} = 28\dfrac{9}{32}$ in.

Third Method

The pipe fitter will often encounter elbows of angles other than 45°. For such, the distance between elbow centers (points *A* and *C*) can easily be calculated with the following procedure:

1. Determine the angle of the elbow.
2. Determine the elbow constant equivalent to its angle.
3. Multiply the elbow constant by the known offset.

In Fig. 8-10, even though only offset *AB* is known, it is possible to determine the length of the other offset *(BC)* and the distance

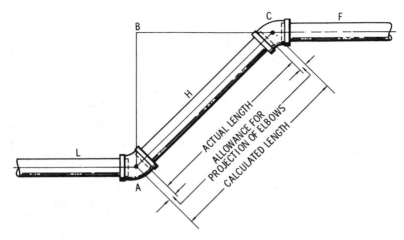

Fig. 8-9. Second method of calculating pipe length.

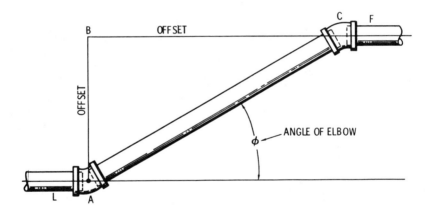

Fig. 8-10. Diagram for elbow constant as given in Table 8-18.

between elbow centers *(AC)*. In order to do this, the following equations must be used:

1. AC = offset AB × constant for AC.
2. BC = offset AB × constant for AB.

Assume that the distance between pipelines *L* and *F* (offset *AB)* in Fig. 8-10 is 20 in. and the angle of elbow is 22½°. In Table 8-18 you will find that for a 22½° elbow, the elbow constant for

Table 8-18. Elbow Constants

Angle of Elbow	Elbow Centers AC	Offset AB
60°	1.15	.58
45°	1.41	1.00
30°	2.00	1.73
22½°	2.61	2.41
11¼°	5.12	5.02
5⅜°	10.20	10.15

AB is 2.41. Substituting values in the second equation above, you have the following:

$$BC = 20 \times 2.41 = 48.2 \text{ in.}$$

The constant for the elbow centers of 22½° elbows is 2.61. Substituting values in the first equation above gives the following results:

$$AC = 20 \times 2.61 = 52.2 \text{ in.}$$

Fourth Method

Offsets may also be calculated by using basic trigonometry. Using the example given in Fig. 8-10, determine the length of the offset *AB* if *AC* is 8 ft. and the angle $\phi = 60°$. From Table 8-19, Sine 60° = .866.

$$\text{Length of offset AB} = .866 \times 8 = 6.93$$

PIPE SUPPORTS

If piping is to be run along the wall or ceiling, it should be attached to the surface with pipe supports (e.g., hangers, straps, clamps). The type of pipe supports used and their spacing will be regulated in accordance with approved local standards.

Pipe straps (perforated metal straps) are used to support small size pipes (Fig. 8-11). Larger pipes require various types of hangers (e.g., rod, spring), chains, or other devices capable of supporting the heavier weight.

Table 8-19. Natural Trigonometrical Functions

Deg.	Sin	Cos	Tan	Sec.	Deg.	Sin	Cos	Tan	Sec.
0	.00000	1.0000	.00000	1.0000	46	.7193	.6947	1.0355	1.4395
1	.01745	.9998	.01745	1.0001	47	.7314	.6820	1.0724	1.4663
2	.03490	.9994	.03492	1.0006	48	.7431	.6691	1.1106	1.4945
3	.05234	.9986	.05241	1.0014	49	.7547	.6561	1.1504	1.5242
4	.06976	.9976	.06993	1.0024	50	.7660	.6428	1.1918	1.5557
5	.08716	.9962	.08749	1.0038	51	.7771	.6293	1.2349	1.5890
6	.10453	.9945	.10510	1.0055	52	.7880	.6157	1.2799	1.6243
7	.12187	.9925	.12278	1.0075	53	.7986	.6018	1.3270	1.6618
8	.1392	.9903	.1405	1.0098	54	.8090	.5878	1.3764	1.7013
9	.1564	.9877	.1584	1.0125	55	.8192	.5736	1.4281	1.7434
10	.1736	.9848	.1763	1.0154	56	.8290	.5592	1.4826	1.7883
11	.1908	.9816	.1944	1.0187	57	.8387	.5446	1.5399	1.8361
12	.2079	.9781	.2126	1.0223	58	.8480	.5299	1.6003	1.8871
13	.2250	.9744	.2309	1.0263	59	.8572	.5150	1.6643	1.9416
14	.2419	.9703	.2493	1.0300	60	.8660	.5000	1.7321	2.0000
15	.2588	.9659	.2679	1.0353	61	.8746	.4848	1.8040	2.0627
16	.2756	.9613	.2867	1.0403	62	.8820	.4695	1.8807	2.1300
17	.2924	.9563	.3057	1.0457	63	.8910	.4540	1.9626	2.2027
18	.3090	.9511	.3249	1.0515	64	.8988	.4384	2.0503	2.2812
19	.3256	.0455	.3443	1.0576	65	.9063	.4226	2.1445	2.3662
20	.3420	.9397	.3640	1.0642	66	.9135	.4067	2.2460	2.4586
21	.3584	.9336	.3839	1.0711	67	.9205	.3907	2.3559	2.5593
22	.3746	.0272	.4040	1.0785	68	.9272	.3746	2.4751	2.6695
23	.3907	.9205	.4245	1.0664	69	.9330	.3586	2.0051	2.7904
24	.4087	.9135	.4452	1.0846	70	.9397	.3420	2.7475	2.9238
25	.4220	.9063	.4663	1.1034	71	.9455	.3256	2.9042	3.0715
26	.4386	.8938	.4877	1.1126	72	.9511	.3090	3.0777	3.2361
27	.4540	.8910	.5095	1.1223	73	.9563	.2924	3.2709	3.4203
28	.4695	.8829	.5317	1.1326	74	.9613	.2756	3.4874	3.6279
29	.4848	.8746	.5543	1.1433	75	.9650	.2588	3.7321	3.8637
30	.5000	.8660	.5774	1.1547	76	.9703	.2419	4.0108	4.1336
31	.5150	.8572	.6009	1.1666	77	.9744	.2250	4.3315	4.4454
32	.5200	.8480	.6249	1.1792	78	.9781	.2079	4.7046	4.9007
33	.5446	.8387	.6494	1.1924	79	.9816	.1908	5.1446	5.2406
34	.5592	.8290	.6745	1.2062	80	.9848	.1736	5.6713	5.7588
35	.5736	.8192	.7002	1.2208	81	.9877	.1564	6.3128	6.3924
36	.5878	.8090	.7265	1.2361	82	.9903	.1392	7.1154	7.1853
37	.6018	.7936	.7536	1.2521	83	.9925	.12187	8.1443	8.2055
38	.6157	.7880	.7613	1.2690	84	.9945	.10453	9.5668	—
39	.6293	.7771	.8098	1.2867	85	.9962	.08716	11.4301	11.474
40	.6428	.7660	.8391	1.3054	86	.9976	.06976	14.3007	14.335
41	.6561	.7547	.8693	1.3230	87	.9986	.05384	10.0811	19.107
42	.6691	.7431	.9004	1.3456	88	.9954	.03490	28.6363	28.654
43	.6820	.7314	.9325	1.3673	89	.9903	.01745	57.2900	57.299
44	.6947	.7193	.9657	1.3902	90	1.0000	Inf.	Inf.	inf
45	.7071	.7071	1.000	1.4142	—	—	—	—	—

Vertical pipe is best supported with a shoulder clamp attached to the flooring at the point through which the pipe passes, or by clamps attached to adjacent walls or columns.

Pipe hangers and anchors are also used for supporting suspended piping or securing it (as in the case of anchors) to adja-

Fig. 8-11. Pipe strap.

cent surfaces. Hangers are similar in appearance and function to pipe straps (see above).

JOINT COMPOUND

Joint compound (also referred to as *pipe dope)* is a substance applied to the male thread when making up screwed joints. The purpose of applying a joint compound is to lubricate the threads so that tightening is made easier. By lubricating the threads, the friction and heat produced by the tightening operation are greatly reduced. Moreover, the joint compound forms a seal inside the screwed joint which prevents leakage and ensures a tight joint.

Joint compounds are commercially available, or they may be made on the job from a variety of different materials. Red lead, white lead, or graphite has frequently been used as a joint compound. Red lead produces a very tight joint, but it hardens to such an extent that it is difficult to unscrew the joint for repairs.

A tape material has also been developed for use in making up screwed joints. It functions in the same manner as a joint compound. The tape is made of teflon and is so thin that it will sink into the threads when wrapped around them.

An old toothbrush is an excellent tool for applying joint compound to the thread. It is important to remember that the joint compound *must* be applied to the *male* thread only. If it is ap-

401

plied to the female thread, some of it will be forced into the pipe where it will lodge as a contaminating substance.

PIPE FITTING WRENCHES

The numerous wrenches used in pipe fitting may be listed as follows:

1. Monkey wrench.
2. Pipe wrench.
   a. Stillson wrench.
   b. Chain wrench.
   c. Strap wrench.
3. Open wrench.

The important point to remember in pipe fitting is to select a suitable wrench for the job at hand. Each of the wrenches listed above is designed for one or more specific tasks. No wrench is suitable for every task encountered in pipe fitting.

A *monkey wrench* has smooth parallel jaws that are especially adapted for hexagonal valves and fittings (Fig. 8-12). Not only does it fit better on the part to be turned, it also does not have the crushing effect of a pipe wrench.

The operating principle of a *pipe wrench* is simple. The harder you pull, the tighter it squeezes the pipe. The pipe wrench was designed for use on pipe and screw fittings only. On parallel-sided objects, its efficiency is not up to that of a monkey wrench, and its squeezing action can do a great deal of damage.

Many inexperienced fitters have learned from experience that using a pipe wrench too large for the job can cause the fitting to stretch or crack. The result is a leaking joint that will require a new fitting to remedy the damage.

A Stillson wrench has serrated teeth jaws that enable it to grip a pipe or round surface in order to turn it against considerable resistance. The correct method for using a Stillson wrench is illustrated in Fig. 8-13. Adjust the wrench so that the jaws will take hold of the pipe at about the middle part of the jaws. To support the wrench and prevent unnecessary lost motion when the wrench engages the pipe, hold the jaw at *A*, with the left hand

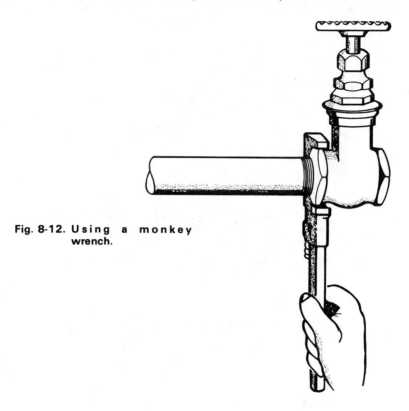

Fig. 8-12. Using a monkey wrench.

pressing it against the pipe. At the beginning of the turning stroke *B*, with the jaw held firmly against the pipe with the left hand, the wrench will at once bite or take hold of the pipe with only the lost motion necessary to bring jaw *C* in contact with the pipe.

Fig. 8-14 shows a *chain wrench* (or *pipe tongs*) and the method in which it is used. Although they are made for small sizes up, they are generally used for 6-in. pipe and larger.

A *strap wrench* (Fig. 8-15) is used when working with plated or polished finish piping in order not to mar the surface. It also comes in handy in tight places where you cannot insert a Stillson wrench.

Open-end wrenches (Fig. 8-16) are used for making up flange couplings. The right size should be used in order to prevent wear-

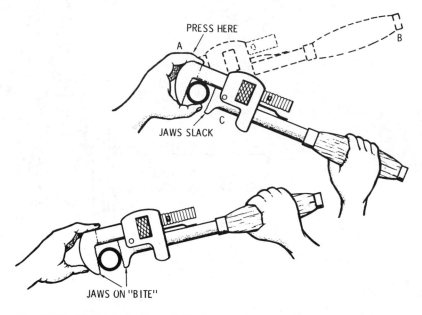

Fig. 8-13. Method of using a Stillson wrench.

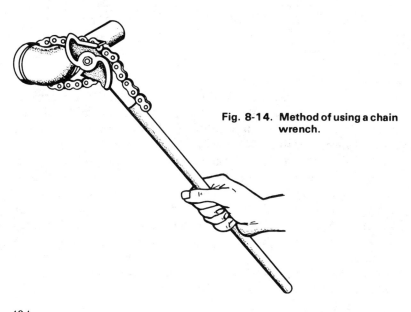

Fig. 8-14. Method of using a chain wrench.

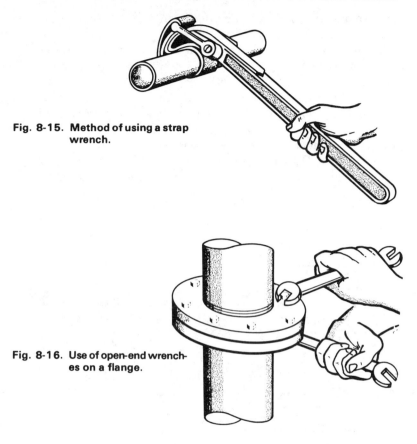

Fig. 8-15. Method of using a strap
wrench.

Fig. 8-16. Use of open-end wrenches on a flange.

ing the bolt heads round or slippage that can cause bruised
knuckles.

PIPE VISE

Either a pipe vise or a machinist vise (Fig. 8-17) can be used in
pipe fitting. The pipe vise is used for pipe only. The machinist
vise, on the other hand, has square jaws or a combination of

405

square and gripper jaws, making it suitable for pipe as well as other work.

Several precautions should be taken when using a pipe vise. It exerts a powerful force at the jaws, which in some cases can do damage to the pipe. That is why experienced fitters never put a valve or fitting into a vise when making up a joint at the bench. There is too much danger of oversqueezing the part as to distort it, or of putting the working parts of a valve out of line. The correct and incorrect methods of connecting a valve to a pipe are illustrated in Fig. 8-18. Always hold a valve between lead-or copper-covered machinist vise jaws while unscrewing the bonnet (Fig. 8-18).

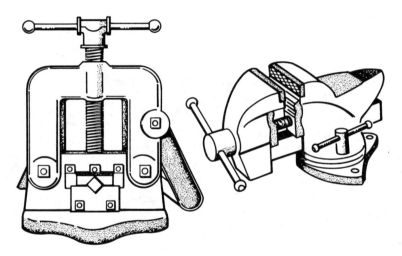

Fig. 8-17. A pipe and machinist vise.

INSTALLATION METHODS

Pipe fitting may be defined as the operations that must be performed in installing a pipe system as made up of pipe and fittings. These pipe fitting operations can be listed as follows:

1. Cutting.
2. Threading.
3. Reaming.
4. Cleaning.
5. Tapping.
6. Bending.
7. Assembling.
8. Make-up.

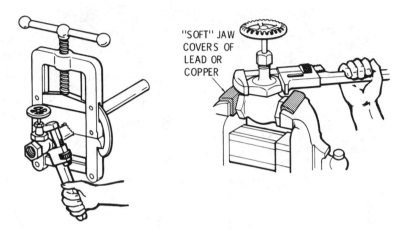

"SOFT" JAW
COVERS OF
LEAD OR
COPPER

Fig. 8-18. Correct method of using the pipe and machinist vise.

Fig. 8-19 illustrates the principal operations in pipe fitting. After being marked to length by nicking with a file, the pipe is put in a vise and cut with a pipe cutter (or hacksaw) as shown in Fig. 8-19A. Any external enlargement is removed with a metal file (Fig. 8-19B). The thread is next cut with stock and dies as in Fig. 8-19C. After carefully cleaning the thread with a hard toothbrush and applying red lead or pipe cement to the freshly cut thread, the joint is made up with a Stillson wrench (Fig. 8-19D).

PIPE CUTTING

Pipe is manufactured in different lengths varying from 12 to 22 ft. Accordingly, it must be cut to required length for the pipe installation. This may be done with either a hacksaw or a pipe cutter. The latter method is generally quicker and more convenient.

When a length of pipe is being cut or threaded, it is held firmly in a pipe vise. The pipe vise should be adjusted just tight enough to prevent the pipe from slipping, but not so tight as to cause the jaw teeth to unduly dig into the pipe.

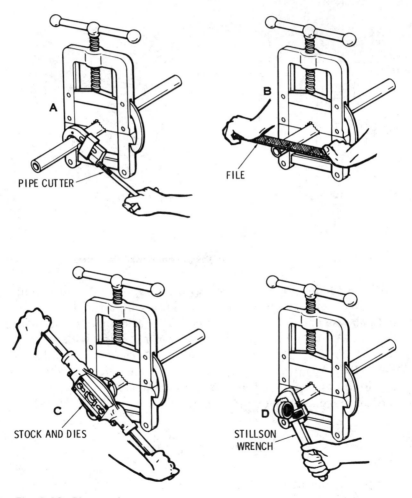

Fig. 8-19. Pipe cutting.

A *pipe cutter* is a tool usually consisting of a hoop-shaped frame on whose stem a slide can be moved by a screw. On the side and frame, several cutting dies or wheels are mounted. In cutting, the pipe cutter is placed around the pipe so that the wheels contact the pipe. The tool is rotated around the pipe, tightening up with the screw stem each revolution until the pipe is cut.

The operating principles of a three-wheel cutter and a combined wheel and roller cutter are illustrated in Fig. 8-20A. The cuts show the comparative movements necessary with the two types of cutters to perform their functions. The three-wheel cut-

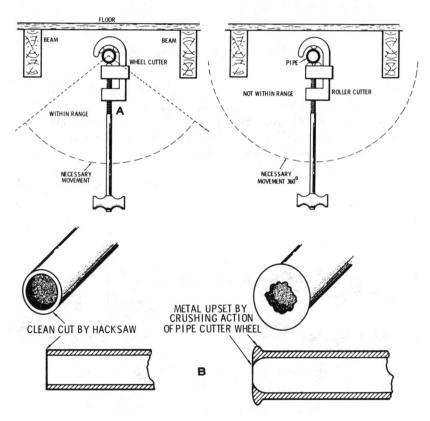

Fig. 8-20. The appearance of a pipe cut by hacksaw and a pipe cut by a cutter wheel.

ter requires only a small arc of movement and is recommended for cutting pipe in inaccessible locations. The wheel cutter has a greater range than the roller cutter and is therefore preferred for general use.

The major disadvantage of a pipe cutter is that it does not cut but *crushes* the metal of the pipe, leaving a shoulder on the outside and a burr on the inside. This does not apply to the knife-type pipe cutter designed to actually cut (not crush) the pipe.

Fig. 8-20B shows the appearance of pipe when cut by a hacksaw or knife cutter and when cut by a wheel pipe cutter. When the latter is used, the external enlargement must be removed by a file and the internal burr by a pipe reamer.

PIPE THREADING

A pipe thread is cut with stock and dies. Adjustable dies are used in pipe threading because of slight variations in fittings, especially cast-iron fittings. Fig. 8-21 illustrates an adjustable pipe stock and dies for double-ended dies. As shown, each pair of dies has one size thread at one end and another size at the other end. Thus, the two dies in the stock are in position for cutting ½-in. thread and by reversing them they will cut ¾-in. thread. The cut shows plainly the reference marks, which must register with each

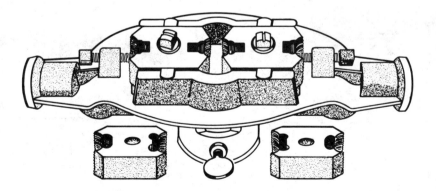

Fig. 8-21. Adjustable pipe stock and dies for double-end dies.

other in adjusting the dies by means of the end setscrews to standard size.

A vise is used in conjunction with the pipe stock and dies when threading a pipe. After securing the pipe in the vise, use plenty of oil in starting and cutting the thread. In starting, press the dies firmly against the pipe until they take hold. After a few turns, blow out the chips and apply more oil. This should be done several times before completing the cut. When complete, blow out the chips as clean as possible and back off the dies. When drawing in your breath preliminary to blowing out the chips, turn your head away from the die to avoid drawing the chips into your lungs.

A nipple is short piece of pipe 12 in. in length or less and threaded at both ends. Nipples are properly cut by using a nipple holder designed for use with hand stock and dies. The holder is double ended and holds two sizes of nipples, one being for ½- and ¾-in. nipples. In construction, there is a pin inside the holder having a fluted end which digs into the nipple end when pressed forward by driving down the wedge. In operation, the nipple is screwed by hand into the holder as far as it will go, then the wedge is driven down sufficiently to firmly secure the nipple. The holder is so arranged that when the thread is cut, the nipple can be removed by simply starting back the wedge, which loosens the inner part of the holder and allows the nipple to be easily unscrewed by hand. The holder can be used for making either right, or right and left, nipples.

PIPE REAMING

The burrs should be removed with a reamer to avoid future trouble with clogged pipes. This is a job that should be done thoroughly.

The correct way of removing the shoulder (i.e., external enlargement) left on a pipe end after cutting it with a pipe cutter is by using a flat file. Obviously at each stroke, the file should be given a turning motion, removing the excess metal through an arc of the circumference. The position of the pipe is changed in the vise from time to time until the excess metal is removed all around the pipe. When the operation is done by moving the file in a straight line, it will result in a series of flat places on the surface. A good pipe threader will also remove the external enlargement of the pipe end caused by using a pipe cutter.

PIPE CLEANING

Dirt, sand, metal chips, and other foreign matter should be removed from pipelines to prevent future problems. When a new pipeline is installed, flush it out completely with water to remove any loose scale or foreign matter.

Check the threads for any dirt or other foreign matter. It is important to be thorough about this because dirt in the threads can also get into the lines when the joints are made up. Dirt can also cause tearing of the metal when screwing up a connection. It increases friction and interferes with making a tight joint.

Flanged faces should also be cleaned thoroughly. Manufacturers usually coat flanges with a heavy oil or grease to prevent rusting. A solvent will easily remove this coating. Special precaution should be taken against dirt on gaskets. Dirt on any part of a flanged joint tends to cause leaks.

PIPE TAPPING

An internal or female thread is cut by means of a pipe tap, a conical screw made of hardened steel and grooved longitudinally. A pipe tap and pipe reamer are illustrated in Fig. 8-22.

Table 8-20 gives drill sizes that permit direct tapping without reaming the holes beforehand. Table 8-21 gives drill sizes for both the Briggs, or American Standard, and the Whitworth, or British Standard.

PIPE BENDING

With the proper tools, pipes may be bent within certain limits without difficulty. An example of a pipe-bending tool is illustrated in Fig. 8-23.

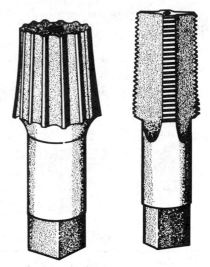

Fig. 8-22. A typical pipe tap and
pipe reamer.

Table 8-20. Drill Sizes for Briggs Standard Pipe Taps
(for direct tapping without reaming)

Size of pipe	1/8	1/4	3/8	1/2	3/4	1	1 1/4	1 1/2	2	2 1/2	3	3 1/2	4
Size of drill	21/64	7/16	9/16	45/64	29/32	1 9/64	1 31/64	1 47/64	2 13/64	2 5/8	3 1/4	3 47/64	4 15/64

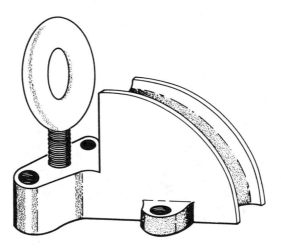

Fig. 8-23. Pipe-bending tool.

413

Pipes can also be bent by hand without the use of special tools. One method involves the complete filling of the pipe with sand and capping both ends so that none of the sand will be lost. Heat the part to be bent and then clamp the pipe in a vise as close to the part to be bent as possible. Now cool the outside with water so that the inside, being hot and plastic, is compressed as the bend is made.

ASSEMBLING AND MAKE-UP

Assembling is the operation of putting together the various lengths of pipe and fittings used in an installation.

Table 8-21. Drill Sizes for Pipe Taps

Size Tap Inches	BRIGGS STANDARD		BRITISH (Whitworth) STANDARD	
	Thread	Drill	Thread	Drill
$\frac{1}{8}$	27	$\frac{21}{64}$	28	$\frac{5}{16}$
$\frac{1}{4}$	18	$\frac{27}{64}$	19	$\frac{7}{16}$
$\frac{3}{8}$	18	$\frac{9}{16}$	19	$\frac{9}{16}$
$\frac{1}{2}$	14	$\frac{11}{16}$	14	$\frac{23}{32}$
$\frac{5}{8}$	—	—	14	$\frac{25}{32}$
$\frac{3}{4}$	14	$\frac{29}{32}$	14	$\frac{29}{32}$
$\frac{7}{8}$	—	—	14	$1\frac{1}{16}$
1	$11\frac{1}{2}$	$1\frac{1}{8}$	11	$1\frac{5}{32}$
$1\frac{1}{4}$	$11\frac{1}{2}$	$1\frac{15}{32}$	11	$1\frac{1}{2}$
$1\frac{1}{2}$	$11\frac{1}{2}$	$1\frac{23}{32}$	11	$1\frac{23}{32}$
$1\frac{3}{4}$	—	—	11	$1\frac{31}{32}$
2	$11\frac{1}{2}$	$2\frac{3}{16}$	11	$2\frac{3}{16}$
$2\frac{1}{4}$	—	—	11	$2\frac{13}{32}$
$2\frac{1}{2}$	8	$2\frac{9}{16}$	11	$2\frac{25}{32}$
$2\frac{3}{4}$	—	—	11	$3\frac{1}{32}$
3	8	$3\frac{3}{16}$	11	$3\frac{8}{32}$
$3\frac{1}{4}$	—	—	11	$3\frac{1}{2}$
$3\frac{1}{2}$	8	$3\frac{11}{16}$	11	$3\frac{3}{4}$
$3\frac{3}{4}$	—	—	11	4
4	8	$4\frac{3}{16}$	11	$4\frac{1}{4}$
$4\frac{1}{2}$	8	$4\frac{11}{16}$	11	$4\frac{3}{4}$
5	8	$5\frac{1}{4}$	11	$5\frac{1}{4}$
$5\frac{1}{2}$	—	—	11	$5\frac{3}{4}$
6	8	$6\frac{5}{16}$	11	$6\frac{1}{4}$
7	8	$7\frac{5}{16}$	11	$7\frac{5}{16}$
8	8	$8\frac{5}{16}$	11	$8\frac{5}{16}$
9	8	$9\frac{5}{16}$	11	$9\frac{5}{16}$
10	8	$10\frac{7}{16}$	11	$10\frac{5}{16}$

If no mistakes have been made in cutting the pipe to the right length or in following the dimensions on the blueprint, the pipe and fittings may be installed without difficulty. In other words, the last joint (either a right or left union, or long screw joint) will come together smoothly or, as they say in the trade, "make up."

Screwed joints are put together with red or white lead pigment mixed with graphite and linseed oil, or with some standard commercial joint compound.

It is unnecessary to put much material on the threads because it will be simply pushed out and wasted when the joint is screwed up. It should be put on evenly and cover all the threads, care being taken not to let any touch the reamed end of the pipe where it may get inside. The red lead is preferably obtained in the powder form and mixed with oil and a little dryer at the time the pipe is to be made up. Get a clean piece of glass on which to prepare the lead. The toothbrush should be laid on the glass after applying the lead in order to avoid getting grit on the brush and paint on the table. When grit becomes mixed with the lead, it prevents close contact of the filling and pipe, thus making the joint less efficient.

The following steps should be taken *before* making up a screwed joint:

1. Ream the pipe ends.
2. Remove burrs from both the inside and outside of the pipe.
3. Thoroughly clean the inside of the pipe.
4. Thoroughly clean the threads.

Clean threads, a suitable joint compound, and proper tightening (neither too much nor too little) are all necessary for a satisfactory screwed joint.

There are four requirements for satisfactorily making up a flanged joint. In the proper order of sequence, these four requirements are:

1. Thorough cleaning.
2. Accurate alignment.
3. Using the proper gasket.
4. Tightening bolts in proper order.

Thoroughly clean the flange face with a solvent to remove any grease, and wipe it dry. The alignment must be accurate for satisfactory make-up. This is particularly important where valve flanges are involved. If the alignment is poor, a severe stress on the valve flanges may distort the valve seats and prevent tight closure (Fig. 8-24).

Selecting a suitable gasket for the service is also very important. Using the wrong gasket may result in a leak.

Fig. 8-25 illustrates the recommended method of tightening the bolts with a wrench. This must not be done in rotation, but in sequence as indicated by the numbers. This is the crossover method. Do not fully tighten on the first round but go over them two or preferably three times to fully tighten.

NONFERROUS PIPES, TUBING, AND FITTINGS

Both brass and copper are used as materials for this type of pipe and tubing. The construction may be either cast or wrought,

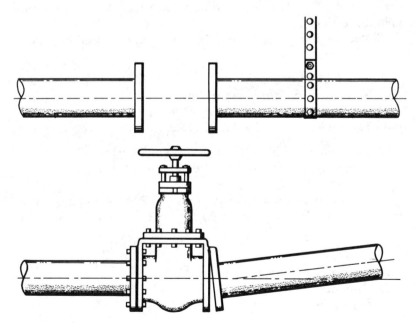

Fig. 8-24. Flange and pipe alignment.

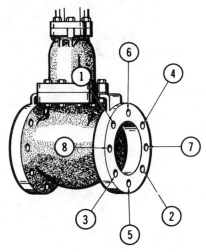

Fig. 8-25. Recommended method of tightening bolts.

and the methods of joining include screwed, flared, and soldered.

The technique used with screwed fittings is the same as with ordinary malleable-iron fittings. This has already been described in the first part of this chapter.

The fittings used for flared soft tube end joints are cast fittings. These are usually used on oil burner construction and supply lines. Flared joint fittings include elbows, tees, couplings, unions, and a full range of reducing and adapter combinations in all standard sizes and combinations of sizes from ⅛ to 2 in. inclusive (Fig. 8-26). A double-seal type of flared joint fitting is illustrated in Fig. 8-27.

The sequence of operations required to make a flared joint are as follows:

1. Cut the tube with a hacksaw to the exact length, using a guide to ensure a square cut.
2. Remove all burrs and irregularities by filing both inside and outside.
3. Slip the coupling nut over the end of the tube and insert the flanging tool.
4. Drive the flanging tool into the tube with a few hammer blows, expanding the tube to its proper flare.

417

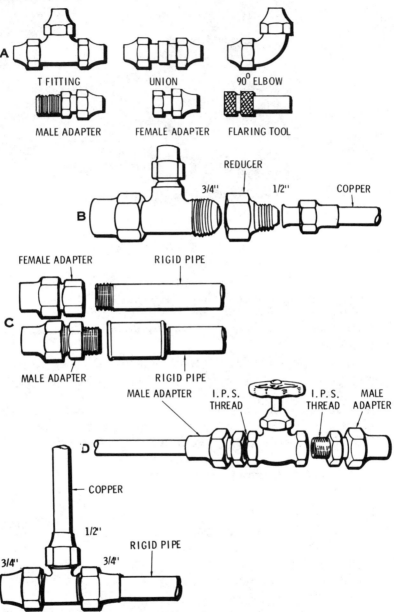

Fig. 8-26. Various flare-type copper tube fittings.

418

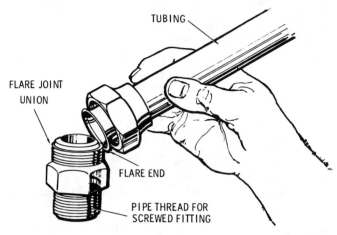

Fig. 8-27. Double-seal flared joint fitting.

5. Assemble the fitting and tighten it by using two wrenches, one on the nut and the other on the body of the fitting.

SOLDERING PIPE

Solder fittings usually come in cast bronze and wrought copper. The two kinds of fittings used are the edge feed fitting and the hole feed fitting (Fig. 8-28).

The basic principle of solder fittings is capillary attraction. Because of capillary attraction, solder can be fed vertically upward between two closely fitted tubes to a height many times the distance required to made a soldered joint, regardless of the size of the fitting.

Either a 50-50 tin-lead solder or a 95-5 tin-antimony solder are recommended for joining copper tube. The former is generally used for moderate pressures with temperatures ranging up to 250°F. The 95-5 tin-antimony solder is used where higher strength is required, but it has the disadvantage of being difficult to handle. Pressure ratings for soldered joints using these two solders are listed in Table 8-22. A suitable paste-type flux is recommended for use with these solders.

419

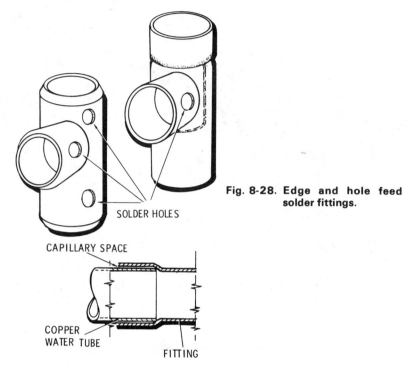

SOLDER HOLES

CAPILLARY SPACE

COPPER
WATER TUBE

FITTING

Fig. 8-28. Edge and hole feed
solder fittings.

The series of operations necessary in making a solder fitting joint are as follows:

1. Measure the tube to proper length so that it will run the full length of the socket of the fitting.
2. Cut the tube end squarely.
3. Clean tube end and socket of fitting.
4. Apply soldering flux to the cleaned areas of tube and fitting socket.
5. Assemble the joint.
6. Revolve the fitting if you can, to spread the flux evenly.
7. Apply heat and solder.
8. Remove residual solder and flux.
9. Allow joint to cool.

With a hole feed fitting that has a feed hole for the solder and a groove inside, the procedure is just the same as for an edge feed fitting except that solder is fed into the feed hole until it appears as a ring at the edge of the fitting. Be sure the hole is kept full of solder, as it shrinks on cooling and solidifying.

Do not select fittings that are oversize because they will result in a loose fit. The capillary action is dependent on a fairly tight fit, although a certain amount of looseness can be tolerated. A loose fit causes the greatest difficulty when working with large-size copper tubes.

A *thorough* cleaning of the tube surface and the fitting socket is absolutely essential for a strong tight and durable joint. *This cannot be emphasized too strongly.*

BRAZING PIPES

Brazing is rapidly taking the place of many operations formerly performed by soldering because it is simpler, quicker, and

Table 8-22. Safe Strength of Soldered Joints
Pressure-Ratings
Maximum Service Pressure, PSI

Solder used in Joints	Service Temperatures Deg. F	Water†		
		¼ to 1 in. Incl.*	1¼ to 2 in. Incl.*	2½ to 4 in. Incl.*
50-50 Tin-Lead**	100	200	175	150
	150	150	125	100
	200	100	90	75
	250	85	75	50
95-5 Tin-Antimony	100	500	400	300
	150	400	350	275
	200	300	250	200
	250	200	175	150
Brazing Filler Metal melting at or above 1000 F‡	250	300	210	170
	350	270	190	155

*Standard copper water tube sizes
**ASTM B32, Alloy grade 50A
†Including refrigerants and other non-corrosive liquids and gases
‡ASTM B260, Brazing Filler Metal

Courtesy Copper & Brass Research Assoc.

421

results in a stronger joint. Like soldering, the brazing alloy is applied at temperatures *below* the melting point of the metal being brazed. In this respect, both brazing and soldering differ from welding, which forms a joint by melting (fusing) the metal at temperatures *above* its melting point.

During the brazing process, the brazing alloy is heated until it adheres to the pipe surface and enters into the porous structure of the metal. The brazed joint is almost always as strong as the brazed metal surrounding it.

Brazing can be used to join nonferrous metals, such as copper, brass, or aluminum, or ferrous metals, such as cast iron, malleable iron, or steel. The brazing alloy used in making the joint depends on the type of metal being brazed. For example, aluminum requires the use of a special aluminum brazing alloy, whereas a low-temperature brazing alloy or a silver alloy is recommended for brazing copper and copper alloys. A local welding supply dealer should be able to provide answers to your questions about which alloy to use.

The procedure for brazing consists essentially of the following operations:

1. Clean both surfaces.
2. Apply a suitable flux.
3. Align and clamp the parts to be joined.
4. Preheat the surface until the flux becomes fluid.
5. Apply a suitable brazing alloy.
6. Allow the surface time to cool.
7. Clean the surface.

The pipe surfaces must be thoroughly cleaned, or the result will be a weak bond or no bond at all. All dirt, grease, oil, and other surface contaminants must be removed, or the capillary attraction so important to the brazing process will not function properly.

Select a flux suited to the requirements of the brazing operation. Fluxes differ in their chemical compositions and the brazing temperature ranges within which they are designed to operate. Apply the flux to the joint surface with a brush.

Align the parts to be joined, securing them in position until after the brazing alloy has solidified. Preheat the metal to the required brazing temperature (indicated by the flux reaching the fluid stage).

After the flux has become fluid, add the brazing alloy. If conditions are right, the brazing alloy will spread over the metal surface and into the joint by capillary attraction. Do *not* overheat the surface. Remove the heat as soon as the entire surface has been covered by the brazing alloy. Allow time for the joint to cool and then clean the surface.

BRAZE WELDING PIPE

Braze welding is another bonding process that does not melt the base metal. In this respect it resembles both soldering and brazing.

Brazing and braze welding operate under essentially the same basic principles. For example, both use nonferrous filler metals that melt above 880°F but below the melting point of the base metal. They differ primarily in application and procedure.

The braze welding process follows most of the steps described for brazing (see above), but will differ principally as follows:

1. Edge preparation is necessary in braze welding and is essentially similar to that employed in gas welding. The edges must be prepared before the surfaces are cleaned.
2. A suitable filler metal is used instead of a brazing alloy.

The filler metal will flow throughout the joint by capillary attraction. Flanges that are to be brazed to copper pipes must be of copper or what is known as brazing metal (98% copper and 2% tin), as gunmetal flange would melt before the brazing alloy ran.

WELDING PIPE

Oxyacetylene welding (gas welding) is probably the most common welding process used for joining pipe and fittings, particularly on smaller installations. Arc welding is also very popular.

Pipe welding should be done only by a skilled and experienced worker. The equipment and the procedure used are much more complicated than those used with soldering, brazing, or braze welding.

Welding forms a joint by melting (fusing) the metal at temperatures above its melting point. The filler metal, electrodes, or welding rods used must be suitable for use with the base metal to be welded, and the procedure used should be such as to ensure complete penetration and thorough fusion of the deposited metal with the base metal. The welding process is employed in many piping installations, but especially those in which large-diameter pipe are used.

Manufactured steel welding fittings are available for almost every conceivable type of pipe connection. These steel welding fittings can be divided into the following two principal categories:

1. Butt welding fittings.
2. Socket welding fittings

Butt welding fittings (Table 8-23) have ends that are cut square or beveled $37\frac{1}{2}°$ for wall thicknesses under $\frac{3}{16}$ in. Wall thicknesses ranging from $\frac{3}{16}$ in. to $\frac{3}{4}$ in. thick are beveled $37\frac{1}{2}°$. For walls ranging from $\frac{3}{4}$ in. to $1\frac{3}{4}$ in. thick, the bevel is U-shaped.

Socket welding fittings (Table 8-24) have a machined recess or socket for inserting the pipe. A fillet weld is made between the pipe wall and the socket end of the fitting. The fillet weld is approximately triangular in cross section, the throat lying in a plane of approximately 45° with respect to the surfaces of the part joined. As shown in Fig. 8-29, the minimum thickness of the socket wall *(L)* is 1.25 times the nominal pipe thickness *(T)* for the designated schedule number of the pipe. Socket welding fit-

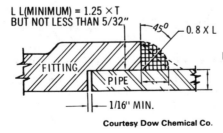

L L(MINIMUM) = 1.25 × T
BUT NOT LESS THAN 5/32″
45°
0.8 X L
FITTING
PIPE
1/16″ MIN.

Fig. 8-29. Fillet weld dimensions.

Courtesy Dow Chemical Co.

tings are generally limited in use to nominal pipe sizes 3 in. and smaller.

In addition to butt and socket welding fittings, flange fittings are also available for welding in sizes ranging from ½ to 24 in.

Table 8-23. Dimensions of Steel Butt Welding Fittings

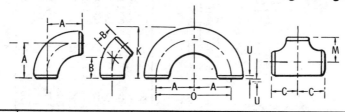

Nominal Pipe Size	Long Radius Elbows			180 Deg Returns			Straight Tees		
	Outside Diameter at Bevel	Center-to-End		Outside Diameter at Bevel	Center-to-Center	Back to Face	Outside Diameter at Bevel	Center-to-End	
		90-Deg Elbows A	45-Deg Elbows B		O	K		Run C	Outlet M
1	1.315	1½	⅞	1.315	3	2³/₁₆	1.315	1½	1½
1¼	1.660	1⅞	1	1.660	3¾	2¾	1.660	1⅞	1⅞
1½	1.900	2¼	1⅛	1.900	4½	3¼	1.900	2¼	2¼
2	2.375	3	1⅜	2.375	6	4³/₁₆	2.375	2½	2½
2½	2.875	3¾	1¾	2.875	7½	5³/₁₆	2.875	3	3
3	3.500	4½	2	3.500	9	6¼	3.500	3⅜	3⅜
3½	4.000	5¼	2¼	4.000	10½	7¼	4.000	3¾	3¾
4	4.500	6	2½	4.500	12	8¼	4.500	4⅛	4⅛
5	5.563	7½	3⅛	5.563	15	10⁵/₁₆	5.563	4⅞	4⅞
6	6.625	9	3¾	6.625	18	12⁵/₁₆	6.625	5⅝	5⅝
8	8.625	12	5	8.625	24	16⁵/₁₆	8.625	7	7
10	10.750	15	6¼	10.750	30	20⅜	10.750	8½	8½
12	12.750	18	7½	12.750	36	24⅜	12.750	10	10
14	14.000	21	8¾	14.000	42	28	14.000	11	
16	16.000	24	10	16.000	48	32	16.000	12	Not
18	18.000	27	11¼	18.000	54	36	18.000	13½	stand-
20	20.000	30	12½	20.000	60	40	20.000	15	ard
24	24.000	36	15	24.000	72	48	24.000	17	

From American Standard for Butt-Welding Fittings, ASA B16.9-1958. All dimensions are in inches. Dimension A is equal to ½ of dimension O.

Courtesy *1960 ASHRAE Guide*

Table 8-24. Dimensions of Socket Welding Fittings

Pipe Size Nominal	Depth of Socket, Min	Center to Bottom of Socket Sched 40 and 80 (A)	Sched 160 (A)	Bore Diameter of Socket, Min (B)	Socket Wall Thickness, Min Sched 40 (C)	Sched 80 (C)	Sched 160 (C)	Bore Diameter of Fitting Sched 40 (D)	Sched 80 (D)	Sched 160 (D)
1/8	3/8	7/16	—	0.420	0.125	0.125	—	0.269	0.215	—
1/4	3/8	7/16	—	0.555	0.125	0.149	—	0.364	0.302	—
3/8	3/8	17/32	—	0.690	0.125	0.158	—	0.493	0.423	—
1/2	3/8	5/8	3/4	0.855	0.136	0.184	0.234	0.622	0.546	0.466
3/4	1/2	3/4	7/8	1.065	0.141	0.193	0.273	0.824	0.742	0.614
1	1/2	7/8	1 1/16	1.330	0.166	0.224	0.313	1.049	0.957	0.815
1 1/4	1/2	1 1/16	1 1/4	1.675	0.175	0.239	0.313	1.380	1.278	1.160
1 1/2	1/2	1 1/4	1 1/2	1.915	0.181	0.250	0.351	1.610	1.500	1.338
2	5/8	1 1/2	1 5/8	2.406	0.193	0.273	0.429	2.067	1.939	1.689
2 1/2	5/8		2 1/4	2.906	0.254	0.345	0.469	2.469	2.323	2.125
3	5/8	2 1/4	2 1/2	3.535	0.270	0.375	0.546	3.068	2.900	2.626

Courtesy 1960 ASHRAE Guide

GAS PIPING

The gas pipe installations in which gas-fired furnaces, boilers, heaters, or other gas appliances are used deserve special attention because of the volatile and highly flammable nature of the fuel. Special attention should be given to the installation of a gas piping to ensure against leakage.

The installation and replacement of gas piping should be done only by qualified workers having the necessary skills and experience.

The following recommendations are offered as a guide for installing and replacing gas piping:

1. All work should be done in accordance with the building, heating, and plumbing codes and standards of the authorities having local jurisdiction. These take precedence over national codes and standards.
2. In an existing installation, the gas supply to the premises and all burners in the system must be shut off before work begins.
3. When installing a system, size the pipes according to the amount of gas to be delivered to each outlet and at the proper pressure. The length of pipe runs and number of outlets are the main determining factors.
4. Use a piping material recommended by the local authorities having jurisdiction. *Never* bend gas piping because it may cause the pipe walls to crack and leak gas. Use fittings for making turns in gas piping. Take all branch connections from the top or side of horizontal pipes (never from the bottom).
5. Locate the gas meter as close as possible to the point at which the gas service enters the structure.
6. Make certain all pipes are adequately supported so that no unnecessary stress is placed on them
7. Offsets should be 45° elbows rather 90° fittings in order to reduce the friction to the flow of gas.
8. Check for gas leaks by applying a soap-and-water solution to the suspected area. *Never* use matches, candles, or any other flame to locate the leak.

INSULATING PIPES

Insulating the *supply* pipes in a low-pressure steam or hot-water space heating system will reduce unwanted heat loss and improve the heating efficiency of the system. The return pipes in a hot-water heating system should also be insulated so that the water reaches the boiler with a minimum of heat loss. Do *not* insulate the return pipes in a steam heating system. Uninsulated pipes will aid in the condensation of any steam that has succeeded in bypassing the thermostatic traps and entering the returns.

A pipe insulation material must be noncombustible, durable, and resistant to moisture. Furthermore, it should be able to retain its original physical shape and insulating properties *after* becoming wet and drying out.

An effective insulation for low-pressure steam and hot-water space heating pipes with temperature up to 300°F consists of alternate layers of plain and corrugated asbestos felts. This material is moisture resistant and nonshrinking.

Fiberglass is also used to insulate steam or hot-water heating pipes. It can be easily applied to the pipe, requiring little more than ordinary cutting shears or a sharp knife.

Another effective pipe insulating material is expanded polystyrene (Fig. 8-30). It has the same noncombustible, moisture-resistant properties as asbestos or fiberglass.

Your local building supply dealer should be able to answer any questions you may have about pipe insulting materials. The manufacturers of these materials also generally provide detailed instruction about how to apply them. You should not have any difficulty if you carefully read and follow these instructions.

PIPING DETAILS

Each piping installation will have design or layout problems that the fitter must solve. Many of these problems, such as installing dirt pockets or siphons, are quite simple for the pipe fitter to handle. More complicated layout problems are solved by calculating and installing lift fittings, swivels and offsets, and drips.

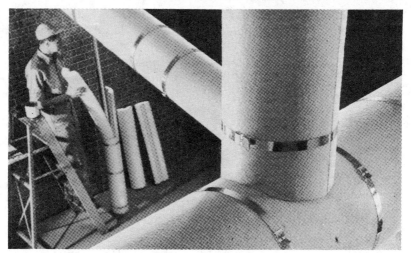

Courtesy Dow Chemical Co.

Fig. 8-30. Expanded polystyrene as a pipe insulating material.

These and other piping details are described in the sections that follow.

Connecting Risers To Mains

A good many steam fitters connect risers directly to mains with a tee; although this method saves on extra labor and expense, it results in a more inefficient operation of the installation. When a tee is used, the condensation falls directly across the path of the steam flowing in the main and will be carried along and finally arrive at the radiator or convector with excess moisture. This problem is avoided with a 45° connection.

Using a 45° connection very effectively drains the condensation from the main, the path of the condensation being along the metal of the pipe and fittings instead of dripping directly into the steam.

The proper method of connecting a riser to a main where the riser has a direct connected drip pipe is by using a 45° connection downward. If the riser has no drip, the riser should be connected to the main with the lead off being 45° upward connecting with a 45° elbow. The runout should pitch ½ in. per foot.

429

Runouts over 8 ft. in length, but with less than ½ in. per foot pitch, should be one size larger than specified in the pipe sizing tables.

If the condensation flows in the opposite direction to the steam, the runouts should be one size larger than the vertical pipe and pitched ¼ in. per foot toward the main. If the runouts are over 8 ft. in length, use a pipe *two* sizes larger than specified.

Connections to Radiators or Convectors

The connections to radiators and convectors must have a proper pitch when installed and be arranged so that the pitch will be maintained under the strains of expansion and contraction. These connections are made by swing joints.

In two-pipe systems, radiators are connected either at the top and bottom opposite end or at the bottom and bottom opposite end. The top connection is not recommended for best performance. Short radiation may be top supply and bottom return connected on the *same* end.

Additional information about radiator and convector connections can be found in Chapter 2 of Volume 3 (Radiators, Convectors, and Unit Heaters).

Lift Fittings

The lift fittings illustrated in Fig. 8-31 are adapted for use on the main return lines of vacuum heating systems at points where it is desired to raise the condensation to a higher level. In operation, the momentum of the water is maintained and assists in making the lift with minimum of loss of vacuum. The lift fitting is constructed with a pocket at the bottom of the lift into which the water drains. As soon as sufficient water accumulates to seal this pocket, it is drawn to the upper portion of the return by the vacuum produced by the pump. The shape of the fitting is such that dirt and scale are usually swept along by the current. Clean-out plugs are provided for use if necessary. A second fitting in a reversed position is recommended for use at the top of the lift to prevent water from running back while the pocket is filled.

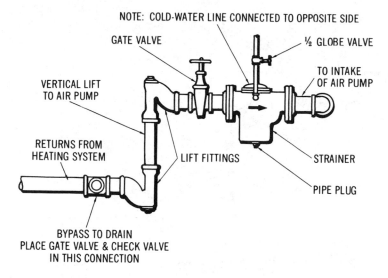

NOTE: COLD-WATER LINE CONNECTED TO OPPOSITE SIDE

GATE VALVE

½ GLOBE VALVE

TO INTAKE
OF AIR PUMP

VERTICAL LIFT
TO AIR PUMP

RETURNS FROM
HEATING SYSTEM

LIFT FITTINGS

STRAINER

PIPE PLUG

BYPASS TO DRAIN
PLACE GATE VALVE & CHECK VALVE
IN THIS CONNECTION

Fig. 8-31. Typical installation of lift fittings.

Drips

A steam main in any steam heating system may be dropped to a lower level without dripping if the pitch is downward with the direction of the steam flow. By the same token, the steam main in any system may be elevated if properly dripped.

Various piping arrangements for dripping the main and riser are illustrated in Figs. 8-32 through 8-39. Fig. 8-39 shows a connection where the steam main is raised and the drain is to a wet return. If the elevation of the low point is above a dry return, it may be drained through a trap to the dry return in a two-pipe vapor, vacuum, or subatmospheric system.

A horizontal steam main can be run over an obstruction if a small pipe is carried below for the condensation with provisions for draining it.

In vacuum steam heating systems, drip traps for steam mains should be either thermostatic or combination float and thermostatic protected by dirt strainer or dirt pockets. Typical methods of making these connections are illustrated in Figs. 8-40 through

431

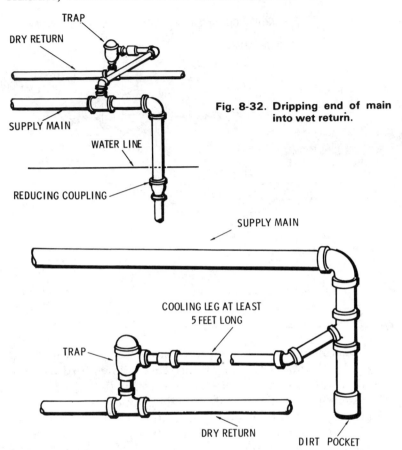

Fig. 8-32. Dripping end of main into wet return.

Fig. 8-33. Dripping end of main into dry return.

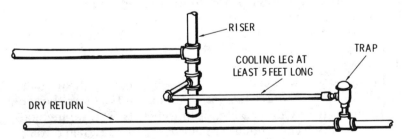

Fig. 8-34. Dripping head of riser into a dry return.

8-43. The bases of supply risers are dripped through drip traps as shown in Figs. 8-44 and 8-45. Methods of connecting return risers are also shown.

In vapor steam heating systems, runouts to supply risers should be dripped separately into a wet return.

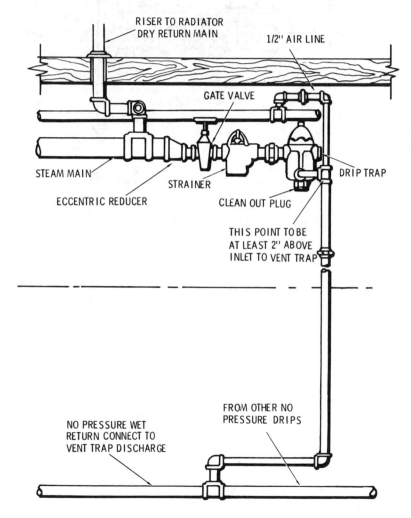

Fig. 8-35. Method of dripping short steam main and discharging conden-sation into pressure wet return near floor.

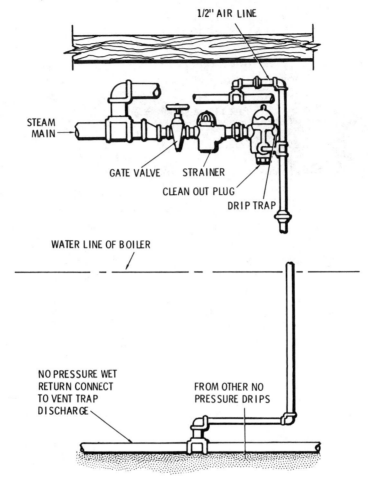

1/2" AIR LINE

STEAM MAIN

GATE VALVE   STRAINER

CLEAN OUT PLUG

DRIP TRAP

WATER LINE OF BOILER

NO PRESSURE WET RETURN CONNECT TO VENT TRAP DISCHARGE

FROM OTHER NO PRESSURE DRIPS

Fig. 8-36. Method of dripping long steam main and discharging condensation into pressure wet return near floor.

Dirt Pockets

On all systems employing thermostatic traps, dirt pockets should be located so as to protect the traps from scale and muck, which will interfere with their operation. Dirt pockets are usually made 8 to 12 in. deep.

434

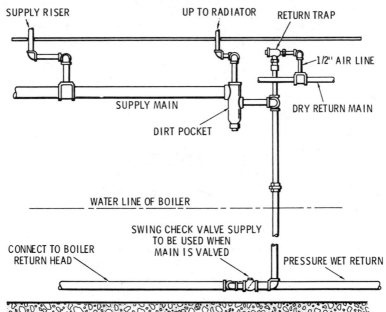

Fig. 8-37. Alternate method of dripping end of supply main through dirt pocket to pressure wet return near floor and venting to overhead dry return.

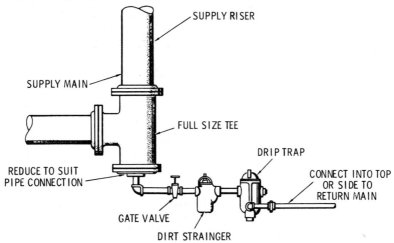

Fig. 8-38. Method of dripping base of main supply riser of downfeed system through dirt strainer and drip trap.

435

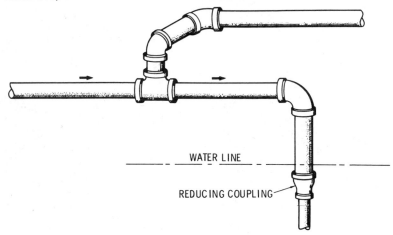

Fig. 8-39. Dripping main where it rises to a higher level.

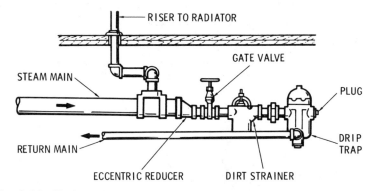

Fig. 8-40. Method of dripping steam mains with drips on end of main.

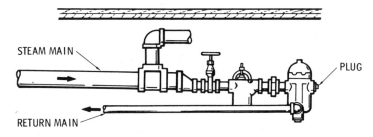

Fig. 8-41. Method of dripping steam mains with relay drip in main.

436

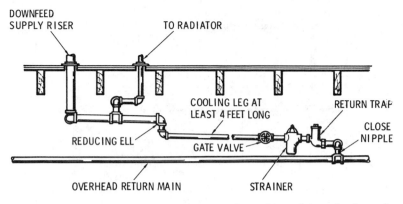

Fig. 8-42. Method of installing drip connection at base of downfeed supply riser to overhead return main through cooling leg, dirt strainer, and return trap.

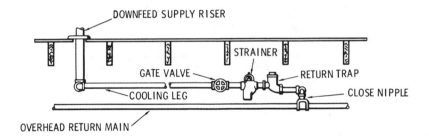

Fig. 8-43. Method of installing drip connection at base of downfeed supply riser to overhead return main.

Siphons

A *siphon* (Fig. 8-46) is used to prevent water from leaving the boiler due to lower pressure in a dry return. Condensation from the drip pipe falls into the loop formed by the siphon and after it is filled, overflows into the dry pipe. The water will rise to different heights *(G* and *H* in the legs of the siphon) to balance the difference in pressure at these points.

If a dry return is used without a siphon, then water would be drawn from the boiler in sufficient amounts to balance the low

437

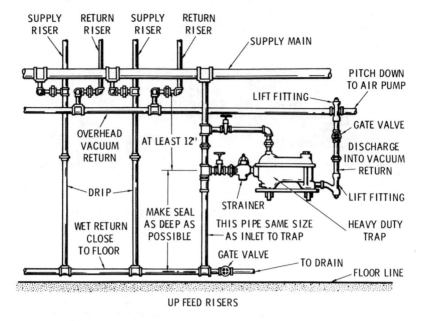

SUPPLY RISER   RETURN RISER   SUPPLY RISER   RETURN RISER

SUPPLY MAIN

PITCH DOWN TO AIR PUMP

LIFT FITTING

OVERHEAD VACUUM RETURN

AT LEAST 12'

GATE VALVE

DISCHARGE INTO VACUUM RETURN

DRIP

LIFT FITTING

MAKE SEAL AS DEEP AS POSSIBLE

STRAINER

WET RETURN CLOSE TO FLOOR

THIS PIPE SAME SIZE AS INLET TO TRAP

HEAVY DUTY TRAP

GATE VALVE

TO DRAIN

FLOOR LINE

UP FEED RISERS

Fig. 8-44. Method of connecting drips from upfeed risers into a wet return and discharging some through a trap into overhead vacuum return main.

pressure in the riser, filling the return and drip to approximately point M (Fig. 8-46).

Hartford Connections

Another method employed to prevent water leaving the boiler is the Hartford connection (or loop) on the wet return (see Chapter 15 of Volume-1, Boilers and Boiler Fittings).

Making Up Coils

In putting together lengths and return bends to form a coil heating unit, there is a right way and a wrong way to do the job. The essential requirement for the satisfactory operation of the coil is providing for proper drainage. To obtain this, the pipes should not be parallel but should have a degree of pitch.

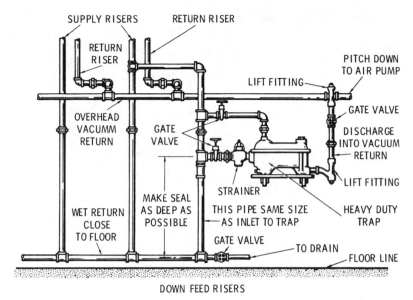

SUPPLY RISERS RETURN RISER

RETURN RISER

PITCH DOWN TO AIR PUMP

LIFT FITTING

GATE VALVE

OVERHEAD VACUMM RETURN

GATE VALVE

DISCHARGE INTO VACUUM RETURN

LIFT FITTING

STRAINER

MAKE SEAL AS DEEP AS POSSIBLE

WET RETURN CLOSE TO FLOOR

THIS PIPE SAME SIZE AS INLET TO TRAP

HEAVY DUTY TRAP

GATE VALVE

TO DRAIN

FLOOR LINE

DOWN FEED RISERS

Fig. 8-45. Method of connecting drips from downfeed risers into wet return and discharging same through a trap into overhead vacuum return main.

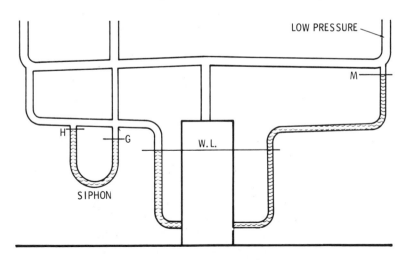

LOW PRESSURE

M

H G

W. L.

SIPHON

Fig. 8-46. Siphon installed on a dry return.

439

A pitch fitting should be used to obtain pitch in the coils rather than the so-called drunken thread method (Fig. 8-47). The drunken thread is obtained by removing the guide bushing from the stock and cutting the thread out of alignment. This gives a poor joint, and one that will eventually break because of corrosion. The corrosion is caused by the deep cut on one side of the pipe resulting from cutting the thread out of alignment.

Relieving Pipe Stress

Long runs of rigidly supported piping carrying steam or hot water, especially when they are at high pressures and temperatures, are often subject to stresses caused by the expansion and contraction of the pipe. These pipe stresses can be relieved in a number of ways, including the installation of either a U-expansion bend or an expansion joint.

Swivels and Offsets

Another method of providing for pipe expansion in steam mains is through the use of swivels and offsets (Fig. 8-48). Allow at least 4 ft. of offset for each inch of expansion to be taken up in the line. The offsets should be place far enough apart to minimize any strain on the threads when expansion or contraction occur.

ELIMINATING WATER POCKETS

Water pockets (i.e., the tendency for water to collect in pipes) are caused by incorrect pipe installation methods (Fig. 8-49 and 8-50). Not only do these water pockets have the potential danger of freezing and damaging the pipes when the boiler is shut down, they also cause that loud and disagreeable hammering in the pipes known as *water hammer*. Water hammer is caused by a sudden rush of steam picking up this undrained water when the radiator or convector is opened and forcing it against any turn in the direction of the main. Hammering in steam lines can be stopped by providing for the proper drainage of the condensa-

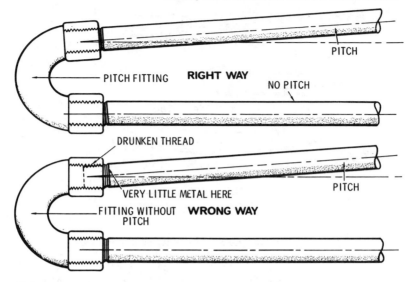

Fig. 8-47. Right and wrong ways of making up coils.

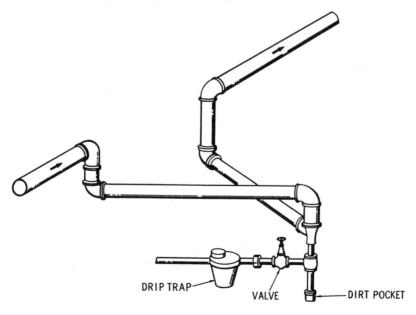

Fig. 8-48. Swivels and offsets.

441

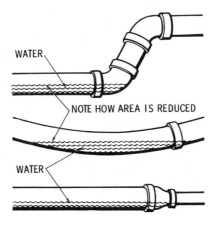

Fig. 8-49. Water pockets resulting from faulty piping methods.

tion. This can be accomplished by installing eccentric fittings or by installing traps of suitable capacity. For example, the water pocket shown in Fig. 8-50 can be avoided by using an eccentric reducing tee (Fig. 8-51).

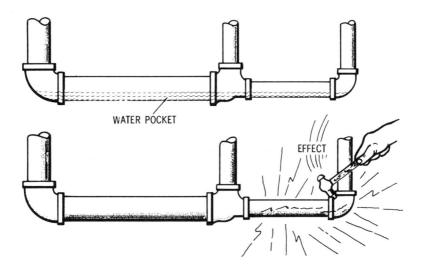

Fig. 8-50. Water hammer caused by using ordinary fittings.

442

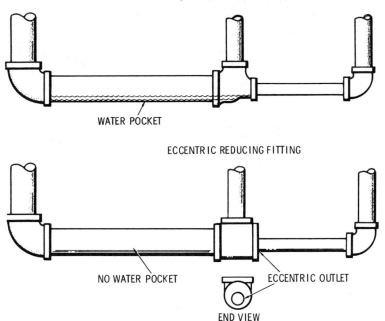

Fig. 8-51. Using an eccentric reducing fitting to eliminate water pockets.

Water trapped in a supply line can sometimes be the cause of radiators failing to heat properly. In this case, the trapped water is usually caused by an improperly pitched supply line, rather than fittings. You can create a slight pitch in the line by slipping wedges under the radiator. Space the wedges so that the radiator remains level (check it with a carpenter's level); otherwise, the radiator will not operate properly.

CHAPTER 9

Valves and
Valve Installation

A valve is a device used for controlling the flow or pressure of a fluid in the pipes of steam and hot-water heating systems. Valves are generally restricted in use to the control function for which they were designed. Specifically, a valve may be used to control a fluid in one of the following ways:

1. Stopping its flow.
2. Checking its flow.
3. Throttling its flow.
4. Reducing its temperature.
5. Relieving its pressure.
6. Reducing or regulating its pressure.

VALVE COMPONENTS AND TERMINOLOGY

The operation of a valve may be controlled either internally or externally, depending upon the type of valve. For example, the operation of a check valve depends on the fluid flow in the pipe. Because the operation of this type of valve is controlled internally and is therefore automatic in nature, no means are provided for external adjustment. Access to the working parts (disc, seat rings, ball, etc.) of a check valve is usually obtained through a cap secured to the valve bolt by bolts or a threaded connection. The operation of most other types of valves is controlled externally, either manually or automatically.

Generally speaking, an externally controlled valve consists of a body to which an extension (a bonnet, yoke, or yoke and bonnet) is attached. The bonnet contains the valve stem, packing nut, and stuffing box. The yoke consists of upright arms mounted on the bonnet or on a gasket placed on the neck of the valve body. When a yoke is used, the stem is threaded and guided through the upper yoke arm.

The valve stem is an adjustable screw that opens or closes the valve by moving a wedge or disc attached to the end of the stem on or off the surface of a valve seat (or seats) contained in the body of the valve.

One method of classifying these externally controlled valves is according to the design of the bonnet. Four common valve bonnet designs are:

1. Union bonnet.
2. Screwed bonnet.
3. Bolted flanged bonnet.
4. Bolted flanged yoke bonnet.

A *union bonnet* (Fig. 9-1A) is secured to the valve body by a heavy ring nut. A bevel on the bottom of the bonnet engages with a corresponding bevel in the body neck. The ring nut fits down over the bonnet, covering the body neck and the bottom portion of the bonnet. A tight seal can be obtained by wrenching down the ring nut. The joint is additionally sealed by the pressure from within the valve body.

446

Fig. 9-1A. Union bonnet.

The advantage of a union bonnet is that its construction provides a quick and easy method of coupling and uncoupling the bonnet and valve body.

A *screwed bonnet* (Fig. 9-1B) is secured to the valve body by a threaded connection. There are two basic types of screwed bonnets. One type has threads on the outside of the base of the bonnet and on the inside of the body neck, and is sometimes referred to as a *screwed-in bonnet*. The other type has threads on the inside of the bonnet and on the outside of the body neck. It is referred to as a *screwed-on bonnet*.

A *bolted flanged bonnet* (Fig. 9-1C) is attached to the top of the valve body neck by bolts. The joint formed by the connection between these two surfaces may be of the following three types: (1) flat-faced joint, (2) male and female joint, or (3) tongue-and-groove joint. Each of these joints is illustrated in Fig. 9-1D.

A flat-faced joint is generally used only in low-pressure service. The possibility of a gasket blowing out is eliminated when a male and female joint is used because it provides correct gasket compression and assures alignment between the bonnet and the valve body. A tongue and-groove joint is used in high-pressure and high-temperature installations. Its construction also assures perfect alignment of the bonnet onto the valve body.

447

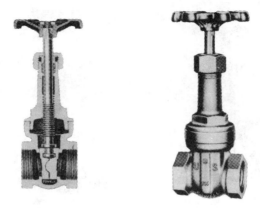

Fig. 9-1B. Screwed bonnet.

Fig. 9-1C. Bolted flange bonnet.

Some valves may also be classified according to the type of valve stem used with the bonnet valves, but sometimes include screwed and union bonnet types. The stems used with these valves include:

1. Outside screw, rising stem with bonnet.
2. Inside screw, rising stem with bonnet.
3. Inside screw, nonrising stem with bonnet.
4. Outside screw, rising stem with yoke.

448

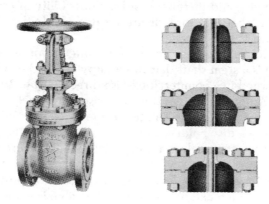

Courtesy Wm. Powell Co.

Fig. 9-1D. Bolted flange bonnet.

The *stem* is the shaft connecting the handwheel at the top of the valve to a wedge or disc in the valve body. The handwheel turns the stem, which changes the position of the wedge or disc and opens or closes the valve.

An *outside screw stem* (Fig. 9-2) is one that has the threaded portion of the stem shaft outside the valve. The advantage of this stem is that the threads do not come in contact with the media

Fig. 9-2. Rising and nonrising stem valves.

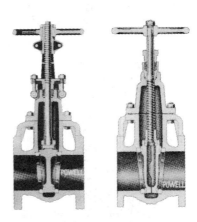

Courtesy Wm. Powell Co.

inside the valve and thereby avoid the possibility of erosion and corrosion. If the stem threads are located inside the valve, it is called an *inside screw stem.*

A stem may also be referred to as a rising or nonrising stem. A *rising stem* is a stem shaft that moves upward as the valve handwheel is turned in a counterclockwise direction (i.e., to the *open* position). A *nonrising stem* does not rise but merely turns with the handwheel. The stem rotates in the bonnet or yoke and is threaded into a disc holder, which is threaded into the wedge or disc. Nonrising stem valves are especially suited where clearance is limited.

A *wedge* (Fig. 9-3) is a device used to control the flow of steam or water through a gate valve. When the valve is wide open, the wedge is lifted entirely out of the main chamber of the valve body, providing a straightway flow area through the valve.

Wedges for gate valves are available in the following types: (1) split wedge, (2) solid wedge, (3) double wedge, and (4) flexible wedge. These wedges are illustrated in Fig. 9-4.

The wedge is attached to the end of the valve stem. When the valve is closed, the wedge is in full contact with the seat.

In globe, angle Y globe and angle, and check valves, a disc is used to control the flow of the steam or water. A *disc* is a flat device attached to the end of the valve stem. The exception to the rule is the swing-check valve, which has no stem. In this valve, the position of the disc is controlled internally by the flow

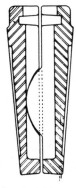

Fig. 9-3. Sectional view of a split wedge.

Courtesy Wm. Powell Co.

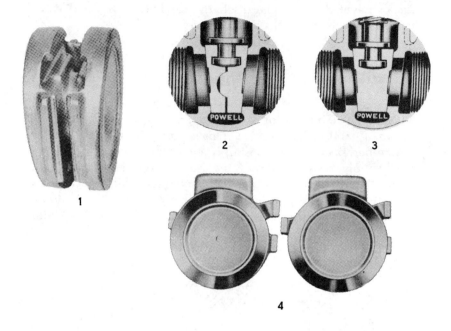

2   3

4

Fig. 9-4. Wedges for gate valves.

of the steam or water. In operation, the disc opens or closes the pathway through the valve depending upon the position of the stem. Some different types of discs and seats are illustrated in Fig. 9-5.

Both composition and metal discs are used in valves. The composition discs are suitable for service other than high temperature. While they do not last as long as metal discs on throttling service, they will enable the seat to last longer by taking most of the wear of wire drawing. They are also easier and cheaper to replace.

Metal discs last longer than composition discs on throttling service and on higher-temperature service. So-called plug-type discs are suitable for close throttling service.

The wedge or disc closes tightly against a seat to close off the valve. As illustrated in Fig. 9-6, the valves and seats may be either

451

flat or beveled. For equal discharge capacity, a beveled valve must be opened more than a flat valve.

VALVE MATERIALS

Valves are made of a number of different metals and metal alloys, and the choice of metal often will mean the difference between good or bad service. This is a *very* important point to remember, because it is not simply a matter of going down to the local supply house with only the valve design and function in mind. You must also consider the *service* for which the valve is intended. Regardless of the type of valve you select, it should be rated for its pressure and service temperature range. You would

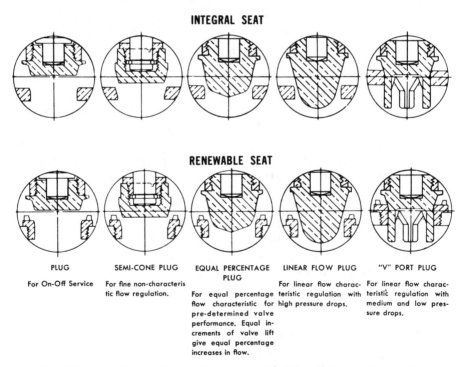

INTEGRAL SEAT

RENEWABLE SEAT

PLUG	SEMI-CONE PLUG	EQUAL PERCENTAGE PLUG	LINEAR FLOW PLUG	"V" PORT PLUG
For On-Off Service	For fine non-characteristic flow regulation.	For equal percentage flow characteristic for pre-determined valve performance. Equal increments of valve lift give equal percentage increases in flow.	For linear flow characteristic regulation with high pressure drops.	For linear flow characteristic regulation with medium and low pressure drops.

Fig. 9-5. Examples of discs and seats provided for globe, angle, and Y valves.

452

not want to select valves rated for 125 psi saturated steam and steam service temperature range to 500°F maximum if your system is capable of producing pressures and temperatures in excess of these limits.

Fig. 9-6. Valve seats.

The standards and specifications prescribing the rules and regulations for valve construction and use will be found in the *latest* publications of the following associations:

1. American Petroleum Institute (API Specifications).
2. American National Standards Institute (ANSI Codes and Standards).
3. American Society of Mechanical Engineers (ASME Boiler Construction and Unfired Pressure Vessel Code).
4. American Society for Testing Materials (ASTM Material Specifications).
5. Manufacturers' Standardization Society of the Valve and Fittings Industry (MSS Standard Practices).

Valves are available in the following metals and metal alloys:

1. Bronzes.
2. Cast iron.
3. Semisteel.
4. Stainless steel.
5. Low carbon and low alloy steel.
6. Special alloys and pure metals.

Bronze valves are suitable for steam pressures of 125 psi and 150 psi at 500°F and 200 psi, 300 psi, and 350 psi at 550°F. For steam pressures of 125 psi and 250 psi at 450°, cast-iron valves are recommended. Steel valves should be used for higher pressures and temperatures. Valves made of pure metals and metal alloys

453

are used in processing systems where resistance to the corrosive and erosive action of the fluids is of prime consideration.

Valve manufacturers are especially willing to help you select the most suitable valve for your needs. They produce excellent informative literature to this end (one of the better examples being *The Powell Handbook of Valve Information* from the Wm. Powell Company of Cincinnati, Ohio) and try to answer all inquiries directed to them.

GLOBE AND ANGLE VALVES

Globe and angle valves are used primarily as throttling devices to control the *rate* of flow. Flow characteristics of both globe and angle valves are illustrated in Fig. 9-7.

A *globe valve* is a valve having a round ball-like shell with a stuffing box extension through which passes the screw spindle, which operates the valve disc. The valve seat is parallel to the line

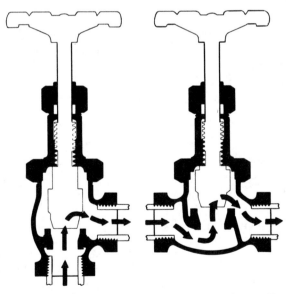

Courtesy Wm. Powell Co.

Fig. 9-7. Sectional view of a globe and angle valve.

of flow, and the inlet and outlet branches are opposite one another.

The design and construction of an *angle valve* is similar to the globe valve, except that the outlet is at right angles to the inlet branch, thus combining in itself a valve and an elbow.

Globe and angle valves are preferred to gate valves (see below) if the valve is to be operated frequently, or if it is to be used as a throttling device (i.e., operated partially open), because either type can open with fewer turns. An additional feature of the angle valve is that it can control both a *change* and the direction of flow. It is designed to create a 90° change in direction. As a result, it saves an elbow and nipple and reduces the pressure loss.

Globe valves are commonly used in pipelines that extend through basement areas or areas of limited access. *Offset* globe valves are often recommended for radiators on second floors or higher. Angle valves are commonly used on first-floor radiators.

Ordinarily, globe and angle valves should be installed with the pressure *under* the disc so that the stuffing box may be repacked without escape of steam or water. Installing these valves in this manner not only promotes easier operation but also provides a certain degree of protection to the packing and reduces the erosive action on the disc and seat faces. High-temperature steam service proves to be the exception. If high-temperature steam is the medium to be controlled, the globe or angle valve should be installed so that the pressure is *above* the valve disc. If the usual installation method (i.e., with the pressure under the disc) is used, then the high-temperature steam will be under the disc when the valve is closed, and the valve stem will be out of the fluid. As a result, the valve stem will cool and contract, causing the disc to lift slightly off the valve seat. The leaks that then occur result in wire drawing on seat and disc faces.

Fig. 9-8 illustrates both the correct and incorrect way to place globe valves on horizontal lines. When the globe valve is placed in an upright position, considerable water will remain in the pipeline and be subject to freezing. When it is placed in a horizontal position, most of this water will drain through the seat opening with less danger of damage by freezing.

Figs. 9-9, 9-10, and 9-11 illustrate a number of commonly used

455

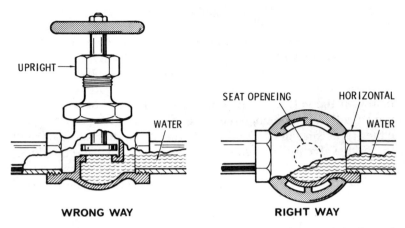

UPRIGHT

WATER

SEAT OPENEING   HORIZONTAL

WATER

WRONG WAY   **RIGHT WAY**

Fig. 9-8. Correct and incorrect way of placing a globe valve on a horizontal line.

globe and angle valves. These illustrations should be referred to when reading the section on troubleshooting valves.

GATE VALVES

A *gate valve* (Fig. 9-12) is one having two inclined seats between which the valve (consisting of a single or double disc) wedges down in closing. In opening, the valve is drawn up into a dome or recess, thus leaving a straight passage the full inside diameter of the pipe. This lifting of the wedges completely out of the waterway, thereby creating an unobstructed passage for the fluid, is a significant feature of the gate valve. As a result, turbulence is minimized, and there is very little pressure drop.

Gate valves are the most frequently used valves and are preferable to globe or angle valves for lines on which it is important to minimize resistance to flow and friction losses. They are also necessary where complete drainage of the pipe must be provided.

The gate valve is primarily suited for locations where valves are to be generally wide open or shut tight. Globe valves are preferable for throttling. When gate valves are used for throt-

456

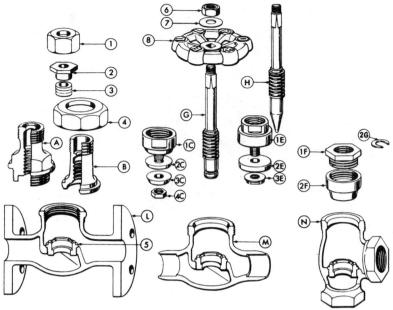

1. PACKING NUT	E. COMPOSITION DISC
2. PACKING GLAND	1E-DISC HOLDER
3. PACKING	2E-NON-METALLIC DISC
4. BONNET RING	3E-DISC LOCKNUT WASHER
5. SEAT RING	F. DISC LOCKNUT
6. HANDWHEEL NUT	1F-DISC NUT
7. IDENTIFICATION PLATE	2F-DISC
8. HANDWHEEL	G. STEM-DISC LOCKNUT
A. SCREWED-IN BONNET	(HORSESHOE RING) TYPE
B. UNION BONNET	2G-HORSESHOE RING
C. HI-LO DISC	L. BODY-GLOBE-FLANGED ENDS
1C-DISC HOLDER	M. BODY-GLOBE-SOLDER JOINT ENDS
2C-NON-METALLIC DISC	N. BODY-ANGLE-THREADED ENDS
3C-DISC PLATE	
4C-DISC NUT	

Courtesy Wm. Powell Co.

Fig. 9-9. Screwed and union bonnet globe and angle valve.

tling, the high velocity of flow tends to damage the surface of the seats.

Wedge gate valves are preferable where the valve is to be installed with the stem extending downward. Nonrising stem gate valves are usually tighter at the stuffing box than those with rising stems and require less head room.

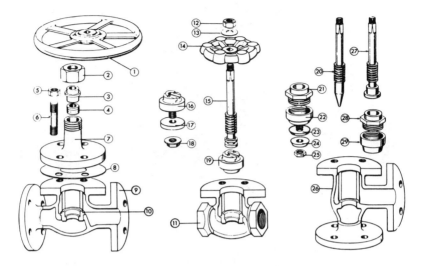

1. HANDWHEEL-ROUND
2. PACKING NUT
3. PACKING GLAND
4. PACKING
5. BODY NUT
6. BODY STUD
7. BONNET
8. GASKET
9. BODY-GLOBE-FLANGED ENDS
10. SEAT RING
11. BODY-GLOBE-THREADED ENDS
12. HANDWHEEL NUT
13. IDENTIFICATION PLATE
14. HANDWHEEL-NON-HEATING
15. STEM-SLIP-ON DISC TYPE
16. COMPOSITION DISC HOLDER
17. NON-METALLIC DISC
18. DISC LOCKNUT
19. SLIP-ON DISC
20. STEM-NEEDLE DISC TYPE
21. HI-LO DISC LOCKNUT
22. DISC HOLDER
23. NON-METALLIC DISC
24. DISC PLATE
25. DISC NUT
26. BODY-ANGLE-FLANGED ENDS
27. STEM-DISC LOCKNUT TYPE
28. DISC LOCKNUT
29. DISC

Courtesy Wm. Powell Co.

Fig. 9-10. Bolted bonnet inside screw globe and angle valve.

The double-disc gate valve enjoys certain advantages over the single-disc type (Fig. 9-13). If the fluid contains foreign matter, the flexibility of operation of the double disc enables one disc to seat tightly if dirt or some other type of foreign matter is present. Moreover, the discs and seats are more readily refaced than those of the single-disc type.

Outside screw and yoke valves should be used when it is necessary to know at a glance whether the valve is open or shut tight.

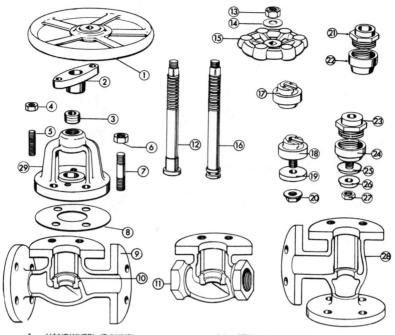

1.	HANDWHEEL (ROUND)	16.	STEM-SLIP-ON TYPE
2.	PACKING GLAND	17.	DISC-ONE-PIECE-SLIP-ON
3.	PACKING	18.	COMPOSITION DISC HOLDER
4.	GLAND STUD NUT	19.	COMPOSITION DISC
5.	GLAND STUD	20.	DISC NUT
6.	YOKE STUD NUT	21.	DISC LOCKNUT
7.	YOKE STUD	22.	DISC
8.	GASKET	23.	DISC LOCKNUT-HI-LO DISC
9.	BODY-GLOBE-FLANGED ENDS	24.	DISC HOLDER
10.	SEAT RING	25.	NON-METALLIC DISC
11.	BODY-GLOBE-THREADED ENDS	26.	DISC PLATE
12.	STEM-DISC LOCKNUT TYPE	27.	DISC NUT
13.	HANDWHEEL NUT	28.	BODY-ANGLE-FLANGED ENDS
14.	IDENTIFICATION PLATE	29.	YOKE
15.	HANDWHEEL		

Courtesy Wm. Powell Co.

Fig. 9-11. Bolted bonnet outside screw and yoke globe and angle valve.

Some examples of gate valves used in heating and cooling installations are illustrated in Figs. 9-14, 9-15, and 9-16. Refer to these illustrations when reading the section on troubleshooting valves in this chapter.

Fig. 9-12. Gate-valve flow characteristics.

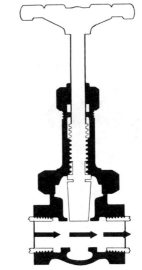

Courtesy Wm. Powell Co.

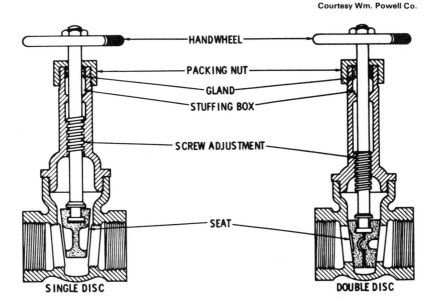

HANDWHEEL

PACKING NUT

GLAND

STUFFING BOX

SCREW ADJUSTMENT

SEAT

SINGLE DISC

DOUBLE DISC

Fig. 9-13. Comparison of single- and double-disc gate valves.

Gate valves are available with either a rising or nonrising stem, the former being more commonly used on smaller gate valves.

CHECK VALVES

A *check valve* is one that automatically opens to permit the passage of liquid in one direction and closes automatically to

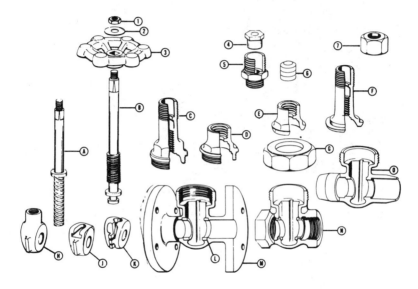

1. HANDWHEEL NUT	D. SCREWED-IN BONNET-NON-RISING STEM VALVES
2. IDENTIFICATION PLATE	
3. HANDWHEEL	E. UNION BONNET-NON-RISING STEM VALVES
4. PACKING GLAND	F. UNION BONNET-RISING STEM VALVES
5. PACKING BOX SPUD (NON-RISING STEM VALVES ONLY)	G. BONNET RING
6. PACKING	H. SOLID WEDGE-NON-RISING STEM VALVES
7. PACKING NUT	J. SOLID WEDGE-RISING STEM VALVES
A. STEM-NON RISING STEM VALVES	K. DOUBLE WEDGE-RISING STEM VALVES
B. STEM-RISING STEM VALVES	L. SEAT RING
C. SCREWED-IN BONNET-RISING STEM VALVES	M. BODY-FLANGED ENDS
	N. BODY-THREADED ENDS
	O. BODY-SOLDER JOINT ENDS

Courtesy Wm. Powell Co.

Fig. 9-14. Screwed and union bonnet rising and nonrising stem gate valve.

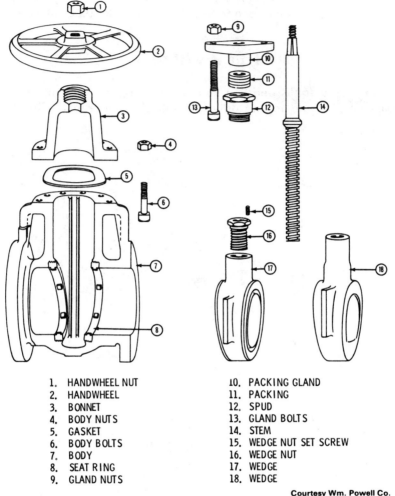

1. HANDWHEEL NUT	10.	PACKING GLAND
2. HANDWHEEL	11.	PACKING
3. BONNET	12.	SPUD
4. BODY NUTS	13.	GLAND BOLTS
5. GASKET	14.	STEM
6. BODY BOLTS	15.	WEDGE NUT SET SCREW
7. BODY	16.	WEDGE NUT
8. SEAT RING	17.	WEDGE
9. GLAND NUTS	18.	WEDGE

Courtesy Wm. Powell Co.

Fig. 9-15. Bolted bonnet inside screw nonrising stem gate valve.

prevent any flow in the opposite direction. In other words, it specifically operates to prevent the reversal of water flow in the line.

A check valve is sometimes placed on the return connection to prevent the accidental loss of boiler water to the returns with

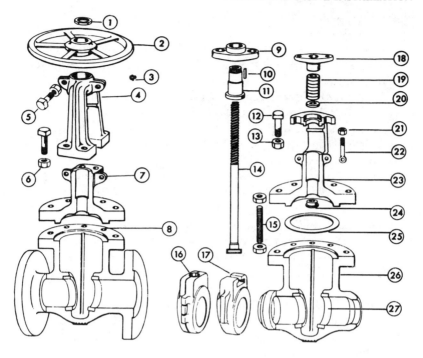

1.	STEM BUSHING NUT	15.	BODY STUD AND NUTS
2.	HANDWHEEL	16.	SPLIT WEDGE
3.	LUBRICANT FITTING	17.	SOLID WEDGE
4.	YOKEARMS	18.	PACKING GLAND
5.	YOKEARM EAR BOLT AND NUT	19.	PACKING
6.	BONNET BOLT AND NUT	20.	PACKING WASHER
7.	TWO-PIECE BONNET	21.	EYEBOLT NUT
8.	BODY-FLANGED ENDS	22.	EYEBOLT
9.	BEARING CAP	23.	ONE-PIECE BONNET
10.	HANDWHEEL KEY	24.	PACK-UNDER-PRESSURE BUSHING
11.	STEM BUSHING	25.	GASKET
12.	BEARING CAP BOLT	26.	BODY-WELDED ENDS
13.	BEARING CAP NUT	27.	SEAT RING
14.	STEM		

Courtesy Wm. Powell Co.

Fig. 9-16. Bolted bonnet outside screw and yoke rising stem gate valve.

consequent danger of boiler damage. When feasible, the Hartford connection is to be preferred over the check valve because the latter is apt to stick or not close tightly. Moreover, the check valve offers additional resistance to the condensation returning to

463

the boiler, which in gravity systems would raise the water line in the far end of the wet return several inches.

The two general types of check valves are: (1) the swing-check valve and (2) the lift check valve. A third type sometimes used is the ball-check valve.

Swing-check valves (Fig. 9-17A) operate best on horizontal lines. If they are to be used on vertical or inclined lines, they should be installed so that the flow will be upward through the valve. When the valve disc is raised in the open position, there is very little obstruction in the flow area. Turbulence is also very low. Swing-check valves are commonly used in combination with gate valves.

Lift check valves (also referred to as *horizontal-lift check valves*) (Fig. 9-17B) have the same flow characteristics as a globe valve. This can be seen by comparing the illustrations. The turbulence and pressure drop is therefore similar for these two valves. Lift check valves are commonly used in combination with globe valves.

A *vertical check valve* should be used on vertical lines only and with the flow upward through the valve.

The *Watts Backflow Preventer* is a safety device containing two check valves, and is designed to prevent backflow in boiler feed lines when the supply pressure falls below system pressure. The primary check valve in the backflow device utilizes a disc

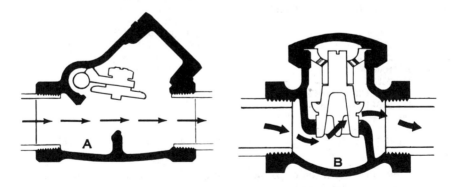

Courtesy Wm. Powell Co.

Fig. 9-17. Flow characteristics of a swing- and lift-check valve.

that seats against a rubber mating part to ensure tight closing. A secondary check valve utilizes a disc-to-metal seating. If the downstream check valve malfunctions (e.g., fouls), leakage will be vented to atmosphere through the vent port, thereby safeguarding the potable water from contamination.

Figs. 9-18 through 9-22 illustrate a number of different check valves used in heating and cooling installations. Refer to these illustrations when reading the section on troubleshooting valves in this chapter.

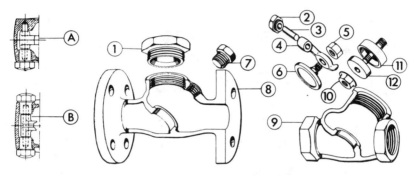

1.	CAP	8.	BODY-FLANGED END
2.	SLIDE PLUG	9.	BODY-THREADED END
3.	CARRIER PIN	10.	DISC LOCKNUT-COMPOSITION DISC
4.	CARRIER	11.	DISC HOLDER
5.	DISC NUT	12.	COMPOSITION DISC
6.	DISC	A.	DETAIL-ONE SIDE PLUG
7.	BUMPER PLUG	B.	DETAIL-TWO SIDE PLUGS

Fig. 9-18. Screwed-cap swing-check valve.

STOP VALVES

Globe, gate, and angle valves are used on steam and hot-(or cold-) water pipelines as *stop valves* (also referred to as *nonreturn valves, nonreturn stop valves,* or *boiler check valves*).

The operating principle of the typical stop valve is shown in Fig. 9-23. It functions as a form of check valve that can be opened or closed by hand control when the pressure in the boiler is greater than in the line, but cannot be opened when the pressure within the boiler is less than that in the line. The counterbalance spring slightly overbalances the weight of the valve and tends to hold the valve open, thus preventing movement of the valve with every fluctuation of pressure.

Stop valves are often used as a safety device in steam power plants where *more* than one boiler is connected to the same header. They are connected directly to the boiler nozzle outlet between the boiler and the header. It functions as a safety device

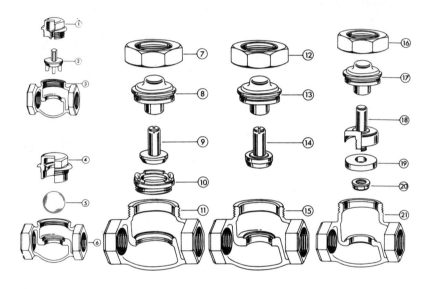

1.	CAP	12.	RING NUT
2.	DISC	13.	DISC GUIDE
3.	BODY-THREADED ENDS	14.	DISC
4.	CAP	15.	BODY-THREADED ENDS
5.	BALL	16.	RING NUT
6.	BODY-THREADED ENDS	17.	DISC GUIDE
7.	RING NUT	18.	DISC HOLDER
8.	DISC GUIDE	19.	DISC-COMPOSITION DISC
9.	DISC	20.	DISC NUT
10.	SEAT RING	21.	BODY-THREADED ENDS
11.	BODY-THREADED ENDS		

Fig. 9-19. Screwed-cap horizontal-lift-check valve.

to prevent backflow of the steam into the boiler. This is particularly important if the boiler is cold.

The stop valve should be installed so that the valve stem is in a vertical or upright position. The pressure of the steam or water should be *under* the disc.

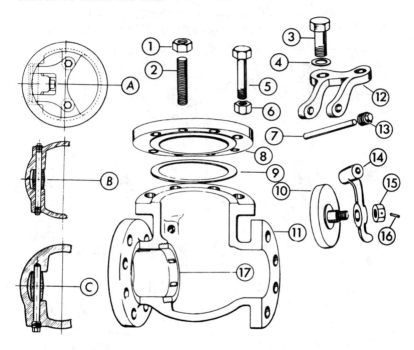

1.	BODY NUT	11.	BODY-FLANGED ENDS
2.	BODY STUD	12.	DISC HOLDER HANGER
3.	CAP SCREW	13.	PIPE PLUG
4.	LOCKWASHER	14.	DISC HOLDER
5.	BODY BOLT	15.	DISC NUT
6.	BODY BOLT NUT	16.	DISC NUT PIN
7.	DISC HOLDER PIN	17.	SEAT RING
8.	CAP	A.	DETAIL-HANGER TYPE DISC
9.	GASKET	B.	DETAIL-PIN TYPE-TWO SIDE PLUGS
10.	DISC	C.	DETAIL-PIN TYPE-ONE SIDE PLUG

Fig. 9-20. Bolted-cap swing-check valve.

THREE-WAY (CROSS) VALVES

The *three-way or cross valve* (Fig. 9-24) is used to control the flow to or from a branch line at the junction of a branch and main line. It is virtually a globe valve having its seat located at the bottom of the body and the passage through the seat connected with a third opening at right angles to the other two openings.

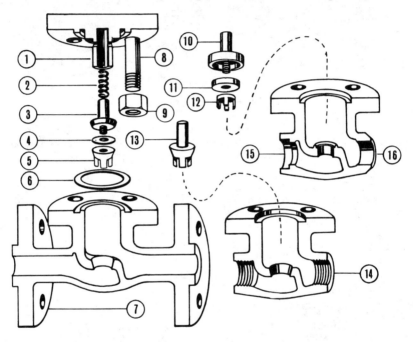

1. CAP - BOLTED	9. CAP BOLT NUT
2. SPRING (OPTIONAL)	10. COMPOSITION DISC HOLDER
3. DISC HOLDER	11. NON-METALLIC DISC
4. TEFLON DISC	12. DISC GUIDE
5. DISC GUIDE	13. METAL DISC
6. GASKET	14. BODY, THREADED ENDS
7. BODY - FLANGED ENDS	15. BODY, SILVER BRAZE ENDS
8. CAP BOLT	16. BODY, SOLDER JOINT ENDS

Courtesy Wm. Powell Co.

Fig. 9-21. Bolted-cap horizontal-lift-check valve.

Y VALVES

A *Y valve* or *Y globe valve* (Fig. 9-25) is designed to combine the operating characteristics of globe and gate valves. In other words, it combines the throttling characteristics of the former with the straightway flow area of the latter. The valve disc can be easily reground and renewed.

TROUBLESHOOTING VALVES

Valve manufacturers generally provide information for servicing and repairing their valves. When the information is not available, as is often the case on older heating and cooling systems, troubleshooting the valve problem generally depends on experience or educated guesswork. Experience has shown that most

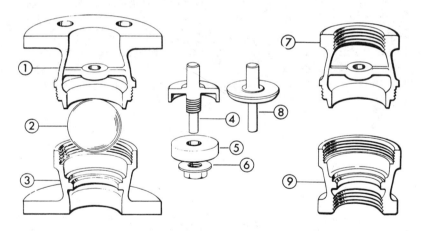

1. TOP BODY - FLANGED ENDS
2. BALL
3. BOTTOM BODY - FLANGED ENDS
4. DISC HOLDER
5. COMPOSITION DISC
6. DISC NUT
7. TOP BODY - THREADED ENDS
8. DISC
9. BOTTOM BODY - THREADED ENDS

Courtesy Wm. Powell Co.

Fig. 9-22. Vertical check valve.

valve problems can be traced to one of the following three sources:

1. Stuffing-box leakage.
2. Seat leakage.
3. Damaged stem.

Valve Stuffing-Box Leakage

Leakage around the valve stuffing box is usually an indication that the stuffing must be adjusted or replaced. This leakage does not occur when the valve is completely opened or closed. Therefore, an absence of leakage is not necessarily an indication that the valve is functioning normally.

Once you have detected leakage, check first to determine whether or not adjusting the packing will stop it. If it is a bolted bonnet valve, turn the packing gland nuts (or gland stud nuts) clockwise alternately with no more than ¼ turn on each until leakage stops. If you are dealing with a screwed and union bonnet valve, turn the packing nut clockwise until the leakage stops. If the leakage will not respond to adjustment, the packing must be replaced.

The procedure for replacing the packing in most valves may be summarized as follows:

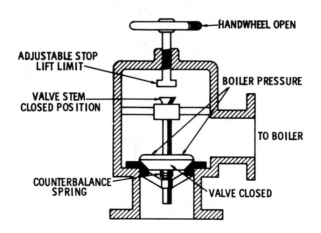

Fig. 9-23. Functions of a stop or nonreturn valve.

470

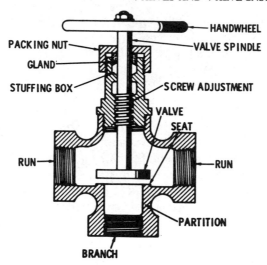

HANDWHEEL

VALVE SPINDLE

PACKING NUT

GLAND

STUFFING BOX

SCREW ADJUSTMENT

VALVE

SEAT

RUN

RUN

PARTITION

BRANCH

Fig. 9-24. A three-way cross valve.

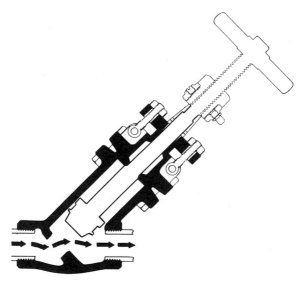

Courtesy Wm. Powell Co.

Fig. 9-25. A Y valve.

471

1. Remove the handwheel nut and the handwheel.
2. Remove the packing nut.
3. Slip the packing gland off of the stem.
4. Replace the packing.
5. Reassemble in reverse order.

The procedure used with bolted bonnet outside screw and yoke valves is a little more complicated. On Y valves of this type, it is necessary to remove the gland flange and gland follower before replacing the packing. On globe and angle valves, the stud nuts and upper valve assembly must be removed.

Because of their design and construction, the problem of stuffing-box leakage does not occur with check valves.

Valve Seat Leakage

Leakage of water from the valve body is usually an indication that the wedge, disc, or seat ring needs replacing. For most valves, the procedure for doing this may be summarized as follows:

1. Open the valve.
2. Remove the bonnet and other components of the upper valve assembly.
3. Run the stem down by turning it in a clockwise direction.
4. Remove the wedge or disc from the stem and replace if necessary.
5. Remove the seat ring with a seat ring wrench and replace if necessary.
6. Reassemble in reverse order.

The disc, disc assembly, or ball are all possible sources of seat leakage in check valves. Access to these components is gained by removing the valve cap (counterclockwise), side plugs, and pins. Reassembly is in reverse order.

Damaged Valve Stems

Sometimes the threads on valve stems become worn or damaged, making the valves inoperable. When this occurs, the stems must be replaced. Before the stem can be replaced, however, all pressure must be removed. Then, with pressure removed, disconnect and remove the bonnet and upper valve assembly. The remainder of the procedure depends upon the type of stem used (i.e., rising or nonrising stem) and other design factors. Basically, the procedure may be summarized as follows:

1. Run the stem down by turning it in a clockwise direction.
2. Rotate the stem in a clockwise direction until the stem threads are completely out of the threaded portion of the upper bushing.
3. Pull the stem out of the stuffing box.
4. Remove the wedge or disc from the stem.
5. Replace the old stem with a new one.
6. Reassemble in reverse order with new packing and gasket (when applicable).

AUTOMATIC VALVES AND VALVE OPERATORS

An *automatic valve* is a controlled device designed to regulate the flow of steam, water, gas, or oil in response to impulses from a controller such as a thermostat. The controller measures changes in the surrounding variable conditions (e.g., temperature, humidity, and pressure) and activates the valve in order to compensate for these changes. The controller, controlled device, and a feedback device constitute a closed-loop automatic control system. A more detailed description of the operating principles of an automatic control system is contained in Chapter 4 (Thermostats and Humidistats).

A number of different automatic valves are used in heating and cooling systems. They are generally available in two-way or three-way models and can be used for either two-position or modulating operation or both. The type of automatic valves used must be carefully sized and selected for the specific application. Valve manufacturers generally provide information to aid you in making a suitable selection.

When selecting an automatic valve, do not confuse the valve body rating with the close-off rating. The former is the actual

473

rating of the valve body, whereas the latter indicates the maximum allowable pressure drop to which the valve may be subjected while fully closed. The close-off rating is a function of the power available from the valve actuator for holding the valve closed against pressure drop (although structural parts such as the valve stem are also sometimes the limiting factor).

A *valve operator* is a controlled device designed to operate an automatic valve. It receives the impulses from the controller and activates the valve stem. There are three types of valve operators:

1. Solenoid operators.
2. Pneumatic operators.
3. Motorized operators.

A *solenoid operator* is used with a valve designed for two-position operation (i.e., fully opened or fully closed). It consists essentially of a magnetic coil, which operates a movable plunger. The plunger controls the opening and closing of the valve. In design, a solenoid-operated valve is less complicated than either the pneumatic or motorized types, but it is limited in size.

The operating principle of a *pneumatic operator* is based on changes in air pressure. Essentially, it consists of a spring-equipped flexible bellows or diaphragm attached to the valve stem. When the air pressure increases, the bellows or diaphragm moves the stem and compresses the spring. A reduction in air pressure results in the operation reversing itself with the spring returning the valve stem to its original position. Pneumatic-operated valves are designed for proportional control of a medium and are available as *normally open* or *normally closed* valves.

Springless pneumatic operators are available with opposing flexible diaphragms located on either side of a single diaphragm. This type of pneumatic operator is limited in use to special applications (e.g., those involving high pressures, large valves).

A *motorized operator* consists of an electric motor that operates the valve stem through a gear train and linkage. Based on their operating principles, electric motors used in motorized operators can be classified as follows:

1. Unidirectional motors.

2. Spring-return motors.

3. Reversible motors.

The reversible motor illustrated in Fig. 9-26 contains both a balancing relay and a feedback potentiometer. The balancing

CONTROLLER
CONNECTIONS

TRANSFORMER
CONNECTIONS

POTENTIOMETER
SLIDING CONTACT

FEEDBACK
POTENTIOMETER

GEAR
TRAIN

OUTPUT
SHAFT
(BOTH
ENDS)

BALANCING
RELAY

RELAY
CONTACTS

CAPACITOR
MOTOR

Courtesy Honeywell

Fig. 9-26. A reversible motor with a balancing relay and a feedback potentiometer.

475

relay controls the motor, which turns the motor drive shaft connected to the gear train.

The balancing relay consists of two solenoid coils with parallel axes, into which are inserted the legs on the U-shaped armature. The armature is pivoted at the center so that it can be tilted by the changing magnetic flux of the two coils to energize the relay. A contact arm is fastened to the armature so that it may touch either of the two stationary contacts as the armature moves back and forth on its pivot. When the relay is balanced, the contact arm floats between the two contacts, touching neither of them. In operation, the motor is started, stopped, and reversed by the single-pole double-throw contacts of the balancing relay.

The feedback potentiometer contained in the motor consists of a coil of wire and a sliding contact. It is similar in construction to the internal view of the Honeywell Q181A Auxiliary Potentiometer illustrated in Fig. 9-27. In operation, the motor shaft moves

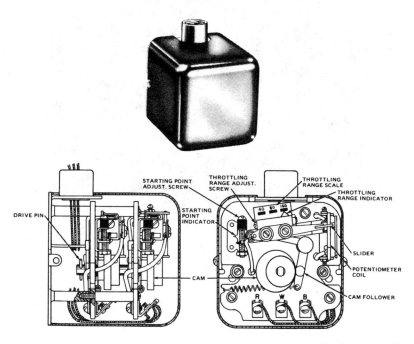

Fig. 9-27. Internal view of a Honeywell Auxiliary potentiometer.

476

the sliding contact of the potentiometer along the coil, establishing contact wherever it touches according to the position of the motor.

Motorized valves are commonly used in residential zoning applications. Examples of valves used for this purpose are illustrated in Figs. 9-28 and 9-29. Both of these valves contain replaceable O-Rings and are equipped with a manual means of opening the valve in case of power failure.

The zone valve shown in Fig. 9-28 is available in either line or low-voltage models. The latter requires a 24-volt power source. One advantage of this zone valve is that the powerhead can be replaced without the removal of the valve body from the pipeline.

Note that the manual opening lever on the powerhead (Fig. 9-28) has two settings: manual open *(man. open)* and automatic *(auto).* Always place the manual opening lever in the automatic

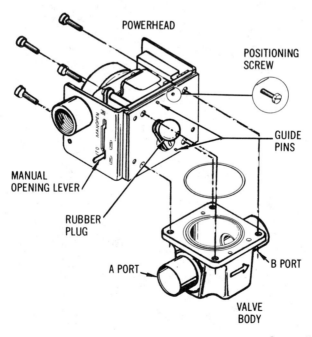

Courtesy Honeywell

Fig. 9-28. A Honeywell zone valve.

477

position after the installation procedure has been completed and *before* running the valve. Always place the manual opening lever in the manual position before removing the valve from the pipeline or replacing a powerhead. *Never* attempt to cycle the valve electrically when the manual opening lever is in this position.

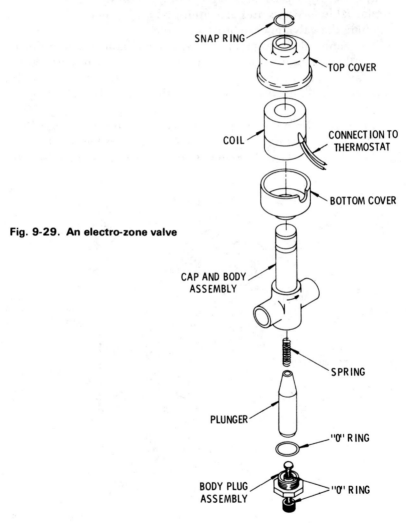

SNAP RING

TOP COVER

COIL

CONNECTION TO THERMOSTAT

BOTTOM COVER

Fig. 9-29. An electro-zone valve

CAP AND BODY ASSEMBLY

SPRING

PLUNGER

"O" RING

BODY PLUG ASSEMBLY

"O" RING

Courtesy Hydrotherm, Inc.

Before attempting to install a new powerhead, drain the water from the system, disconnect the valve from the electrical power source, and remove the conduit connections (if fitted). Then, place the manual opening lever on the *old* powerhead in the manual position and remove the cover. The remainder of the procedure may be summarized as follows:

1. Take out the four screws connecting the powerhead to the valve body and remove the old powerhead.
2. Replace the old O-Ring in the valve body with a new one.
3. Place the manual opening lever on the new powerhead in the manual open position.
4. Align the powerhead on the valve body in the same position as the old one, and screw it down.
5. Reconnect the conduit and electrical connections.
6. Adjust the thermostat or controller so that the valve runs smoothly through its cycle.

The motorized zoning valve shown in Fig. 9-29 is energized by impulses from a standard two-wire thermostat. A wiring diagram for four such valves used in a residential installation is illustrated in Fig. 9-30. Low-voltage DC power and circular-burner control is supplied by a unit containing a transformer, rectifier, and two circuit breakers. When supplied for boilers with self-energizing controls, a built-in power failure manual switch is furnished that permits manual boiler operation.

The valve itself contains a water-damped piston that is lifted by a strong magnetic flux and opens a straight through passage for unrestricted flow. The valve closes slowly under the action of gravity.

The butterfly valve is ideally suited for regulating the flow of water or steam in applications where tight close-off is not required. Typical applications include zone control of gravity hot-water or low-pressure steam heating systems. For tight close-off, a final shutoff valve must also be used.

The motorized valve illustrated in Fig. 9-31 is a single-seated type used for two-position operational control of hot or chilled water, steam radiators, or zoned residential heating and cooling systems. It contains a replaceable composition disc and is designed to operate at fluid temperatures up to a maximum of 250°F or steam pressure up to 15 psi maximum.

Automatic room temperature control on hot-water or two-pipe steam heating systems can be accomplished by using self-actuating thermostatic radiator valves (Fig. 9-32). These valves can be used on freestanding radiators, convectors, and baseboard heating units. Electric power is not required for their operation.

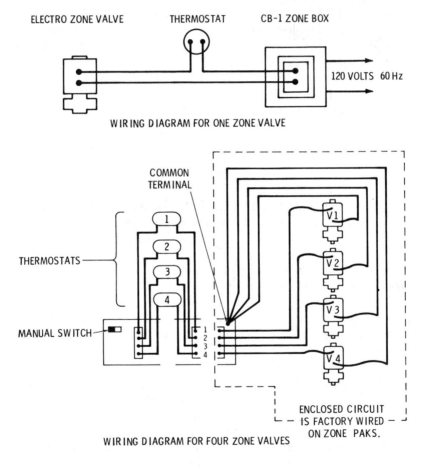

Fig. 9-30. Wiring diagram for the zone valve.

Valve linkage is used to connect an electric motor to an automatic valve. The *type* of valve linkage used will depend upon the type of automatic valve used in the heating or cooling application. The valve manufacturer's recommendations should be followed to avoid problems.

The Honeywell Q601 Valve Linkage illustrated in Fig. 9-33 is an example of the type of linkage used with automatic steam or hot-water valves (Fig. 9-34). Installation and adjustment of this linkage for a two-position valve may be summarized as follows:

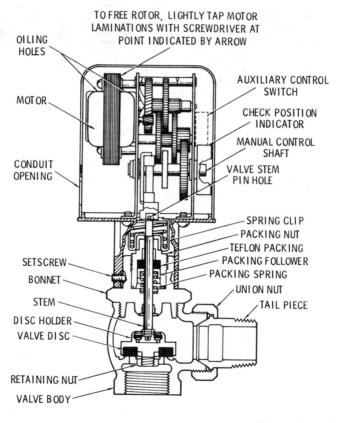

Courtesy Honeywell

Fig. 9-31. Honeywell V5045A valve body and V2045A operator.

481

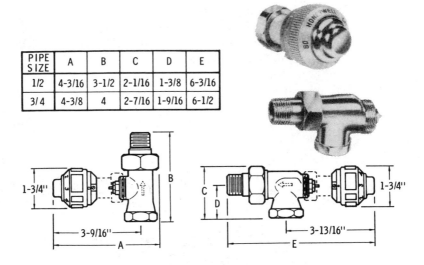

PIPE SIZE	A	B	C	D	E
1/2	4-3/16	3-1/2	2-1/16	1-3/8	6-3/16
3/4	4-3/8	4	2-7/16	1-9/16	6-1/2

Fig. 9-32. Thermostatic radiator valves showing dimensions of straight-through and angle pattern bodies.

1. Follow the manufacturer's instructions and connect the motor and valve to the linkage. Be sure to align the key on the crank arm with the *keyway* on the motor shaft.
2. The lift adjustment on the valve linkage allows the linkage to be used with number of valve sizes. Set the lift adjustment equal to the rated lift of the valve according to the three steps illustrated in Fig. 9-35.
3. Use the adjusting screw to put tension on the valve in the closed position (Fig. 9-36).
4. Run the motor and valve all the way open. Check to see that the motor runs all the way to the end of the stroke without putting tension on the valve stem. The existence of tension is an indication that it may jam open.

Adjusting the same linkage for a three-way valve involves the same steps, particularly the alignment of the key on the crank arm with the keyway on the motor shaft. The lift adjustment, however, should be set ½ division above the rated valve lift (Fig. 9-35).

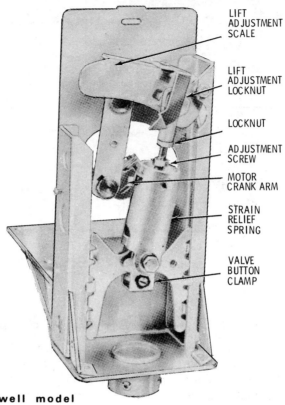

LIFT
ADJUSTMENT
SCALE

LIFT
ADJUSTMENT
LOCKNUT

LOCKNUT

ADJUSTMENT
SCREW

MOTOR
CRANK ARM

STRAIN
RELIEF
SPRING

VALVE
BUTTON
CLAMP

Fig. 9-33. Honeywell model Q601 valve linkage.

Courtesy Honeywell

With the motor closed and the valve at the bottom of its stroke, loosen the locknut and turn the adjustment screw down until the top of the washer is even with the pointer (Fig. 9-36).

When the motor is run all the way open, the washer at the bottom of the strain-relief spring should move upward $1/16$ in. (Fig. 9-37). If it moves more than $1/16$ in., reduce the lift adjustment. Increase the lift adjustment if it moves less than $1/16$ in. Any change in the lift adjustment will require that the close-off tension at the bottom of the stroke be reset before continuing. In operation, the motor should be free to run through its entire stroke.

VALVE SIZE	DIMENSIONS IN INCHES		
	A	B	C
2-1/2	9-1/2	3-1/2	4-5/8
3	11	3-15/16	4-7/8
4	13	4-1/2	5-5/8
5	15	5-1/2	7-3/8
6	16-1/2	6-3/8	7-3/8

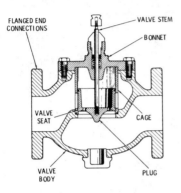

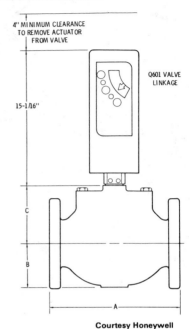

Courtesy Honeywell

Fig. 9-34. Cage-type single-seated valve used for control of steam, liquids, or noncombustible gases.

Another type of valve linkage that requires adjustment during its installation is the Honeywell Q455 Valve Linkage illustrated in Fig. 9-38. This linkage is also used with automatic valves in many different types of heating and cooling applications. The method of connecting a Q455 Valve Linkage to a water or steam valve is illustrated in Fig. 9-38. The adjustment procedure for a two-way automatic valve is as follows:

1. Set the lift adjustment on the crank arm to equal the rated lift of the valve (Fig. 9-39).
2. With the motor running closed, turn the adjustment screw down (Fig. 9-39) until the valve is closed and washer *(A)* (Fig. 9-40) has been pushed down 1/16 in. as indicated on the scale marks.
3. Run the motor all the way open. If the adjustment is correct, the valve will not reach the top of its stroke and jam. A tendency to jam will be indicated by tension in the linkage.

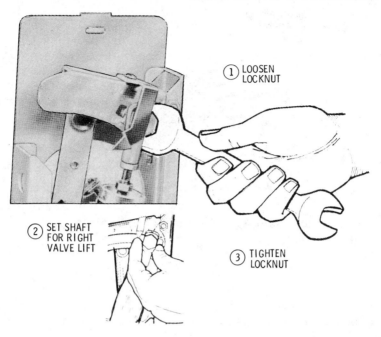

① LOOSEN LOCKNUT

② SET SHAFT FOR RIGHT VALVE LIFT

③ TIGHTEN LOCKNUT

Courtesy Honeywell

Fig. 9-35. Setting the left adjustment.

The procedure used for connecting a motor and an automatic three-way valve to a Honeywell Q455 valve linkage is identical to that used for the two-way type valve. However, there are significant differences in the adjustment procedure.

The lift adjustment should be set at *one mark higher* than the rated lift of the automatic valve. Then, with the motor closed, turn the shaft adjustment screw down until the valve is closed and washer *(A)* has been pushed down $\frac{1}{16}$-in. Now, run the motor all the way open and check washer *(B)* for a $\frac{1}{16}$-in. upward movement (Fig. 9-40).

Any readjustment of the lift requires a readjustment of the shaft adjustment screw for proper close-off force on the valve stem at the bottom of the stroke.

An example of a type of valve linkage that does *not* require adjustment during installation is the Honeywell Q618 valve link-

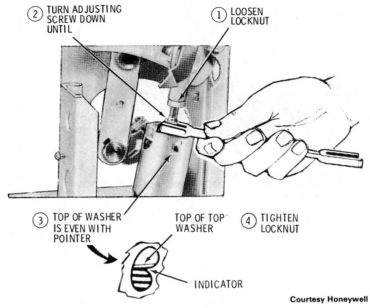

Courtesy Honeywell

Fig. 9-36. Using the adjusting screw to put tension on the valve.

age. This valve linkage can be used in most motorized valve applications

VALVE PIPE CONNECTIONS

The pipe ends or connections of valves are commonly available in the following three types:

1. Threaded ends.
2. Flanged ends.
3. Grooved ends.

Threaded end valves are tapped usually with female threads into which the pipe is threaded. These are the cheapest of the three types to use because less material and finishing is required. On the other hand, it is difficult to remove them without dismantling a considerable portion of the piping (unless extra fittings such as unions are used).

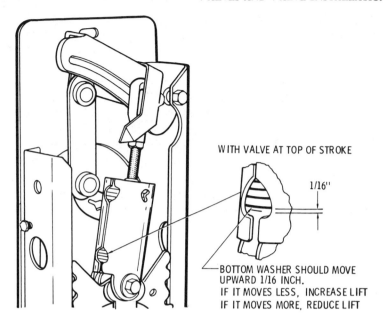

WITH VALVE AT TOP OF STROKE

1/16"

BOTTOM WASHER SHOULD MOVE
UPWARD 1/16 INCH.
IF IT MOVES LESS, INCREASE LIFT
IF IT MOVES MORE, REDUCE LIFT

Courtesy Honeywell

Fig. 9-37. Adjusting linkage for a three-way valve.

Flanged end valves are used where heavy viscous media are to be controlled or where high-pressure steam service is the rule. The different types of flanged ends available for use are shown in Fig. 9-41. Although flanged ends make a stronger, tighter, and more leakproof connection, their initial cost and the cost of installation is higher. This higher cost can be traced to a number of factors. For example, more metal is required in their manufacture. Moreover, flanged ends must be carefully and accurately machined before they are connected. Finally, additional parts such as gaskets, bolts, nuts, and companion flanges (to which the valve end flanges are bolted) must be provided to complete the connection.

Grooved ends are lighter in weight than flanged ends and require only an ordinary socket or speed wrench for installation or removal. The grooved end is connected to groove piping with a coupling patented by the Vitaulic Company of America.

Pipe threads in valve bodies are gauged to standard tolerances. Threaded end valves should be installed with the proper-size wrenches. The wrench should be applied to the pipe side of the valve when installing. Doing so minimizes the possibility of distorting the valve body, particularly when the valve is made of a malleable material such as bronze. Closing the valve tightly before installation will also reduce the possibility of distortion.

Fig. 9-42 illustrates the method of tightening flanged valves and fittings. Note the numbering sequence of the flange bolts in the illustration. The reason for tightening down bolts diametrically opposite one another is to create a gradual and uniform tightness. This results in uniform stress across the entire cross section of the flange, which eliminates the possibility of a leaky gasket.

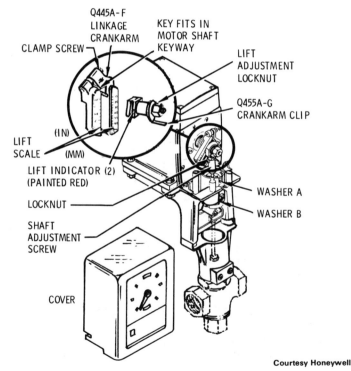

Courtesy Honeywell

Fig. 9-38. Honeywell model Q455 valve linkage connecting a motor to a water or steam valve.

VALVE INSTALLING POINTERS

Unless valves are installed properly, they will not operate efficiently and can cause problems in the system. There are certain precautions you should take when installing valves that will

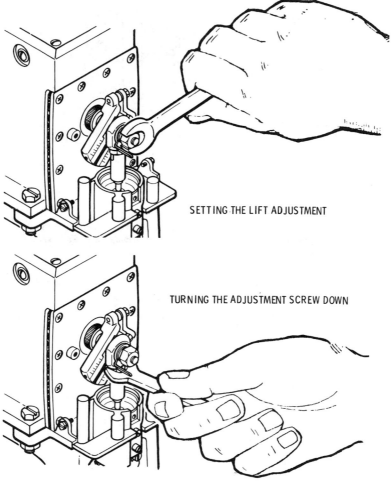

SETTING THE LIFT ADJUSTMENT

TURNING THE ADJUSTMENT SCREW DOWN

Courtesy Honeywell

Fig. 9-39. Setting the lift adjustment to equal the rated lift of the valve it is used with.

489

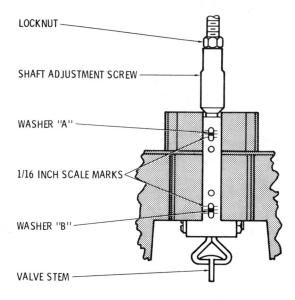

LOCKNUT

SHAFT ADJUSTMENT SCREW

WASHER "A"

1/16 INCH SCALE MARKS

WASHER "B"

VALVE STEM

Courtesy Honeywell

Fig. 9-40. Amount of tension put on the valve stem at the ends of the valve stroke is indicated by the washer inside the strain release mechanism.

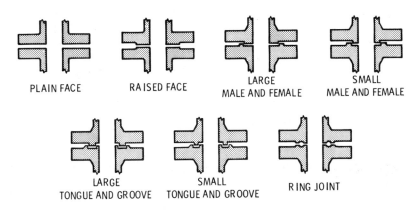

PLAIN FACE

RAISED FACE

LARGE
MALE AND FEMALE

SMALL
MALE AND FEMALE

LARGE
TONGUE AND GROOVE

SMALL
TONGUE AND GROOVE

RING JOINT

Courtesy Wm. Powell

Fig. 9-41. Various types of flanged ends used on valves.

490

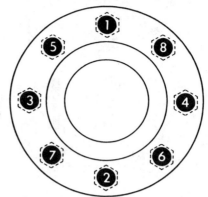

Fig. 9-42. Recommended tightening sequence for flange bolts.

improve their performance and minimize the possibilities of a malfunction. These precautions may be summarized as follows:

1. Always clean out a valve before installation because dirt, metal chips, and other foreign matter can foul it. This can be done by flushing the valve with water or blowing it out with compressed air.
2. Clean the piping before installing the valve. If you cannot flush or blow out the foreign matter, the end of the pipes should be swabbed with a damp cloth.
3. Only apply paint, grease, or joint sealing compound to the pipe threads (i.e., the *male* threads). *Never* apply these substances to the valve body threads because you run the risk of their getting into the valve itself and interfering with its operation.
4. Install valves in a location that can be reached conveniently (Fig. 9-43). If the valve is placed so that it is awkward to reach, it sometimes may not be closed tightly enough. This can eventually cause leaks to develop in the valve.
5. When necessary, support the piping so that additional strain is not placed on the valve. Small or medium-size valves can be supported with hangers placed on either side. Large valves should always be independently suspended.
6. Sufficient clearance is particularly important when rising stem valves are used.

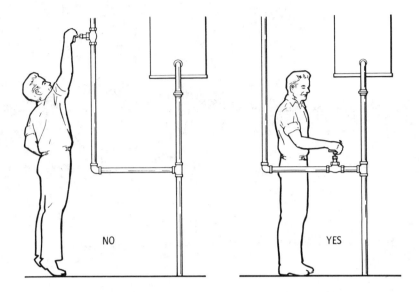

Fig. 9-43. Install valves in a convenient, easily reached location.

SOLDERING, BRAZING, AND WELDING VALVES TO PIPES

The joining of valves to pipes by soldering, brazing, or welding offers certain advantages over threaded and flanged joints in reduced maintenance and repairs. Furthermore, joints of this type enable you to use lighter pipe and require fewer flanges and gaskets. The major disadvantage is that an experienced and skilled worker is required to do the job. This can result in higher labor costs.

Fusion welding is the most commonly used method of welding a valve to a pipe. Essentially, it consists of fusing the metal of the valve and pipe connection by applying heat at a temperature above the melting point of the metal. The following two types of welds can be made: (1) a butt weld, and (2) a socket weld.

Both soldering and silver brazing are procedures used to join two metals, but they differ from fusion welding in that temperatures *below* the melting point of the metal are used. Either a suitable soldering flux or a silver brazing alloy is heated to a temperature at which it will flow into the joint formed by the valve connection and pipe end. Silver brazing is replacing soldering in many applications because it is quicker and results in a stronger joint.

A more detailed description of the procedures used in soldering, brazing, and welding pipe connections is given in Chapter 8 (Pipes, Pipe Fittings, and Piping Details). These procedures are summarized in the following sections.

SOLDERING OR SILVER BRAZING PROCEDURE

The basic procedure for soldering or silver brazing pipe or tubing to valves is as follows:

1. Cut the tube or pipe end square, and make sure the diameter is not undersize or out of round.
2. Remove all burrs with a metal file.
3. Clean the pipe or tubing end (at least to the depth of the socket) and the inside of the valve socket with steel wool and a cloth to wipe away the residue.
4. Clean all surfaces with a suitable solvent and wipe dry.
5. Apply solder flux or silver brazing flux to the inside of the valve socket and the outside of the pipe or tubing.
6. Insert the pipe or tubing into the valve socket until it seats against the shoulder within the socket.
7. Turn the valve and the pipe or tubing once or twice to evenly distribute the flux.
8. Make certain the valve is in open position before applying heat. A nonmetallic disc should be removed before the heat is applied. After removing the disc, the valve bonnet or bonnet ring should be replaced handtight to prevent distortion to the threaded sections when heating the valve.
9. Make certain the valve and pipe or tubing is properly supported during the soldering or silver brazing process. Any strain on the joint while cooling will weaken it.
10. Apply flame evenly around piping or tubing adjacent to

valve ends until solder or brazing alloy suitable for the service flows upon contact.

11. *Soldering:* Apply solder to the joint between the pipe or tubing and the end of the valve socket. Apply the flame toward the bottom of the valve socket until all the solder is absorbed. Control the direction of the flame away from the valve body to avoid excessive heating, which causes distortion and improper functioning of the valve.

12. *Silver brazing:* Apply brazing alloy to the joint between the pipe or tubing and the end of the valve socket. Wave the flame over the valve hexes to draw the metal alloy into the socket leaving a solid fillet of brazing alloy at the joint. Control the direction of the flame away from the valve body to avoid excessive heating which causes distortion and improper functioning of the valve.

13. Remove all excess and loose matter from the surface with a clean cloth or brush.

BUTT WELDING PROCEDURE

The procedure for making a butt weld is as follows:

1. Machine the pipe ends for the butt welding joint.
2. Clean the pipe ends, valve joint, and the inside of the valve socket with a degreasing agent to remove oil, grease, or other foreign materials.
3. Align by means of fixtures, and tack weld in place.
4. Make certain the valve is in the open position before applying heat. The valve bonnet should be handtight to prevent distortion or damage to the threads. Nonmetallic discs should be removed before applying heat.
5. The valve and pipe should be supported during the welding process, and must not be strained while cooling.
6. Preheat the welding area 400°F to 500°F.
7. Depending upon the welding method used (gas, arc, etc.), a butt weld is normally completed in two to four passes. The first pass should have complete joint penetration and be flush with the internal bore of the pipe. Make sure the first

494

pass is clean and free from cracks before proceeding with the second pass. The second pass should blend smoothly with the base metal and be flush with the external diameter. Avoid excessive heat because this can cause distortion and possible malfunctioning of the valve.

8. Use a wire brush and a clean cloth to remove discoloration.

SOCKET WELDING PROCEDURE

The procedure for making a socket weld is as follows:

1. Cut the pipe end square making sure the diameter is not undersize or out of round.
2. Remove all burrs with a metal file.
3. Clan the pipe end, valve joint, and the inside of the valve socket with a degreasing agent to remove any oil, grease, or other foreign material.
4. Insert the pipe end into the valve socket and space by backing off the pipe after it hits against the shoulder inside the spacing collar. Tack weld in place.
5. Make certain the valve is in the open position before applying heat. Valve bonnets should be handtight to prevent distortion or damage to the threads. Nonmetallic discs should be removed before applying heat.
6. The valve and pipe should be supported during the welding process, and must not be strained while cooling.
7. Preheat the welding area 400°F to 500°F.
8. A socket weld can generally be completed in two to four passes, depending upon the welding method used. Make sure the first pass is clean and free from cracks before proceeding with the second pass. Avoid excessive heat as it may cause distortion to the valve bonnet.
9. Use a wire brush and a clean cloth to remove discoloration.

CHAPTER 10

Steam and Hot-Water Line Controls

The pipelines of steam and hot-water heating systems contain a number of different devices or controls designed to maintain the circulation of the steam or water. These steam and hot-water line controls include such devices as pumps, valves, steam traps, air eliminators, and pipeline strainers. Some of these devices, such as pumps and valves, are designed to directly control the rate of flow of the steam or water. Others indirectly influence the flow rate. For example, air eliminators remove air pockets that retard circulation. Steam traps perform a similar function by removing both air and condensation, and returning the latter to the boiler.

Most of the line control devices commonly used in steam and hot-water heating systems are covered in this chapter. However, with the exception of certain specialized applications, valves are described in Chapter 9 (Valves and Valve Installation).

HEATING AND CIRCULATING PUMPS

Heating and circulating pumps are used to maintain the desired flow rate of steam or hot water in a heating system.

The two types of pumps used in *steam* heating systems are: (1) condensation pumps, and (2) vacuum heating pumps. The condensation pump is used in a gravity steam heating system to return the condensation to low- or medium-pressure boilers. Vacuum heating pumps are used in either return-line or variable-vacuum heating systems to return condensation to the boiler *and* to produce a vacuum in the system by removing air and vapor along with the condensation.

Circulating pumps are used in forced hot-water (hydronic) heating systems to maintain the flow rate of the water in the system. They are smaller in size than either condensation or vacuum heating pumps and are usually of the motor-driven centrifugal type.

CONDENSATION PUMPS

The design of a *standard* gravity steam heating system is such that the heat-emitting units cannot be placed at a level lower than the water level in the boiler. The movement of the condensation depends upon the gravity flow principle and therefore must move from a higher to a lower level until it reaches the boiler. This places certain limitations on the design of the heating system unless mechanical means are used to compensate for the lack of gravity. A condensation pump serves this purpose.

Many different condensation pumps have been developed for use in steam heating systems. Screw, rotary, turbine, reciprocating, and centrifugal pumps are some of the types used for this purpose. One of the most common used condensation pumps found in low-pressure steam heating systems is the motor-driven centrifugal pump equipped with receiver (tank) and float-control automatic switch.

In operation, condensation enters the receiver and fills the tank. A float connected to an automatic switch rises with the water until the tank is almost full. At that point, the float closes

498

the switch and starts the pump motor. The water is pumped from the receiver and the float drops, causing the switch to open and shut off the pump motor.

The centrifugal pump shown in Fig. 10-1 is used to pump condensation from a lower level return line to one at a higher level, or against a higher pressure. These units are also used to pump condensation from a flash tank to a boiler (Fig. 10-2) and for other special applications.

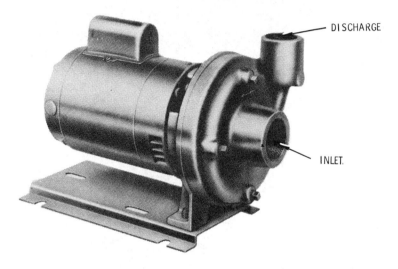

Courtesy Sarco Co.

Fig. 10-1. Typical centrifugal pump.

As shown in Fig. 10-3, the basic components of this pump consist of an impeller *(A)* with an inlet at its center rotating on a shaft *(D)*. The condensation enters the inlet orifice and flows radially through vanes to the outer periphery *(F)* of the impeller; it has approximately the same velocity as the periphery. The head of pressure developed by the pump is the result of the velocity imparted to the condensation by the rotating impeller.

When the condensation leaves the outer periphery of the impeller, it flows around the volute casing *(B)* and through the discharge orifice *(E)* of the pump. A wear ring *(C)* is provided to prevent bypassing of the condensation.

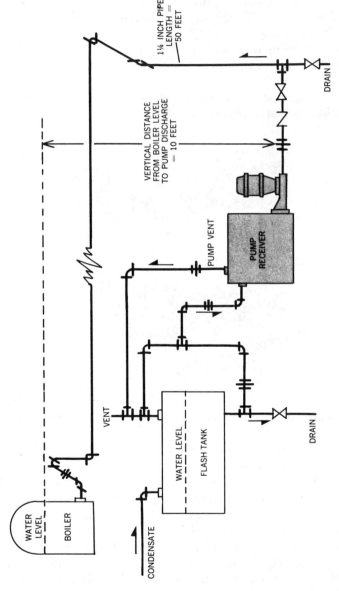

1¼ INCH PIPE LENGTH = 50 FEET

VERTICAL DISTANCE FROM BOILER LEVEL TO PUMP DISCHARGE = 10 FEET

DRAIN

PUMP RECEIVER

PUMP VENT

VENT

WATER LEVEL

FLASH TANK

DRAIN

CONDENSATE

WATER LEVEL

BOILER

Fig. 10-2. Pumping condensation from flash tank to boiler.

500

Condensation pumps are available in either single or duplex units (Fig. 10-4). The latter are used in installations where it is necessary to have a pump available for use at all times. A duplex unit is actually two condensation pumps fitted with a mechanical

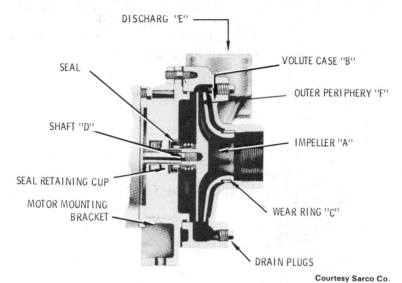

DISCHARG "E"

SEAL

VOLUTE CASE "B"

OUTER PERIPHERY "F"

SHAFT "D"

IMPELLER "A"

SEAL RETAINING CUP

MOTOR MOUNTING BRACKET

WEAR RING "C"

DRAIN PLUGS

Courtesy Sarco Co.

Fig. 10-3. Basic components of a centrifugal pump.

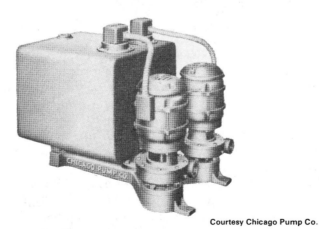

Courtesy Chicago Pump Co.

Fig. 10-4. Condensation pump and receiver for low- and medium-pressure steam heating system.

501

alternator. Both pumps feed into the same receiver. If one pump malfunctions, the other starts automatically and continues to provide uninterrupted pumping service for the system.

Vertical condensation pumps (Fig. 10-5) are available for use in installations where the space for the pump is limited, where the returns run below the floor, or where it is undesirable to place a horizontal pump in a sunken area. Vertical condensation pumps are also available in both single and duplex units.

Fig. 10-5. Vertical condensation pump and receiver.

Courtesy Chicago Pump Co.

The use of a condensation pump in a gravity steam heating system is shown in Fig. 10-6. Note the arrangement of gate and check valves on the discharge side of the pump. This type of system is commonly referred to as a *condensate-return steam heating system.* Using a condensation pump to return the condensation provides greater design flexibility for the system. The major disadvantage is that larger steam traps and piping must be used than in vacuum heating systems.

Condensation pumps can also be used as mechanical lifts in vacuum steam heating systems (Fig. 10-7). By connecting the vent outlet of the condensation pump to a return line above the level of the vacuum heating pump, the same vacuum return condition is maintained in the piping below the water level as in the

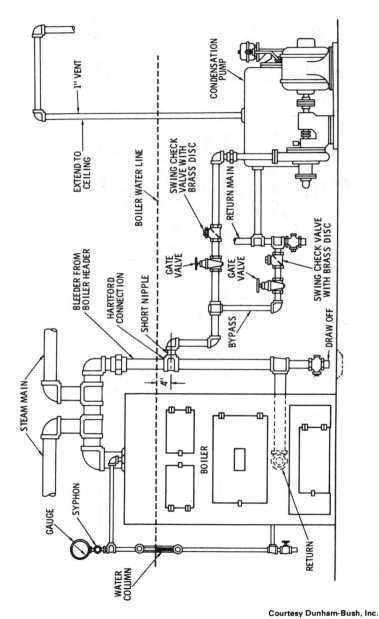

Fig. 10-6. Piping of a condensation-return steam heating system.

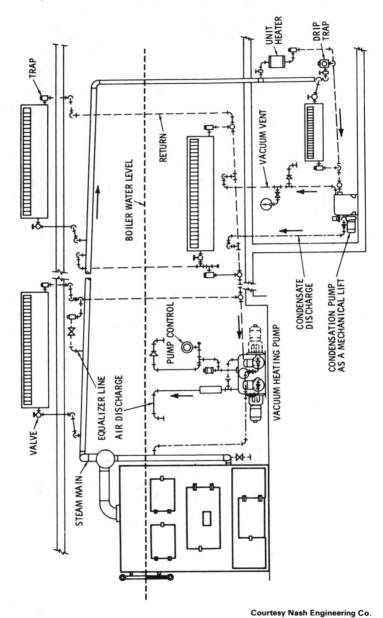

Fig. 10-7. A condensation pump used as a mechanical lift.

rest of the system. In this arrangement, the only purpose of the condensation pump is to lift the condensation from the lower level to the higher one without reducing the capacity of the vacuum heating pump.

VACUUM HEATING PUMPS

Vacuum heating pumps are used to maintain the vacuum in mechanical vacuum heating systems by removing air, vapor, and condensation from the lines. Many vacuum pumps are available in both single and duplex units, and are designed to automatically adjust themselves to the varying conditions of the system. The duplex units have the advantage over the single pumps because they provide automatic standby service. If one of the pumps in a duplex unit should happen to malfunction, the other one cuts in and picks up the load.

A vacuum heating pump is operated either by steam or electricity. Steam-driven vacuum pumps are sometimes used in high-pressure steam systems, but these pumps have been generally replaced by automatic motor-driven return-line pumps. Examples of vacuum heating pumps are shown in Figs. 10-8 and 10-9.

Courtesy Nash Engineering Co.

Fig. 10-8. Return-line vacuum heating pump and receiver.

Fig. 10-9. Duplex return-line vacuum and boiler feed pump for a vacuum
steam heating system.

In operation, the pump is started before the steam enters the system. When the pump removes the air from the lines, steam quickly fills the radiators of the system. The radiators remain full of steam because the air is automatically removed as fast as it accumulates. By quickly exhausting the air and condensation from the system, the vacuum pump causes the steam to circulate more rapidly resulting in faster warm-up time and quieter operation.

The condensation of steam in the lines creates the vacuum, and the pump maintains it by continuing to pump air from the system. The vacuum maintained by the pump is only a *partial* one because it is not possible with this device to extract all the air. Each stroke of the pump piston or plunger removes only a fraction of the air, depending on the percentage of clearance in the pump cylinder, the resistance of valves, and other factors; hence theoretically an infinite number of strokes would be necessary to obtain a perfect vacuum, not considering line and pump resistance.

Vacuum pumps designed to remove only air from a system are referred to as *dry* pumps. Those that remove both air and condensation are called *wet* pumps. When a wet pump is used, the condensation is pumped back to the boiler. In operation, the air, being heavier than steam, passes off through thermostatic retainer valves to the pump. When the steam reaches the retainer valves, they close automatically to prevent the steam passing into the dry return line to the pump and breaking the vacuum. The air from the pump is passed into a receiver where it is discharged through an air vent. The condensation is pumped back to the boiler generally by means of a centrifugal pump.

In most vacuum systems, the pump is controlled by a vacuum regulator and a float control. The vacuum regulator cuts in when the vacuum drops to a preset level and cuts out when the vacuum reaches its highest point. The float control operates independently from the vacuum regulator, starting the pump when condensation reaches a certain level in the receiver.

Two typical installations in which vacuum heating pumps are used are illustrated in Figs. 10-10, 10-11, and 10-12. In the vacuum air line heating system, shown in Fig. 10-10, thermostatic type air line valves are used instead of radiator air vents. The primary purpose of the vacuum heating pump is to expel air from the system.

The vacuum return-line system is very similar to a condensation return steam heating system except that a vacuum pump is used to provide a low vacuum in the pipes and to return the condensation to the boiler (Fig. 10-12). Because of the vacuum condition, smaller steam traps and piping can be used.

WATER CIRCULATING PUMPS

Forced hot-water heating systems (i.e., hydronic heating systems) use small compact pumps to provide the motive force for the water in the pipes. They are usually referred to as *circulators* or *water circulating pumps*. An example of one of these pumps, the Sundstrand Model W406 water circulating pump, is shown in Fig. 10-13. The cutaway of this pump in Fig. 10-14 shows some of the essential parts. The impeller is mounted on a steel shaft and both the shaft and the rotor are plated to resist corrosion. The use of a "wet" rotor in this pump model eliminates the need for shaft seals, thereby avoiding seal leakage and wear. The unit is powered by a split-capacity motor, and no starting switches or

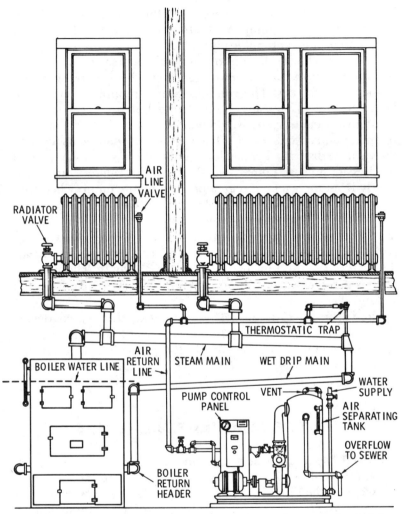

Fig. 10-10. An air line pump installed on a one-pipe steam heating system to create a vacuum air line system.

relays are necessary. The Sundstrand pump is a hermetically sealed and self-lubricating unit. These design elements eliminate some of the major problems of older pump models such as: (1) seal leakage, (2) bearing deterioration and seizure, (3) noise, and (4) vibration.

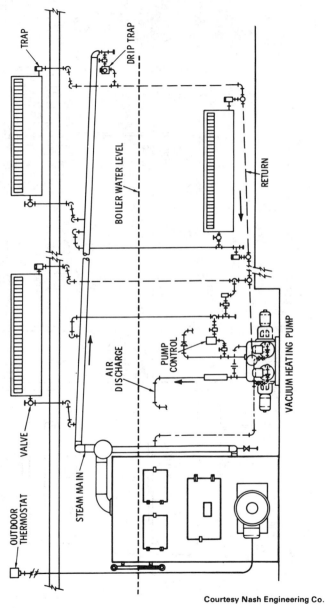

Fig. 10-11. Steam heating system that utilizes the thermodynamic properties of steam.

509

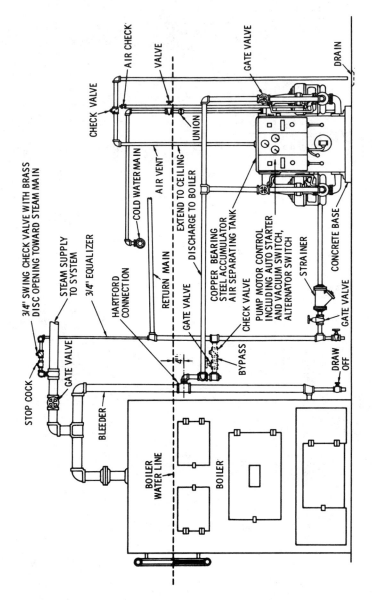

Fig. 10-12. Piping of a vacuum return-line steam heating system.

510

Fig. 13. Model W406 water
circulating pump.

Fig. 10-13. Model W406 water circulating pump.

511

MOTOR WINDING

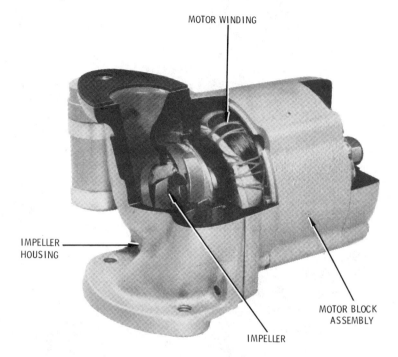

IMPELLER
HOUSING

MOTOR BLOCK
ASSEMBLY

IMPELLER

Courtesy Sundstrand Corp.

Fig. 10-14. Cutaway view of model W406 water circulating pump

Motor-driven centrifugal pumps are the most commonly used in hot-water heating systems. The smaller systems use small centrifugal circulators that can be mounted in standard fittings at a location in the return line as close as possible to the boiler. These smaller circulators are described in this chapter.

A number of different factors are important in selecting water circulating pumps for both large and small heating systems. Among the factors that must be considered when selecting a pump are:

1. Amount of water to be handled.
2. Temperature of the water to be handled.
3. Head against which the pump must operate.
4. Working head of the system.
5. Pump suction head.

The temperature of the water handled by the pump will determine the type of pump packing selected. The working head of the system is the sum of the static head and the friction. The pump will have either a positive or negative suction head. Data necessary for selecting a suitable water circulating pump are usually supplied by pump manufacturers.

The water circulating pumps used in small residential hot-water heating systems are installed directly in the return line. These are commonly single-suction pumps in which both the motor and the pump share a common shaft. The major objection to installing a pump directly in the return line is that the pipe connections must be broken in order to service the rotor. Some pumps, such as the Sundstrand model W406, are designed to be serviced without removing the entire unit. The operating head of water circulating pumps used in smaller hot-water heating systems is limited, and it is the general practice to size the pipelines of the system *after* selecting a pump capable of meeting the requirements of the system.

In residential hot-water heating systems, the water circulating pump should be mounted on the return line as close to the boiler as possible. This point cannot be overemphasized. *Never* mount these circulators at the highest point in the system.

Hermetically sealed, self-lubricating pumps should never be oiled or lubricated. It is not only unnecessary, it could also damage the pump. Pumps designed for forced hot-water heating should not be used on domestic-service water systems.

The following recommendations for installing a water circulating pump specifically apply to the one illustrated in Fig. 10-13.

1. Mount the circulator so that the motor shaft is in the horizontal plane.
2. Connect the hot line to the black motor lead.
3. Connect the ground wire to the white motor lead.
4. Connect the safety mechanical ground to the ground screw located inside the junction box (Fig. 10-15).
5. Turn on the water supply and open the isolating gate valves. Make sure the water fills *above* the level of the circulator.
6. Run the circulator at least 15 minutes, cycling on and off several times.

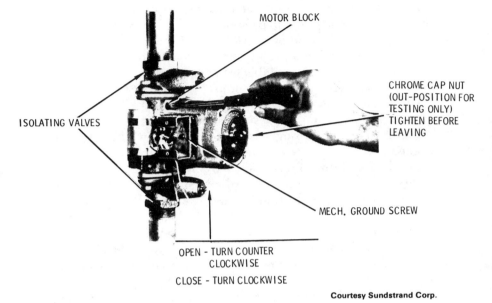

MOTOR BLOCK

CHROME CAP NUT
(OUT-POSITION FOR
TESTING ONLY)
TIGHTEN BEFORE
LEAVING

ISOLATING VALVES

MECH. GROUND SCREW

OPEN - TURN COUNTER
CLOCKWISE
CLOSE - TURN CLOCKWISE

Courtesy Sundstrand Corp.

Fig. 10-15. Installing and operating a pump.

7. Purge the air from the system after installation.

Sometimes the failure of a forced hot-water heating system to produce heat can be traced to a water circulating pump malfunction. Before attempting to repair or replace the unit, check the electric power to the pump from the system controls. The problem may be due to an electrical failure, rather than a mechanical failure of the pump itself. *Always check fuses.*

Suggestions for servicing a Sundstrand water circulating pump are illustrated in Figs. 10-15 and 10-16, and may be listed as follows:

1. Shut off the electric power.
2. Turn the chrome cap nut counterclockwise to "out-position," engaging the pump shaft (Fig. 10-15).
3. Turn on the electric power and set the controls to start the pump. The chrome cap nut shown in Fig. 10-15 must spin in the same direction as the arrow on the nameplate when the pump is running.

4. Check the water supply and bleed the system with the pump running.

5. If the chrome cap nut does not spin or cannot be turned manually, remove the motor block assembly.

6. Before attempting to remove the motor block assembly, shut off the electric power and disconnect the wires from the unit.

7. Shut off the water supply by turning the stems of the isolating valves (gate valves) clockwise as far as possible (Fig. 10-15).

8. Remove the four motor block screws (Fig. 10-16), and pull the motor block assembly with the O-Ring from the impeller housing (Fig. 10-16).

9. If the motor cannot be repaired, it can be exchanged for a new motor block assembly.

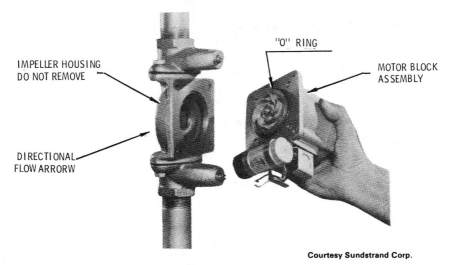

Courtesy Sundstrand Corp.

Fig. 10-16. Removing motor block assembly.

STEAM TRAPS

A steam trap is a device that opens to expel air and condensation from steam lines, and closes to prevent the escape of steam.

All steam traps operate on the fundamental principle that the pressure within the trap at the time of discharge will be slightly in excess of the pressure against which the trap must discharge. This includes the friction head, the velocity head, and the static head on the discharge side of the trap. The steam trap cannot operate unless the excess pressure of discharge is greater than the total back pressure.

Each steam trap used in steam heating is designed for a specific range of applications for which its operating characteristics best suit it. Although there is no universal steam trap per se, the many different types can be grouped into the following three classes on the basis of their operating characteristics:

1. Separating traps.
2. Return traps.
3. Air traps.

Separating traps are designed to release condensation but close against steam. They are float-operated, thermostatically operated, or float and thermostatically operated. Thermostatic traps are designed to release air and condensation but close against them.

Return traps may be operated to receive condensation under a vacuum and return it to atmosphere or a higher pressure. *Air traps* are generally operated by a float.

SIZING STEAM TRAPS

Selecting the correct size steam trap (or traps) for a system is an important factor in its operational efficiency. For example, an oversized trap will operate less efficiently than a correctly sized one, and will tend to create abnormal back pressure. Moreover, the installation cost will be higher and the operational life expectancy will be reduced.

Manufacturers of steam traps generally provide information in the form of capacity ratings and related data to make these selections easier. Data should be based on hot condensation under actual operating conditions rather than cold-water ratings. If possible, always try to determine the basis for a manufacturer's ratings. Table 10-1 and the sizing example were provided by Sarco Company, Inc., a manufacturer of steam traps.

Table 10-1. Sizing Traps for Steam Mains

Condensation Load in Pounds per Hour

Per 1000 Feet of Insulated Steam Main*—

Ambient Temperature 70°—Insulation 80% Efficient

Steam Pressure (PSIG)	\multicolumn{13}{c}{MAIN SIZE}												0°F. Correction Factor†
	2"	2½"	3"	4"	5"	6"	8"	10"	12"	14"	16"	18"	20"	24"	
10	6	7	9	11	13	16	20	24	29	32	36	39	44	53	1.58
30	8	9	11	14	17	20	26	32	38	42	48	51	57	68	1.50
60	10	12	14	18	24	27	33	41	49	54	62	67	74	89	1.45
100	12	15	18	22	28	33	41	51	61	67	77	83	93	111	1.41
125	13	16	20	24	30	36	45	56	66	73	84	90	101	112	1.39
175	16	19	23	26	33	38	53	66	78	86	98	107	119	142	1.38
250	18	22	27	34	42	50	62	77	92	101	116	126	140	168	1.36
300	20	25	30	37	46	54	68	85	101	111	126	138	154	184	1.35
400	23	28	34	43	53	63	80	99	118	130	148	162	180	216	1.33
500	27	33	39	49	61	73	91	114	135	148	170	185	206	246	1.32
600	30	37	44	55	68	82	103	128	152	167	191	208	232	277	1.31

*Chart loads represent losses due to radiation and convection for standard steam.
†For outdoor temperature of 0°F, multiply load value in table for each main size by correction factor corresponding to steam pressure.

AUTOMATIC HEAT-UP

Select a steam trap with a pressure rating equal to or greater than the pressure in the steam supply main, but with a capacity based on the estimated pressure at the trap inlet. The pressure at the inlet of the steam trap can be considerably less than the pressure in the steam supply main.

If the steam trap is connected into a common piping return system, it may have to operate against a certain amount of static pressure. This static (back) pressure can cause a reduction in the operating capacity of the steam trap. Table 10-2 illustrates the effect of back pressure on steam trap capacity.

The safety factor for a steam trap is the ratio between its maximum discharge capacity and the condensation load it is ex-

517

Table 10-2. Effect of Back Pressure on Steam Trap Capacity
(Percentage Reduction in Capacity)

% Back Pressure	Inlet Pressure PSIG			
	5	25	100	200
25	6	3	0	0
50	20	12	6	5
75	38	30	25	23

Courtesy Sarco Company

pected to handle. The actual safety factor to use for any particular application will depend upon the accuracy of the estimated condensation load, the accuracy of the estimated pressure conditions at trap inlet and outlet, and the operational characteristics of the trap.

The application for which a steam trap is to be used is also an important factor in its selection. For example, a float and thermostatic trap is recommended for use as a steam line drip trap at pressures ranging from 16 psig to 125 psig. For the same application at pressures of 126 psig or above, an inverted bucket trap is suggested. Unusual operating conditions may also influence the choice of a steam trap for a particular application. A careful reading of the selection guide and related literature provided by the manufacturer will greatly reduce the possibility of error in choosing a suitable steam trap.

INSTALLING STEAM TRAPS

The installation of steam traps requires the following modifications in the piping:

1. Install a long vertical drip and a strainer between the trap and the apparatus it drains. The vertical drip should be as long as the installation design will permit. *Exception:* Thermostatic traps in radiators, convectors, and pipe coils are attached directly to the unit without a strainer.
2. A gate valve should be installed on each side of a trap, along with a valved bypass around the traps if continuous service is required. This permits removal of the trap for servicing, repair, or replacement without interrupting service.

3. A check valve and gate valve should be installed on the discharge side of a trap used to discharge condensation against back pressure or to a main located above the trap (as for lift service).

FLOAT TRAPS

A *float trap* (Fig. 10-17) is operated by the rise and fall of a float connected to a discharge valve. The change of condensation level in the trap determines the level of the float. When the trap is empty, the float is at its lowest position and the discharge valve is closed. As the condensation level in the trap rises, the float also rises and gradually opens the valve. The pressure of the steam then pushes the condensation out of the valve. Because the opening of the valve is portioned to the flow of condensation through the trap, the discharge of condensation from the trap is generally continuous. On some float traps, a gauge glass is used to indicate the height of the condensation in the trap chamber.

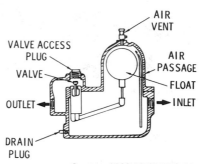

Fig. 10-17. Float trap.

Courtesy *1960 ASHRAE Guide*

One of the principal disadvantages of a float trap is the tendency of the valve to malfunction. Valve malfunctions can result from the sticking of moving parts or excess steam leakage due to unequal expansion of the valve and seat.

Float traps are designed for steam pressures ranging from vacuum conditions to 200 psig and are used to drain condensation from heating systems, steam headers, steam separators, laundry equipment, and other steam process equipment. When used in heating systems, a float trap should be equipped with a thermostatic air vent (see "Float and Thermostatic Traps").

THERMOSTATIC TRAPS

The operation of a *thermostatic trap* (Fig. 10-18) is based on the expansion or contraction of an element under the influence of heat or cold.

Thermostatic traps are of the following two types: (1) those in which the discharge valve is operated by the relative expansion of metals, and (2) those in which the action of the liquid is utilized for this purpose. The latter is probably the most commonly used thermostatic trap found in modern steam heating systems. Thermostatic traps of large capacity for draining blast coils or very large radiators are called *blast traps*.

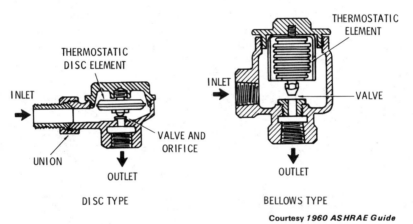

Courtesy *1960 ASHRAE Guide*

Fig. 10-18. Thermostatic trap.

Modern thermostatic traps consist of thin corrugated metal bellows or discs enclosing a hollow chamber which is filled with a liquid or partially filled with a volatile liquid. When steam comes in contact with the expansive element, the liquid expands or becomes a gas and thereby creates a certain amount of pressure. The element expands as a result of this pressure and closes the valve against the escape of the steam.

BALANCED-PRESSURE
THERMOSTATIC TRAPS

As shown in Fig. 10-19, the principal parts of a *balanced-pressure thermostatic trap* consist of a flexible bellows, a valve head, and a valve seat. The bellows is partially filled with a

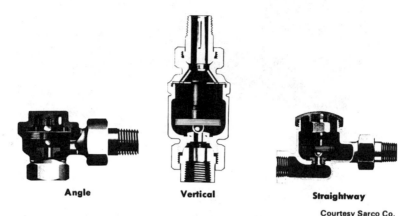

Angle   Vertical   Straightway

Courtesy Sarco Co.

Fig. 10-19. Balanced-pressure, bellows-type thermostatic steam traps.

volatile fluid and hermetically sealed. The fluid sealed in the bellows has a pressure-temperature relationship that closely parallels, but is approximately 10°F below, that of steam. When the condensation surrounding the bellows reaches approximately 10°F below saturated steam pressure, the fluid *inside* the bellows begins to build up pressure. When the temperature of the condensation approaches that of steam, the pressure inside the bellows exceeds the external pressure. This pressure imbalance causes the bellows to expand, driving the valve head to its seat and closing the trap (Fig. 10-20). When the condensation surrounding the bellows cools, the vaporized fluid condenses and reduces the internal pressure. The reduction of internal pressure causes the bellows to contract, opening the trap for discharge.

A balanced-pressure thermostatic steam trap is vulnerable to water hammer and corrosive elements in the condensation. The latter problem can be handled by fitting the trap with anticorrosive internal components. On the plus side, this type of trap has

521

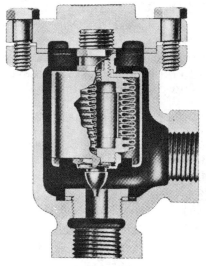

Fig. 10-20. Balanced-pressure thermostatic steam trap.

Courtesy Sarco. Co.

relatively large capacity and high air venting capability. It is completely self-adjusting within its pressure range.

FLOAT AND THERMOSTATIC TRAPS

A *float and thermostatic trap* (Fig. 10-21) has both a thermo-static element to release air and a float element to release the condensation. As such, it combines features of both the float trap and the thermostatic trap.

These traps are recommended for installations in which the volume of condensation is too large for an ordinary thermostatic trap to handle. Float and thermostatic traps are also used in low-pressure steam heating systems to drain the bottom and end of steam risers (Figs. 10-22 and 10-23). Other applications include the draining of condensation from unit heaters, preheat and reheat coils in air conditioning systems, steam-to-water heat exchangers, blast coils, and similar types of process equipment.

If the trap is operating properly, it will immediately and con-tinuously discharge condensation, air, and noncondensable gases from the system that enter the inlet orifice of the trap.

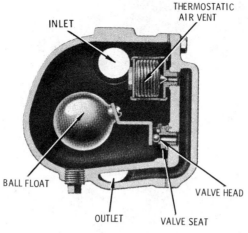

INLET

THERMOSTATIC
AIR VENT

BALL FLOAT

VALVE HEAD

OUTLET

VALVE SEAT

Courtesy Sarco. Co.

Fig. 10-21. Float and thermostatic steam trap.

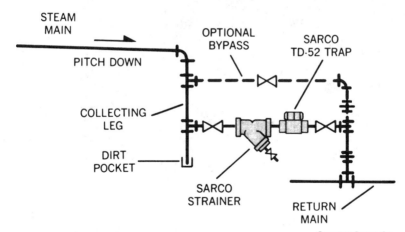

STEAM
MAIN

PITCH DOWN

OPTIONAL
BYPASS

SARCO
TD-52 TRAP

COLLECTING
LEG

DIRT
POCKET

SARCO
STRAINER

RETURN
MAIN

Courtesy Sarco. Co.

Fig. 10-22. Draining end of low-pressure steam risers.

The condensation is handled by the ball float, which is connected by a level assembly to the main valve head. Condensation entering through the trap inlet causes the ball float to rise, moving the level assembly and opening the valve for discharge.

Air and noncondensable gases are discharged through the

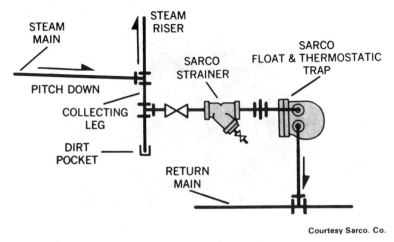

Fig. 10-23. Draining bottom of low-pressure steam riser.

thermostatic air vent. The thermostatic element is also designed to prevent the flow of steam around the float valve.

Float and thermostatic traps operate under pressures ranging from vacuum to a maximum pressure of 200 psig; however, the great majority of them are designed for 40 psig or less.

These traps have limited resistance to water hammer. They are also vulnerable to corrosive elements in the condensation unless fitted with anticorrosive internal components.

THERMODYNAMIC STEAM TRAPS

A thermodynamic steam trap (Figs. 10-24 and 10-25) contains only one moving part, a hardened stainless-steel disc that functions as a valve. Because of its construction simplicity, this is an extremely rugged trap and is especially well suited for service on medium- and high-pressure steam lines operating under pressures up to 600 psig. The minimum operating pressure for some makes of these traps is as low as 3.5 psig.

Thermodynamic steam traps are small, unaffected by water hammer, and can be mounted in any position. The operating principles of a thermodynamic steam trap are illustrated in Fig. 10-26.

BUCKET TRAPS

Bucket traps are either of the upright or inverted design and are used in both low- and high-pressure steam heating systems.

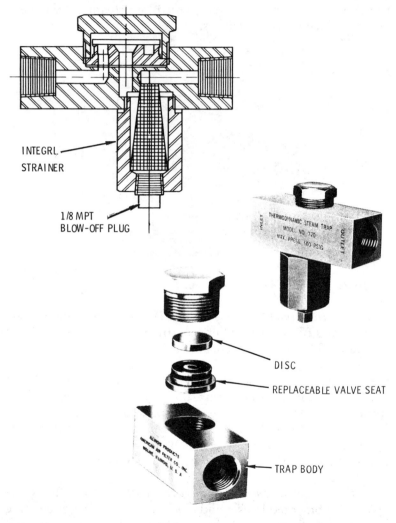

INTEGRL. STRAINER

1/8 MPT BLOW-OFF PLUG

DISC

REPLACEABLE VALVE SEAT

TRAP BODY

Courtesy American Air Filter

Fig. 10-24. Thermodynamic steam trap.

525

Both types of bucket traps are designed to respond to the difference in density between steam and condensation. The construction of a bucket trap is such that it has good resistance to water hammer. On the other hand, most bucket traps, unless modified, have limited air venting capabilities. Bucket traps also have a tendency to lose their waterseal and blow steam continuously during sudden pressure changes.

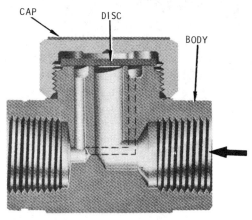

Courtesy Sarco Co.

Fig. 10-25. Thermodynamic steam trap.

In an *upright bucket trap* (Fig. 10-27), the condensation enters the trap and fills the space between the bucket and the walls of the trap. This causes the bucket to float and forces the valve against its seat, the valve and its stem usually being fastened to the bucket. When the water rises above the edges of the bucket, it floats into it and causes it to sink, thereby withdrawing the valve from its seat. This permits the steam pressure acting on the surface of the water in the bucket to force the water to a discharge opening. When the bucket is emptied, it rises and closes the valve and another cycle begins. The discharge from this type of trap is intermittent.

In the *inverted bucket trap* (Fig. 10-28) steam floats the inverted submerged bucket and closes the valve. Water entering the trap fills the bucket, which sinks and through compound leverage opens the valve, and the trap discharges.

526

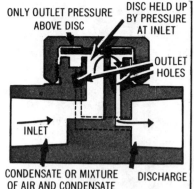

A high-velocity jet of flash steam reduces pressure under the disc and at the same time, by recompression, builds up pressure in the control chamber above the disc. This drives th disc to the seats assuring tight closure withuot steam loss.

Presure of condensate or air lifts the disc off its seats. Flow is across the underside of the disc to the three outlet holes. Discharge continues until the flashing condensate approaches steam temperature.

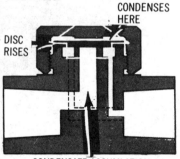

Steam pressure in the control chamber, acting over the total disc area, holds the disc closed against inlet pressure acting over the smaller inlet seat area.

As soon as condensate, even at steam temperature, collects it reduces heat transferred to the control chamber. Pressure in the chamber decreases as steam trapped there condenses. The disc is lifted by inlet pressure and condensate is discharged.

Courtesy Sarco Co.

Fig. 10-26. Operating principles of a thermodynamic steam trap.

527

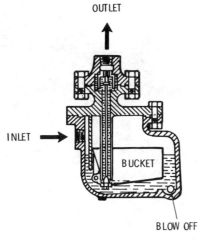

Fig. 10-27. Upright bucket trap.

Courtesy *1960 ASHRAE Guide*

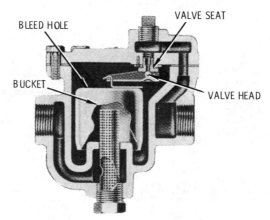

Courtesy Sarco. Co.

Fig. 10-28. Inverted bucket trap.

FLASH TRAPS

A *flash trap* (Fig. 10-29) is used to drain condensation from steam lines; steam, water, and oil heaters; unit heaters; and other

528

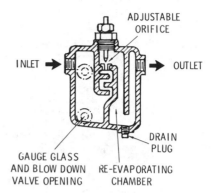

ADJUSTABLE
ORIFICE

INLET ➡

OUTLET

Fig. 10-29. Flash trap.

DRAIN
PLUG

GAUGE GLASS
AND BLOW DOWN
VALVE OPENING

RE-EVAPORATING
CHAMBER

Courtesy 1960 ASHRAE Guide

equipment in which the pressure differential between the steam supply and condensation return is 5 psig or more.

The operation of a flash trap depends upon the property of condensation at a high pressure and temperature to flash into steam at a lower pressure. The condensation flows freely through the trap due to the pressure difference between the inlet and outlet orifices. The free flow of the condensation is interrupted by the introduction of steam into the inlet chamber where it mixes with the remaining condensation. The steam heats the condensation and causes it to flash, thereby temporarily halting its flow through the orifice and allowing it to accumulate in the trap.

Except for an adjustable orifice used for adjusting the pressure differential, a flash trap contains no other moving parts. Flash traps operate intermittently. They are generally available for pressures ranging from vacuum to 450 psig.

IMPULSE TRAPS

An *impulse trap* (Fig. 10-30) operates with a moving valve actuated by control cylinder. When the trap is handling condensation, the pressure required to lift the valve is greater than the reduced pressure in the control cylinder, and consequently the valve opens allowing a free discharge of condensation. As the

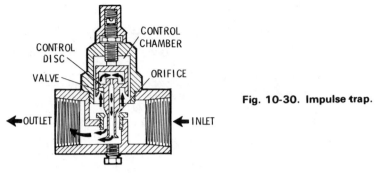

CONTROL
CHAMBER

CONTROL
DISC

VALVE

ORIFICE

OUTLET

INLET

Fig. 10-30. Impulse trap.

remaining condensation approaches steam temperature, flashing results, flow through the valve orifice is choked, and the pressure builds up in the control chamber, closing the valve.

TILTING TRAPS

The operation of a *tilting trap* (Fig. 10-31) is intermittent in nature. With this type of trap, condensation enters a bowl and rises until its weight overbalances that of a counterweight, and the bowl sinks to the bottom. As the bowl sinks, a valve is opened, thus admitting live steam pressure on the surface of the water, and the trap then discharges. After the water is discharged, the counterweight sinks and raises the bowl, which in turn closes the valve, and the cycle begins again.

LIFTING TRAPS

A *lifting trap* (Fig. 10-32) is an adaptation of the upright bucket trap and is available for pressures ranging from vacuum to 150 psig.

Condensation in the chamber of the trap accumulates until it reaches a level high enough to cause the steam valve in the high-pressure inlet to open. Steam then enters the auxiliary high-pressure inlet on the top of the trap at a pressure higher than the

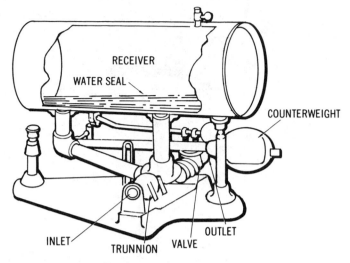

RECEIVER

WATER SEAL

COUNTERWEIGHT

INLET

TRUNNION

VALVE

OUTLET

Courtesy *1960 ASHRAE Guide*

Fig. 10-31. Tilting trap.

trap inlet pressure. This high-pressure steam forces the condensation to a point above the trap, and against a back pressure higher than that which is possible with normal steam pressure. As the condensation is pushed out of the trap, the float or bucket descends to the bottom and causes the valve in the high-pressure inlet to close, shutting off the steam supply. Condensation then begins to refill the float chamber, and the cycle is repeated.

BOILER RETURN TRAPS

A *boiler return trap* (or *alternating receiver*) is a device used in some vapor steam heating systems to return condensation to the boiler under varying pressure conditions of operation up to the working limit of the boiler. A vapor steam heating system in which a boiler return trap is used is sometimes referred to as a *return-trap system*. A typical installation in which a boiler return trap is used is shown in Fig. 10-33.

As shown in Fig. 10-34, a boiler return trap consists of a chamber containing a float, which is linked to two valves. These

531

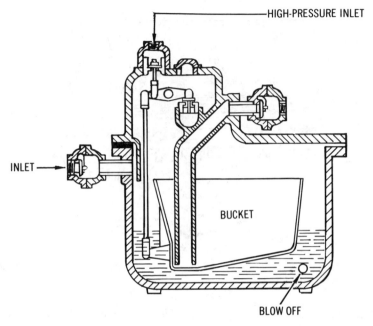

Courtesy *1960 ASHRAE Guide*

Fig. 10-32. Lifting trap.

valves control the openings to two connections on the top of the trap. One of these connections (the steam inlet) is connected to the steam header and direct boiler pressure. The other connection is vented to the atmosphere.

Condensation returning from the radiators is unable to enter the boiler by ordinary gravity flow because the higher pressure of the boiler keeps the check valve closed. As a result, the condensation is forced to back up in the vertical pipe connected to the boiler return trap. As the condensation rises, it fills the bottom of the trap and lifts the float. At a certain level, the float causes the air valve to close and the steam valve to open, allowing steam at boiler pressure to enter the top of the trap. This steam at boiler pressure plus the gravity head (a boiler return trap *must* be located at least 6 in. above the water level in the boiler) is sufficient to force the condensation back down the pipe, through the check valve, and into the boiler.

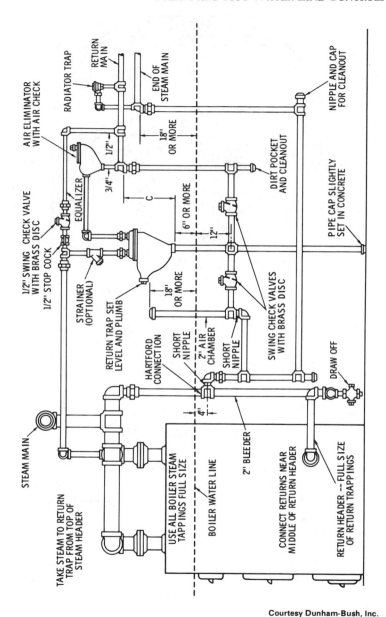

Fig. 10-33. Boiler piping of a return-trap steam heating system.

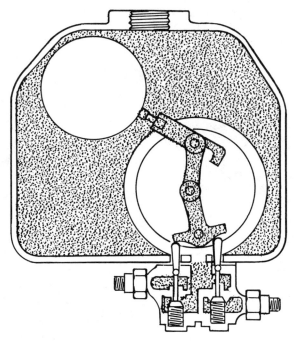

Courtesy Sarco Co.

Fig. 10-34. Boiler return trap.

EXPANSION TANKS

Expansion tanks are installed in hot-water space heating systems to limit increases in pressure to the allowable working pressure of the equipment, and to maintain minimum operating pressures.

When the temperatures rise during the operation of the system, the water volume also increases and builds up pressure. The pressure in the system is relieved to a certain extent by the storage of the excess water volume in the expansion tank. When temperatures drop, there is a corresponding drop in water volume and the water returns to the system.

Maximum pressure at the boiler is maintained by an ASME

pressure-relief valve. Minimum pressure in the system is generally maintained by either an automatic or manual water-fill valve.

The two basic types of expansion tanks are: (1) an *open* expansion tank (Fig. 10-35) and (2) a *closed* expansion tank (Fig. 10-36). A system in which the former is used is referred to as an *open tank system*. A closed expansion tank is used in a *closed tank system*.

Many problems are caused by using an expansion tank of inadequate size. Table 10-3 lists recommended sizes for expansion tanks in both open and closed tank systems. Additional information about expansion tanks and how they operate can be found in Chapter 7 of Volume 1 (Hot-Water Heating Systems).

Table 10-3. Recommended Sizes for Expansion Tanks

Open System	
Nominal Capacity—Gallons	Square Feet of Radiation
10	300
15	500
20	700
26	950
Closed System	
Nominal Capacity—Gallons	Square Feet of Radiation
18	350
21	450
24	650
30	900
35	1100

TROUBLESHOOTING EXPANSION TANKS

An undersized expansion tank or one that is completely filled up with water will cause the boiler pressure to increase when the water heats. Because the expansion tank is too small or too filled with water to absorb the excess pressure, the relief valve will begin to drip. The dripping relief valve is only symptomatic of the real problem, and replacing the valve will in no way solve it.

There is not much you can do about an undersized expansion tank except replace it. As a rule-of-thumb, expansion tanks

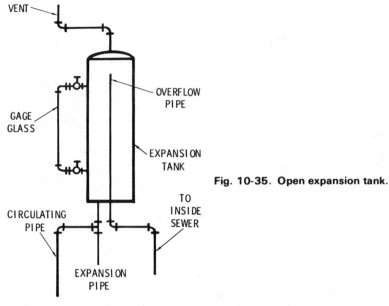

Fig. 10-35. Open expansion tank.

should be sized at 1 gal. for every 23 sq. ft. of radiation, or 1 gal. for every 3500 Btu of radiation installed on the job. In Table 10-3, the allowance is slightly higher.

If the problem is a completely filled tank, it should be partially drained so that there is enough space to permit future expansion under pressure. The first step in draining an expansion tank is to open the drain valve. The water will gush out at first in a heavy flow and then tend to "gurgle" out because a vacuum is building up inside the tank. Inserting a tube into the drain valve opening will admit air and break the vacuum, and the water will return to its normal rate of flow. After a sufficient amount of water has been removed, the drain valve can be closed.

AIR ELIMINATORS

Sometimes air pockets will form in the pipelines of steam or hot-water heating and cooling systems and retard circulation.

536

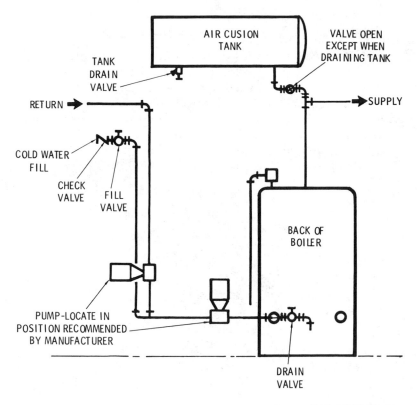

Fig. 10-36. Closed expansion tank.

One method of eliminating these air pockets is to install one or more air eliminators at suitable locations in the pipeline.

An *air eliminator* (or air vent) is a device designed to permit automatic venting of air. These automatic venting devices are available in a number of sizes, shapes, and designs. Not only are air eliminators used for venting convectors, baseboard radiators, and other heat-emitting devices; they are also frequently used for this purpose on overhead mains and circulating lines.

Three types of air eliminators (air vents) used in steam or hot-water heating and cooling systems are:

1. Float-type air vents.

537

2. Thermostatic air vents.
3. Combination float and thermostatic air vents.

A *float-type air vent* (Fig 10-37) consists of a chamber (body) containing a float attached to a discharge valve by a lever assembly. The float-controlled discharge valve vents air through the large orifice at the top. The float action prevents the escape of any fluid because the float closes the valve tightly when it rises. When the float drops, the lever assembly pulls the valve from its seat, and the unit discharges air.

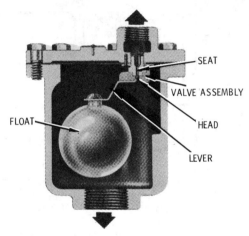

Fig. 10-37. Float-type air vent.

Float-type air vents are available for hot-water heating and cooling systems to 300 psi, and low-pressure steam heating systems to 15 psi. A float-type air vent used in a steam heating system should be equipped with a check valve, which prevents air return under vacuum.

Steam cannot be maintained at its saturated temperature when air is present in the system. As shown in Table 10-4, the temperature of the steam decreases as the percentage of air increases. A *thermostatic air vent* is specifically designed for removing air from a steam system. The one shown in Fig. 10-38 consists of a valve head attached to a bellows, operating in conjunction with a

Table 10-4. Effects of Air on Temperature of a Steam and Air Mixture

Mixture Pressure psig	Pure Steam	5% Air	10% Air	15% Air
2	219°	216°	213°	210°
5	227°	225°	222°	219°
10	239°	237°	233°	230°
20	259°	256°	252°	249°

Courtesy Sarco Company

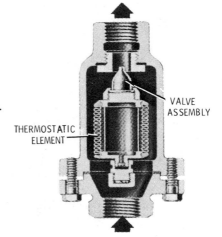

Fig. 10-38. Thermostatic air vent.

VALVE ASSEMBLY

THERMOSTATIC ELEMENT

Courtesy Sarco Co.

thermostatic element. Its operating principle resembles that of a thermostatic steam trap. When air is present, the temperature of the steam drops. The temperature drop is sensed by the thermostatic element, which causes the valve in the vent to open and discharge the air. When the air has been discharged, the temperature of the steam rises, and the valve closes tightly.

In some special applications, it is necessary to use an air eliminator that will close when the vent body contains steam or water, and opens when it contains air or gases. Combination float and thermostatic air vents have been designed for this purpose.

A *combination float and thermostatic air vent* (Fig. 10-39) consists of a vent body or chamber containing a float attached to a valve assembly. The float rests on a thermostatic element which responds to the temperature of the steam. The operation of this element is similar to the one used in a thermostatic steam trap. When the vent body is filled with air or gas, the float is at its lowest point, causing the thermostatic bellows to contract. Because the float is at a low point in the vent body, the head is moved off the valve seat and the vent discharges the air or gas. The head moves up and closes the valve when either water or steam enters the vent body. The entry of water into the vent body forces the float upward and eventually closes off the valve. The entry of steam, on the other hand, causes the thermostatic bellows to expand and force the float upward, closing the valve.

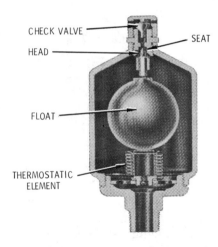

CHECK VALVE

HEAD

SEAT

FLOAT

THERMOSTATIC ELEMENT

Fig. 10-39. Combination float and thermostatic air vent.

PIPELINE VALVES

Pipeline valves are used to regulate the temperature, pressure, or flow rate of the steam or water in the lines. Some valves (e.g., check valves) deal with only one of these functions; other valves are designed to handle more than one function.

Details about the design and construction of valves and the methods used for servicing, repairing, and installing them are found in Chapter 9 (Valves and Valve Installation). This chapter is limited to a description of the function of the following four control valves commonly found in steam and hot-water space heating systems:

1. Regulating valves.
2. Water-tempering valves.
3. Heating-control valves.
4. Balancing valves.

STEAM-PRESSURE-REGULATING VALVES

Steam-pressure-regulating valves (Fig. 10-40) are used for main line and high-capacity process regulation service, heating applications, and dead-end service. They are recommended for institutional, industrial, and commercial applications where it is necessary to reduce and maintain pressures in a line at the required service pressure level. The Watts No. 127 steam-pressure-regulator shown in Fig. 10-40 is a single-seated, remote-control, diaphragm-type regulator that may be installed with the diaphragm above or below the line (Fig. 10-41).

Before installing the Watts No. 127 pressure regulator, the pipe ends should be reamed and threads cut to size. After this is done, blow out the lines to remove any loose scale or other foreign matter that could damage the valve seat of the regulator.

Install the Watts regulator with the arrow on the body pointing in the direction of the steam flow. Connect a 1/4-in. or 3/8-in. equalizer pipe (control pipe) between the diaphragm chamber on the pressure regulator and a point in the reduced pressure line *at least* 18 in. from the regulator (distances greater than 18 in. are preferred when possible) (Fig. 10-42). Always connect the equalizer pipe to a straight run of the line. When the pressure regulating valve is installed above a horizontal line or on a vertical line, on steam service, an equalizer pipe with a waterseal loop should be installed (Fig. 10-43). A steam safety-relief valve should be installed on the reduced pressure line *just beyond* the connection with the equalizer pipe. As shown in Fig. 10-42, the safety-relief valve should always be installed in an upright (i.e., vertical) position.

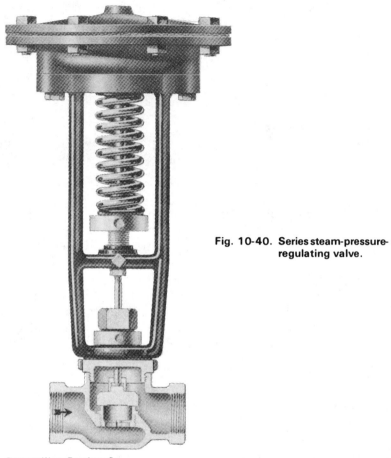

Fig. 10-40. Series steam-pressure-regulating valve.

Sometimes uneven pressure will cause the regulator to pulsate. The regulator can be stabilized by throttling a needle valve installed in the equalizer pipe (Figs. 10-42 and 10-43).

Adjustment of the steam-pressure-regulating valve depends upon the location of the diaphragm. If it is above the line, the adjustment collar should be turned from left to right to increase the reduced pressure. Turning it the opposite direction will decrease the reduced pressure. When the diaphragm is located below the line, the adjustment collar should be turned right to left.

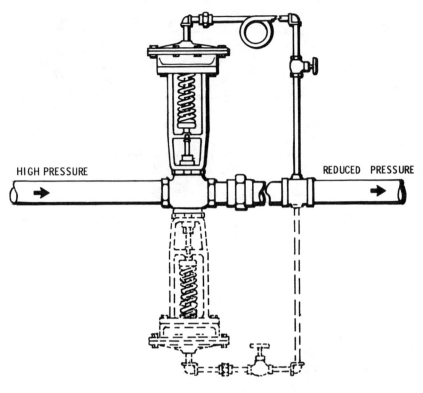

HIGH PRESSURE

REDUCED PRESSURE

Fig. 10-41. Installation of a pressure-regulating valve above and below the line.

Maintenance procedures for these regulating valves are relatively simple. Springs and diaphragm assemblies can be easily changed by removing the two bolts securing the diaphragm chamber (Fig. 10-44). Access to the valve disc and seat is gained by removing the topwork of the regulator as shown in Fig. 10-45. A hammer and blunt tool are applied to the lugs on the bonnet (Fig. 10-46).

As with all valves, a leak may occur around the valve stem of the regulator. Sometimes tightening the packing nut lightly will stop the leak. If light nut pressure will not stop the leak, the stem should be repacked (see Chapter 9, Valves and Valve Installation for packing instructions). *Never* tighten the packing nut enough

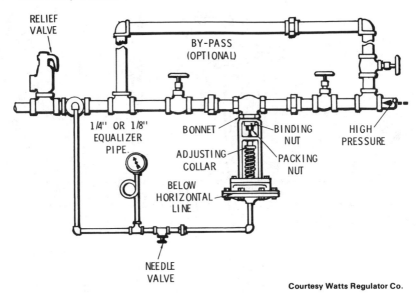

RELIEF
VALVE

BY-PASS
(OPTIONAL)

1/4" OR 1/8"
EQUALIZER
PIPE

BONNET

ADJUSTING
COLLAR

BELOW
HORIZONTAL
LINE

BINDING
NUT

PACKING
NUT

HIGH
PRESSURE

NEEDLE
VALVE

Fig. 10-42. Piping installation with an equalizer pipe.

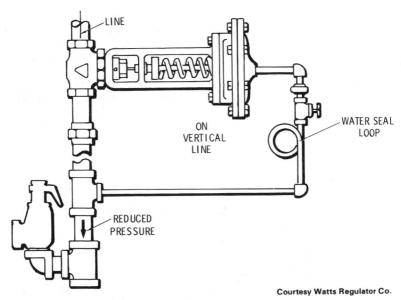

LINE

ON
VERTICAL
LINE

WATER SEAL
LOOP

REDUCED
PRESSURE

Fig. 10-43. Equalizer pipe with a waterseal loop.

544

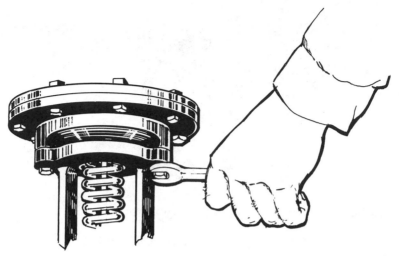

Fig. 10-44. Removing bolts securing the diaphragm chamber.

Fig. 10-45. Removing topwork.

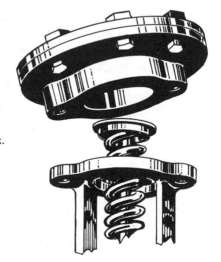

545

Courtesy Watts Regulator Co.

Fig. 10-46. Applying hammer and blunt tool to lugs.

to bind the stem. This will cause uneven regulation and excessive stem wear.

TEMPERATURE REGULATORS

Temperature regulators are used for many heating and cooling applications including small-flow instantaneous heaters or coolers (shell and tube or shell and coil heat exchangers), small storage or tank heaters, and similar installations.

The Sarco Type 25T temperature regulator shown in Fig. 10-47 is a diaphragm-operated valve used for regulating temperature in a variety of different process applications.

Before startup, the main valve is normally in a closed position and the pilot valve is held open by spring force. The steam enters the orifice inlet, passes through the pilot valve and into the diaphragm chamber, and out the control orifice. Control pressure builds up in the diaphragm chamber when the flow through the pilot valve exceeds the flow through the control orifice. This buildup in pressure opens the main valve.

The bulb of the temperature regulator is immersed in the medium being heated. At a predetermined temperature setting, the liquid in the bulb expands through the capillary tubing into the bellows and throttles the pilot valve. The main valve will deliver the required steam flow as long as the control pressure is maintained in the diaphragm chamber. The main valve closes when heat is no longer required.

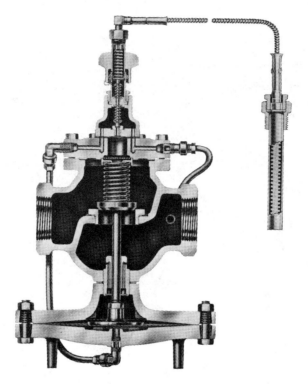

Courtesy Sarco Co.

Fig. 10-47. Temperature regulator.

ELECTRIC CONTROL
VALVES (REGULATORS)

An *electric control valve* (regulator) is designed to provide remote electric on-off control in heating systems and steam process applications (Fig. 10-48).

The solenoid pilot at the top of the valve is connected to a

547

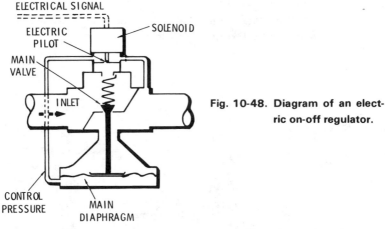

Fig. 10-48. Diagram of an electric on-off regulator.

Courtesy Sarco Co.

room thermostat, automatic time clock, or some similar device from which it can receive an electrical signal. When the solenoid pilot is electrically energized, the pilot valve opens, and pressure builds up in the diaphragm chamber. As a result, control pressure is applied to the bottom of the main valve diaphragm and the main valve is opened. The pilot valve closes when the solenoid pilot is deenergized, and control pressure is relieved through the bleed orifice. Steam pressure acting in conjunction with the force of the main valve return spring combine to close the main valve.

WATER TEMPERING VALVES

Water tempering valves are used in hot-water space heating systems where it is necessary to supply a domestic hot water supply at temperatures considerably lower than those of the water in the supply mains. The water tempering valve automatically mixes hot and cold water to a desired temperature, thus preventing scalding at the fixtures. These valves are designed for use with hot-water space heating boilers equipped with tankless heaters, boiler coils, or high-temperature water heaters.

The A.W. Cash Type TMA-2 valve illustrated in Fig. 10-49 is a thermostatic water tempering valve that is shipped from the factory preset to operate at 140°F. These valves can also be field adjusted to change the temperature of the mixed water leaving the valve by loosening the adjustment nut and turning the adjustment screw either clockwise (for colder water) or counterclockwise (for hotter water).

If turning the adjustment screw does not produce the desired mixed water temperature, carefully touch the hot-water inlet on the valve to make sure hot water is being delivered. If you are certain hot water is getting into the valve and a temperature adjustment still fails to produce the desired results, the problem most likely lies inside the valve. Problems with these valves can usually be caused by one of the following:

1. Binding of bonnet to push-rod.
2. Sticking of push-rod to O-ring.

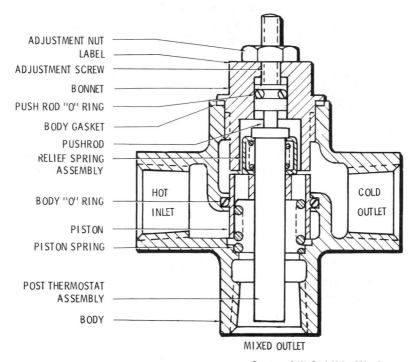

Courtesy A.W. Cash Valve Mfg. Corp.

Fig. 10-49. Diagram of a model TMA-2 water tempering valve.

3. Binding of piston.
4. Sticking of body O-Ring.

Before attempting to service or repair these valves, close off the hot, cold, and mixed water connections. Water must not be allowed to enter the valve when the bonnet has been removed.

Access to the internal parts of the hot-water tempering valve illustrated in Fig. 10-49 is gained by unscrewing the bonnet (i.e., turning it counterclockwise) and removing it. If the bonnet is binding to the push-rod, pull the push-rod out of the bonnet and wipe it off with a crocus cloth. Do the same with the inside of the bonnet. If the push-rod O-Ring is sticking, it should be removed and replaced. Reassembly is in reverse order; first place the push-rod O-ring, reinsert the push-rod, and then screw the bonnet back on.

A binding piston should be removed, cleaned (with a crocus cloth), and lubricated. Access to the piston is also gained by unscrewing the bonnet. A body O-Ring that is sticking should be removed and replaced. Access to the body O-Ring is gained by unscrewing and removing the bonnet (leaving the push-rod in the bonnet). Push the piston and piston spring up and out through the top of the tempering valve. Lift out the post thermostat assembly. When reassembling, be sure to lubricate both the body O-Ring and piston.

The design of the Watts No. N170 Series water tempering valve differs from the one described above in that the discharge or mixed water orifice is located in the bonnet (Figs. 10-50 and 10-51). The water temperature can be changed by loosening the locknut and turning the adjustment screw. Each *full* turn of the adjustment screw is equal to approximately a 10°F change in temperature.

A typical installation in which a Watts No. N170 valve is used is shown in Fig. 10-52. Tempered water at 140°F can be delivered to the system. The thermostat in the valve makes trapping unnecessary except in extreme cases.

A two-temperature recirculating hot-water supply system is shown in Fig. 10-53. In this system, a water tempering valve and recirculating line are used to maintain approximate fixture water temperatures of 140°F in the mains at all times. A relatively small

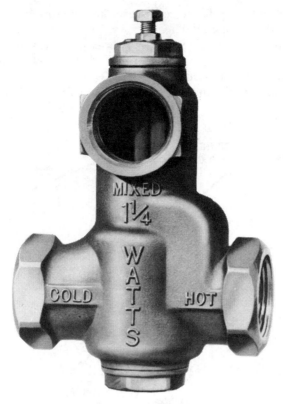

Fig. 10-50. Model N170 water tempering valve.

capacity recirculator is used in the recirculating return piping because very little hot water is required to maintain the low temperature in the mains. Long runs of recirculating piping should be insulated to reduce the heat loss from the piping.

Tempering valves cannot compensate for rapid pressure fluctuations in the system. Where such water pressure fluctuations are expected to occur, a pressure equalizing valve should be installed.

The Watts 70A Series tempering valve shown in Fig. 10-54 is designed for small domestic water supply systems and tankless heater installations. Piping connections for these applications are shown in Fig. 10-55. A balancing valve should be installed below the tempering valve in the cold-water line to compensate for the pressure drop through the heater.

This valve is available with both threaded and sweat connections. It is also available in both high- (120° to 160°F) and low- (100° to 130°F) temperature models. Temperature changes are made by turning the dial-type adjustment cap on the valve (Fig. 10-56).

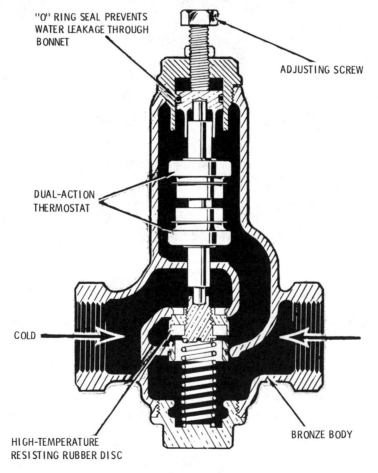

"O" RING SEAL PREVENTS WATER LEAKAGE THROUGH BONNET

ADJUSTING SCREW

DUAL-ACTION THERMOSTAT

COLD

HIGH-TEMPERATURE RESISTING RUBBER DISC

BRONZE BODY

Fig. 10-51. Components of a water tempering valve.

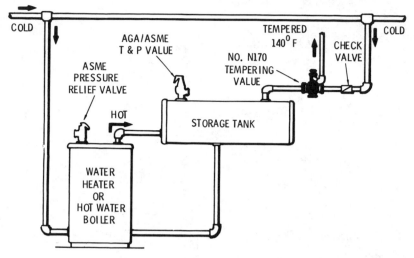

Courtesy Watts Regulator Co.

Fig. 10-52. Basic hot-water supply system using model N170 tempering valve.

The Sarco Type MB water blender (Fig. 10-57) has a 55°F adjustment range for supplying tempered water to a system. It is a three-way double-ported balancing valve, essentially resembling the Watts and A.W. Cash valves in construction except for an extended "bonnet" containing spirals. This type of construction allows a certain degree of pressure fluctuation between the hot- and cold-water inlets without disturbing the control of the tempered water.

HOT-WATER HEATING CONTROL

A *hot-water heating control* consists of an outdoor liquid expansion-type bulb connected by a capillary system to a double-ported three-way automatic mixing valve (Fig. 10-58). It is designed to blend the hot water from the boiler with the cooler return water in inverse proportion to the outside water and deliver the blended water to the circulating system.

In operation, the outdoor bulb reacts to changes in temperature and creates pressure. This pressure is transferred through a

553

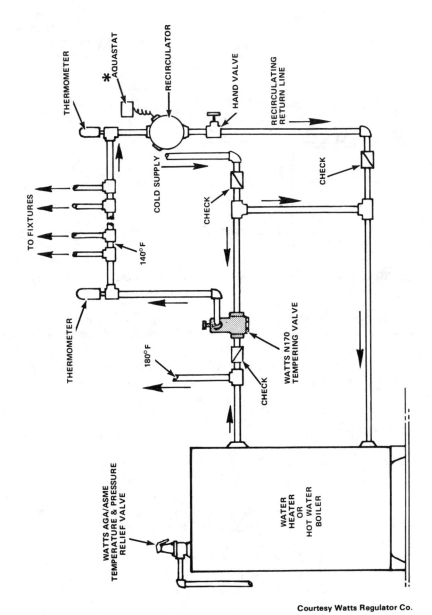

Fig. 10-53. Two-temperature hot-water supply system.

capillary system to the indoor bulb and then to the mixing valve, which is positioned to increase or decrease the amount of hot water from the boiler. Temperature-range adjustments can be made by turning an adjustment on top of the valve. A typical installation in which a hot-water heating control is used is shown in Fig. 10-59.

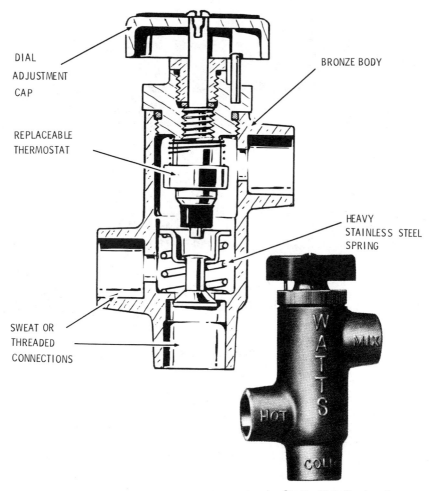

DIAL
ADJUSTMENT
CAP

BRONZE BODY

REPLACEABLE
THERMOSTAT

HEAVY
STAINLESS STEEL
SPRING

SWEAT OR
THREADED
CONNECTIONS

Courtesy Watts Regulator Co.

Fig. 10-54. Model N70A series water tempering valve.

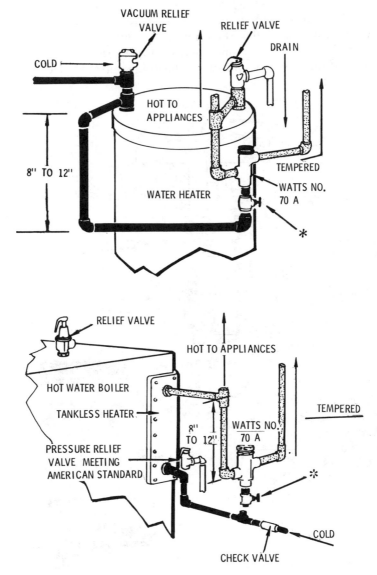

Fig. 10-55. Piping connections.

556

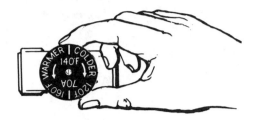

Fig. 10-56. Dial-type adjustment cap.

Courtesy Watts Regulator Co.

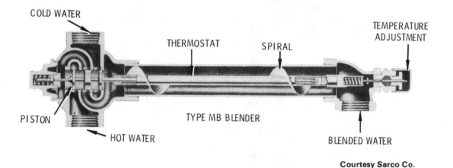

Courtesy Sarco Co.

Fig. 10-57. Water blender.

BALANCING VALVES,
VALVE ADAPTERS, AND FILTERS

Hot-water heating and cooling systems often require additional balancing not foreseen in the preliminary planning. An
effective method of balancing a heating or cooling system is to
install balancing valves, valve adapter units, or balancing fittings
at suitable locations in the pipelines in order to regulate water
flow through the radiators, convectors, baseboard panels, radiant
coils, return mains, and branches.

A *balancing valve* is a control device that functions as a combination balancing, indicating, and shutoff valve. This valve is used
to balance a hot-water heating or cooling system and at the same
time indicate the percent of flow through the valve. An example
of one of these balancing valves is shown in Fig. 10-60.

557

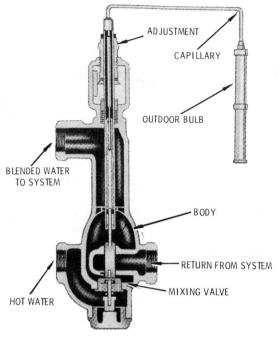

Fig. 10-58. Hot-water heating control.

These valve are available in many body patterns and connection types, the selection depending on the requirements of the installation. Among the body patterns available are angle, angle union, globe, and globe union. The different connections include screwed, sweat, male, or female unions.

The balancing valve shown in Fig. 10-60 is fitted with a balancing yoke that fits over the bonnet. The stem has a stop washer that shoulders on the balancing yoke. The yoke is rotated until the indicator on the calibrated dial points to the percent of flow required at a particular setting. At this point, it is locked in place with an Allen setscrew. The valve can then be opened until the stop washer contacts the top of the yoke, thereby permitting correct percentage of flow. For service work, the valve can be shut and, when opened, will never open beyond its set point. In addition, the valve is equipped with an O-Ring that seals off

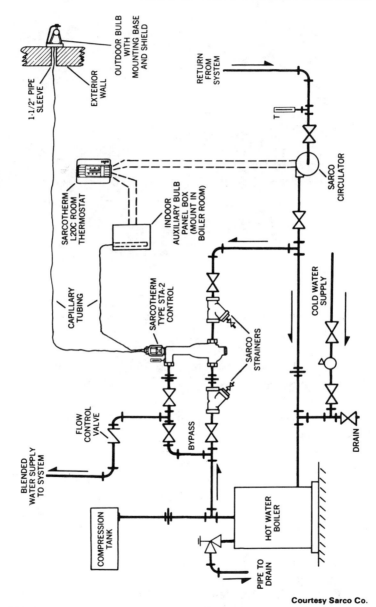

1-1/2" PIPE SLEEVE

EXTERIOR WALL

OUTDOOR BULB WITH MOUNTING BASE AND SHIELD

RETURN FROM SYSTEM

SARCOTHERM L20C ROOM THERMOSTAT

INDOOR AUXILIARY BULB PANEL BOX (MOUNT IN BOILER ROOM)

SARCO CIRCULATOR

CAPILLARY TUBING

SARCOTHERM TYPE STA-2 CONTROL

COLD WATER SUPPLY

SARCO STRAINERS

FLOW CONTROL VALVE

BYPASS

DRAIN

BLENDED WATER SUPPLY TO SYSTEM

COMPRESSION TANK

HOT WATER BOILER

PIPE TO DRAIN

Courtesy Sarco Co.

Fig. 10-59. Hot-water heating control used to control heating system water temperature in accordance with outdoor temperature.

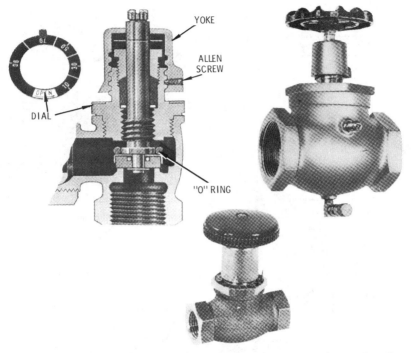

YOKE

ALLEN
SCREW

DIAL

"O" RING

Fig. 10-60. Balancing valves.

against the bonnet to form a positive back seat, and the valve can be repacked under full line pressure.

Balanced fittings, such as the ones shown in Fig. 10-61, are used for balancing branch or circuit resistance of radiators, convectors, heating or cooling coils, unit heaters, and other heat transfer equipment employing hot or chilled water. Some are available with an integral manual air vent. They can also be obtained with a thread connection or with a sweat connection for nominal copper tube.

When a balancing fitting is used, a stop valve must be installed with it to allow shutdown for necessary service.

Valve adapters are devices used to convert copper, bronze, cast brass, or cast-iron tees to balancing valves. Some typical examples of valve adapters are shown in Fig. 10-62.

560

Fig. 10-61. Balancing valve fittings.

Courtesy Sarco Co.

Courtesy Maid-O'-Mist, Inc.

Fig. 10-62. Valve adapters.

These adapter devices can be threaded into cast-iron tees, or soldered or sweat-fitted into copper, bronze, or brass tees. They can also be inserted in a side outlet or run of tee of the same size to complete either a straightway or angle balancing valve. Because there is no inside reduction of pipe diameter, there is no water restriction except for the balancing required. Balancing is accomplished by using a screwdriver to the adjustment screw at the top of the adapter device.

PIPELINE STRAINERS

Rust, dirt, metal chips, scale, and other impurities are commonly found in both new and old pipelines. Unless these impurities are captured and removed from the pipes, they can damage valves, traps, and other equipment. Protection against these potentially damaging impurities is provided by installing a *strainer* or *scrape strainer* ahead of each mechanical device in the

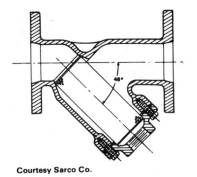

Fig. 10-63. Flanged pipeline strainer.

Courtesy Sarco Co.

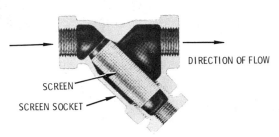

DIRECTION OF FLOW

SCREEN

SCREEN SOCKET

Courtesy Sarco Co.

Fig. 10-64. Pipeline strainer with thread connections.

pipeline. Strainers are constructed with a screen socket placed at an angle to the normal direction of flow. The impurities are captured by the screen.

Two typical examples of pipeline strainers are shown in Figs. 10-63 and 10-64. These devices are constructed from bronze, semisteel, and steel, and are available in a variety of sizes. The screen socket is tapered and positioned to collect the impurities suspended in the steam or hot water. Both standard and specially designed screens are available from manufacturers.

APPENDIX A

Professional and Trade Associations

Many professional and trade associations have been organized to develop and provide materials and/or services in connection with heating, ventilating, and air conditioning. These materials and services include:

1. Formulating and establishing specifications and professional standards.
2. Certifying equipment and materials.
3. Conducting product research.
4. Promoting interest in the product.

Much useful information can be obtained by contacting these associations. With that in mind, the names and addresses of the principal organizations have been included in this appendix. They are listed in alphabetical order.

AIR-CONDITIONING AND REFRIGERATION
INSTITUTE (ARI)
1815 North Fort Meyer Drive
Arlington, VA 22209
Manufacturers of air conditioning, warm-air heating, and commercial and industrial refrigeration equipment.

AIR-CONDITIONERS AND REFRIGERATION
WHOLESALERS (ARW)
P.O. Box 640
1351 South Federal Highway
Deerfield Beach, FL 33441
Air conditioning and refrigeration equipment wholesalers.

AIR CONDITIONING CONTRACTORS
OF AMERICA (ACCA)
1228 17th Street, NW
Washington, DC 20036
Formerly (until 1978) National Environmental Systems Contractors Association. Heating, air conditioning, and refrigeration systems contractors.

AIR DIFFUSION COUNCIL (ADC)
435 North Michigan Avenue
Chicago, IL 60611
Manufacturers of registers, grilles, diffusers, and related equipment.

AIR MOVEMENT AND CONTROL ASSOCIATION (AMCA)
30 West University Drive
Arlington Heights, IL 60004
Formerly (until 1977) Air Moving and Conditioning Association. Manufacturers of air moving and conditioning equipment.

AIR FILTER INSTITUTE. *See* Air Conditioning
and Refrigeration Institute.

AMERICAN GAS ASSOCIATION (AGA)
1515 Wilson Boulevard
Arlington, VA 22209
Residential gas operating and performance standards from distributors and transporters of natural, manufactured, and mixed gas.

AMERICAN NATIONAL STANDARDS INSTITUTE
(ANSI)
1430 Broadway
New York, NY 10018
Clearinghouse for nationally coordinated safety, engineering, and industrial standards.

AMERICAN SOCIETY FOR TESTING
AND MATERIALS (ASTM)
1916 Race Street
Philadelphia, PA 19103
Engineering standards for materials.

AMERICAN SOCIETY OF HEATING,
REFRIGERATION AND AIR-CONDITIONING
ENGINEERS (ASHRAE)
345 East 47th Street
New York, NY 10017
Professional association.

AMERICAN SOCIETY OF MECHANICAL
ENGINEERS (ASME)
345 East 47th Street
New York, NY 10017
Develops safety codes and standards for various types of equipment (boiler and pressure vessel codes, etc.)

AMERICAN VENTILATION ASSOCIATION (AVA)
Box 7464
Houston, TX 77008
Association of residential ventilating equipment manufacturers and dealers.

BETTER HEATING-COOLING COUNCIL. *See* Hydronics
Institute

EDISON ELECTRIC INSTITUTE (EEI)
1111 19th Street, NW
Washington, DC 20036
Association of investor-owned electric utility companies.

ELECTRICAL ENERGY ASSOCIATION. *See* Edison Electric
Institute

FIREPLACE INSTITUTE. Merged with Wood Energy Institute
in 1980 to form Wood Heating Alliance.

HEATING AND PIPING CONTRACTORS NATIONAL
ASSOCIATION. *See* Mechanical Contractors Association of
America

HOME VENTILATING INSTITUTE (HVI)
4300-L Lincoln Avenue
Rolling Meadows, IL 60008
Develops performance standards for residential ventilating
equipment.

HYDRONICS INSTITUTE
35 Russo Place
Berkley Heights, NJ 07922
Manufacturers, installers, etc., of hot-water and steam heating
and cooling equipment. Formed by a merger of the Better
Heating-Cooling Council (1956-1970) and the Institute of Boiler
and Radiator Manufacturers (1915-1970).

INSTITUTE OF BOILER AND RADIATOR
MANUFACTURERS. *See* Hydronics Institute

566

INTERNATIONAL SOLAR ENERGY SOCIETY
National Science Center
P.O. Box 52
Parkville, Victoria 3052
Australia.
Educational and research organization.

MECHANICAL CONTRACTORS ASSOCIATION
OF AMERICA (MCAA)
Suite 750
5530 Wisconsin Avenue, NW
Washington, DC 20015
Contractors of piping and related equipment used in heating, cooling, refrigeration, ventilating, and air conditioning.

NATIONAL ASSOCIATION OF PLUMBING-
HEATING-COOLING CONTRACTORS
1016 20th Street, NW
Washington, DC 20005
Local plumbing, heating, and cooling contractors association.

NATIONAL BOARD OF FIRE UNDERWRITERS
(NBFU)
85 John Street
New York, NY 10036
Safety standards.

NATIONAL BUREAU OF STANDARDS
U.S. Department of Commerce
Washington, DC 20025
Government regulatory agency.

NATIONAL ENVIRONMENTAL SYSTEMS
CONTRACTORS ASSOCIATION. See
Air Conditioning Contractors of America

NATIONAL FIRE PROTECTION ASSOCIATION (NFPA)
Batterymarch Park
Quincy, MA 02269
Fire safety standards.

NATIONAL LP-GAS ASSOCIATION (NLPGA)
1301 West 22nd Street
Oak Brook, IL 60521

NATIONAL OIL FUEL INSTITUTE. *See* National Oil Jobbers
  Council

NATIONAL OIL JOBBERS COUNCIL
1707 H Street, NW
Washington, DC 20006
Absorbed National Oil Fuel Institute in 1974. Independent
  wholesale petroleum marketers and retail fuel oil dealers.

NATIONAL WARM AIR HEATING AND AIR CONDITION-
  ING ASSOCIATION.
See Air Conditioning Contractors of America

PLUMBING-HEATING-COOLING INFORMATION
  BUREAU (PHCIB)
35 East Wacker Drive
Chicago, Il 60601
Association of plumbing, heating, and cooling manufactur-
  ers, contractors, wholesalers.

REFRIGERATION AND AIR CONDITIONING CONTRAC-
  TORS ASSOCIATION. *See* Air Conditioning Contractors
  of America

REFRIGERATION SERVICE ENGINEERS SOCIETY (RSES)
960 Rand Road
Des Plaines, IL 60018
Association of refrigeration, air conditioning, and heating
  equipment installers, servicemen, and salesmen.

SHEET METAL AND AIR CONDITIONING
CONTRACTORS NATIONAL ASSOCIATION (SMACNA)
8224 Old Courthouse Road
Vienna, VA 22180
Sheet-metal contractors who install ventilating, warm-air heating,
and air-handling equipment and systems.

STEAM HEATING EQUIPMENT MANUFACTURERS
ASSOCIATION. Defunct.

STEEL BOILER INSTITUTE. Defunct.

UNDERWRITERS' LABORATORIES
333 Pfingsten Road
Northbrook, IL 60062
Testing laboratory. Promotes safety standards for equipment.

WOOD ENERGY INSTITUTE. Merged with Fireplace Institute
in 1980 to form Wood Heating Alliance.

WOOD HEATING ALLIANCE (WHA)
Suite 700
1101 Connecticut Avenue, SW
Washington, DC 20036
Manufacturers, dealers, and suppliers of fireplaces, fireplace
components, and related equipment.

APPENDIX B

Psychrometric Charts

The atmosphere around us is essentially made up of dry air and water vapor in various percentages, each with its own characteristics. The water vapor is not dissolved in the air in the sense that it loses its own individuality, but merely serves to moisten the air.

Psychrometry is that branch of physics concerned with the measurement or determination of atmospheric conditions, particularly the moisture content of air. These measurements are obtained with a psychrometer and are graphically represented on a *psychrometric chart*. Insofar as air conditioning is concerned, psychrometry is specifically concerned with the thermodynamic properties of moist air. The technical application of thermodynamics to air conditioning is called *psychrometrics*.

The typical *sling psychrometer* consists of a handle and an attached unit containing the wet-bulb and dry-bulb thermometers. The thermometer unit fits inside the handle when not in use. The wet-bulb, dry-bulb, and relative humidity scales are read

571

from the handle. Temperature readings are obtained by twirling the thermometer unit around the handle and then allowing the thermometers time to stabilize.

The readings given on a psychrometric chart are represented by the relative positions of a number of different lines running vertically, horizontally, and diagonally. These lines are simplified for purposes of explanation in Figs. B-1, B-2, B-3, and B-4.

In Fig. B-1, the horizontal distances on the chart are a measure of sensible heat as obtained from dry-bulb temperatures. The vertical distances are a measure of latent heat as obtained from the dew point temperatures. The inclined (solid) lines are a measure of the total heat (not including the heat of liquid) and are constant for a given wet-bulb temperature.

The curved lines indicate the relative humidity between the limiting conditions of dry and saturated air. As shown in Fig. B-2, the grains in water vapor per pound of dry air in the mixture can be obtained by proceeding to the left through the dew point temperature to the scale of grains at the left of the chart.

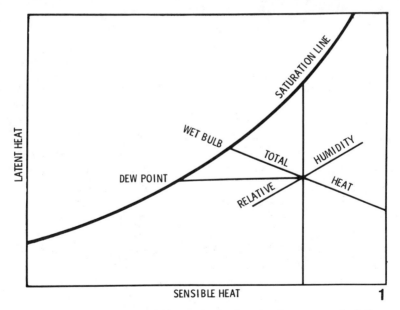

Fig. B-1. Horizontal distances as measure of sensible heat; vertical distances of latent heat.

572

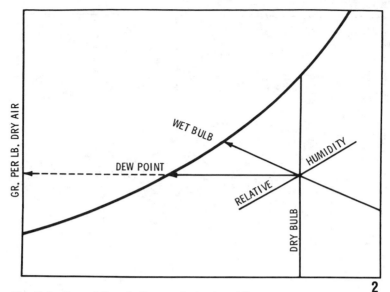

Fig. B-2. Curved lines indicate relative humidity.

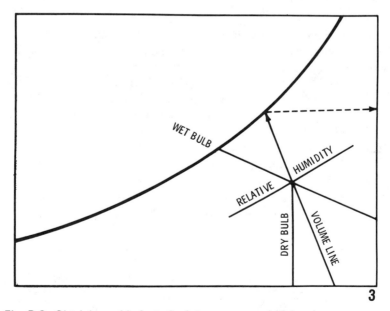

Fig. B-3. Obtaining cubic feet of mixture per pound of dry air.

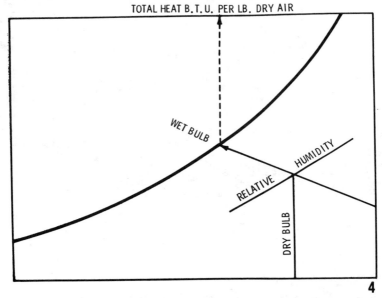

Fig. B-4. Obtaining total heat per pound of dry air.

The cubic feet of mixture per pound of dry air in the mixture can be obtained by proceeding from the intersection of dry-bulb, wet-bulb, and dew point temperature upward and parallel with the inclined volume lines (shown dotted) to the saturation curve and then directly to the volume scale at the right of the chart (Fig. B-3). The grains of water vapor per cubic feet of mixture is obtained by dividing the reading obtained through Fig. B-2 by that obtained through Fig. B-3.

The total heat per pound of dry air in the mixture can be obtained by following up along the inclined wet-bulb temperature line to the saturation curve and then vertically upward to the total heat scale at the top of the chart (Fig. B-4).

Index

The Audel® Mail Order Bookstore

Here's an opportunity to order the valuable books you may have missed before and to build your own personal, comprehensive library of Audel books. You can choose from an extensive selection of technical guides and reference books. They will provide access to the same sources the experts use, put all the answers at your fingertips, and give you the know-how to complete even the most complicated building or repairing job, in the same professional way.

Each volume:

- ● **Fully illustrated**
- ● **Packed with up-to-date facts and figures**
- ● **Completely indexed for easy reference**

APPLIANCES

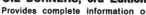

HOME APPLIANCE SERVICING, 3rd Edition

A practical book for electric & gas servicemen, mechanics & dealers. Covers the principles, servicing, and repairing of home appliances. 592 pages; 5¼ x 8¼; hardbound. **Price: $12.95**

REFRIGERATION AND AIR CONDITIONING
LIBRARY—2 Vols. Price: $21.95

REFRIGERATION: HOME AND COMMERCIAL

Covers the whole realm of refrigeration equipment from fractional-horsepower water coolers, through domestic refrigerators to multi-ton commercial installations. 656 pages; 5½ x 8¼; hardbound. **Price: $12.95**

AIR CONDITIONING: HOME AND COMMERCIAL

A concise collection of basic information, tables, and charts for those interested in understanding, troubleshooting, and repairing home air conditioners and commercial installations. 464 pages; 5½ x 8¼; hardbound. **Price: $10.95**

OIL BURNERS, 3rd Edition

Provides complete information on all types of oil burners and associated equipment. Discusses burners—blowers—ignition transformers—electrodes—nozzles—fuel pumps—filters—Controls. Installation and maintenance are stressed. 320 pages; 5½ x 8¼; hardbound. **Price: $9.95**

Use the order coupon on the back page of this book.

All prices are subject to change without notice.

AUTOMOTIVE

AUTOMOBILE REPAIR GUIDE, 4th Edition
A practical reference for auto mechanics, servicemen, trainees, and owners Explains theory, construction. and servicing of modern domestic motorcars. 800 pages; 5½ x 8¼; hardbound. **Price: 14.95**

CAN-DO TUNE-UP™ SERIES
Each book in this series comes with an audio tape cassette. Together they provide an organized set of instructions that will show you and talk you through the maintenance and tune-up procedures designed for your particular car. All books are softcover.

AMERICAN MOTORS CORPORATION CARS
(The 1964 thru 1974 cars covered include: Matador. Rambler. Gremlin. and AMC Jeep (Willys).) 112 pages; 5½ x 8½; softcover. **Price: $7.95**

CHRYSLER CORPORATION CARS
(The 1964 thru 1974 cars covered include: Chrysler, Dodge, and Plymouth.) 112 pages; 5½ x 8½; softcover. **Price: $7.95**

FORD MOTOR COMPANY CARS
(The 1954 thru 1974 cars covered include: Ford, Lincoln, and Mercury.) 112 pages; 5½ x 8½; softcover. **Price: $7.95**

GENERAL MOTORS CORPORATION CARS
(The 1964 thru 1974 cars covered include: Buick, Cadillac, Chevrolet, Oldsmobile, and Pontiac.) 112 pages; 5½ x 8½; softcover. **Price: $7.95**

PINTO AND VEGA CARS
1971 thru 1974. 112 pages. 5½ x 8½; softcover. **Price: $7.95**

TOYOTA AND DATSUN CARS
1964 thru 1974. 112 pages; 5½ x 8½; softcover. **Price: $7.95**

VOLKSWAGEN CARS
(The 1964 thru 1974 cars covered include: Beetle. Super Beetle. and Karmann Ghia.) 96 pages; 5½ x 8½; softcover. **Price: $7.95**

AUTOMOTIVE AIR CONDITIONING
You can easily perform most all service procedures you've been paying for in the past. This book covers the systems built by the major manufacturers, even after-market installations. Contents: introduction—refrigerant—tools—air conditioning circuit—general service procedures—electrical systems—the cooling system—system diagnosis—electrical diagnosis—troubleshooting. 232 pages; 5½ x 8½; softcover. **Price: $7.95**

Use the order coupon on the back page of this book.

All prices are subject to change without notice.

DIESEL ENGINE MANUAL, 3rd Edition

A practical guide covering the theory, operation, and maintenance of modern diesel engines. Explains diesel principles—valves—timing—fuel pumps—pistons and rings—cylinders—lubrication —cooling system—fuel oil and more. 480 pages; 5½ x 8¼; hardbound. **Price: $10.95**

GAS ENGINE MANUAL, 2nd Edition

A completely practical book covering the construction, operation, and repair of all types of modern gas engines. 400 pages; 5½ x 8¼; hardbound. **Price: $9.95**

BUILDING AND MAINTENANCE

ANSWERS ON BLUEPRINT READING, 3rd Edition

Covers all types of blueprint reading for mechanics and builders. This book reveals the secret language of blueprints, step-by-step in easy stages. 312 pages; 5½ x 8¼; hardbound. **Price: $9.95**

BUILDING MAINTENANCE, 2nd Edition

Covers all the practical aspects of building maintenance. Painting and decorating; plumbing and pipe fitting; carpentry; heating maintenance; custodial practices and more. (A book for building owners, managers, and maintenance personnel.) 384 pages; 5½ x 8¼; hardbound. **Price: $9.95**

COMPLETE BUILDING CONSTRUCTION

At last—a *one-volume* instruction manual to show you how to construct a frame or brick building from the footings to the ridge. Build your own garage, tool shed, other outbuilding—even your own house or place of business. Building construction tells you how to lay out the building and excavation lines on the lot; how to make concrete forms and pour the footings and foundation; how to make concrete slabs, walks, and driveways; how to lay concrete block, brick and tile; how to build your own fireplace and chimney: It's one of the newest Audel books, clearly written by experts in each field and ready to help you every step of the way. 800 pages; 5½ x 8¼; hardbound. **Price: 19.95**

GARDENING & LANDSCAPING

A comprehensive guide for homeowners and for industrial, municipal, and estate groundskeepers. Gives information on proper care of annual and perennial flowers; various house plants; greenhouse design and construction; insect and rodent controls; and more. 384 pages; 5½ x 8¼; hardbound. **Price: $9.95**

CARPENTERS & BUILDERS LIBRARY, 4th Edition (4 Vols.)

A practical, illustrated trade assistant on modern construction for carpenters, builders, and all woodworkers. Explains in practical, concise language and illustrations all the principles, advances, and shortcuts based on modern practice. How to calculate various jobs. **Price: $35.95**

> Vol. 1—Tools, steel square, saw filing, joinery cabinets. 384 pages; 5½ x 8¼; hardbound. **Price: $10.95**
> Vol. 2—Mathematics, plans, specifications, estimates 304 pages; 5½ x 8¼; hardbound. **Price: $10.95**
> Vol. 3—House and roof framing, laying out foundations. 304 pages; 5½ x 8¼; hardbound. **Price: $10.95**
> Vol. 4—Doors, windows, stairs, millwork, painting. 368 pages; 5½ x 8¼; hardbound. **Price: $10.95**

Use the order coupon on the back page of this book.

All prices are subject to change without notice.

CARPENTRY AND BUILDING

Answers to the problems encountered in today's building trades. The actual questions asked of an architect by carpenters and builders are answered in this book. 448 pages; 5½ x 8¼; hardbound. **Price: $10.95**

WOOD STOVE HANDBOOK

The wood stove handbook shows how wood burned in a modern wood stove offers an immediate, practical, low-cost method of full-time or part-time home heating. The book points out that wood is plentiful, low in cost (sometimes free), and nonpolluting, especially when burned in one of the newer and more efficient stoves. In this book, you will learn about the nature of heat and its control, what happens inside and outside a stove, how to have a safe and efficient chimney, and how to install a modern wood burning stove. You will also learn about the different types of firewood and how to get it, cut it, split it, and store it. 128 pages; 8½ x 11; softcover. **Price: $7.95**

HEATING, VENTILATING, AND AIR CONDITIONING LIBRARY (3 Vols.)

This three-volume set covers all types of furnaces, ductwork, air conditioners, heat pumps, radiant heaters, and water heaters, including swimming-pool heating systems. **Price: $32.95**

Volume 1

Partial Contents: Heating Fundamentals . . . Insulation Principles . . . Heating Fuels . . . Electric Heating System . . . Furnace Fundamentals . . . Gas-Fired Furnaces . . . Oil-Fired Furnaces . . . Coal-Fired Furnaces . . . Electric Furnaces. **Price: $11.95**

Volume 2

Partial Contents: Oil Burners . . . Gas Burners . . . Thermostats and Humidistats . . . Gas and Oil Controls . . . Pipes, Pipe Fitting, and Piping Details . . . Valves and Valve Installations 560 pages; 5½ x 8¼; hardbound. **Price: $11.95**

Volume 3

Partial Contents: Radiant Heating . . . Radiators, Convectors, and Unit Heaters . . . Stoves, Fireplaces, and Chimneys . . . Water Heaters and Other Appliances . . . Central Air Conditioning Systems . . . Humidifiers and Dehumidifiers. 544 pages; 5½ x 8¼; hardbound. **Price: $11.95**

HOME MAINTENANCE AND REPAIR: Walls, Ceilings, and Floors

Easy-to-follow instructions for sprucing up and repairing the walls, ceiling, and floors of your home. Covers nail pops, plaster repair, painting, paneling, ceiling and bathroom tile, and sound control. 80 pages; 8½ x 11; softcover. **Price: $6.95**

HOME PLUMBING HANDBOOK, 2nd Edition

A complete guide to home plumbing repair and installation. 200 pages; 8½ x 11; softcover. **Price: $7.95**

MASONS AND BUILDERS LIBRARY—2 Vols.

A practical, illustrated trade assistant on modern construction for bricklayers, stonemasons, cement workers, plasterers, and tile setters. Explains all the principles, advances, and shortcuts based on modern practice—including how to figure and calculate various jobs. **Price: $17.95**

Vol. 1—Concrete, Block, Tile, Terrazzo. 368 pages; 5½ x 8¼; hardbound. **Price: $9.95**

Vol. 2—Bricklaying, Plastering, Rock Masonry, Clay Tile. 384 pages; 5½ x 8¼; hardbound. **Price: 9.95**

Use the order coupon on the back page of this book.

All prices are subject to change without notice.

PLUMBERS AND PIPE FITTERS LIBRARY—3 Vols.

A practical, illustrated trade assistant and reference for master plumbers, journeymen and apprentice pipe fitters, gas fitters and helpers, builders, contractors, and engineers. Explains in simple language, illustrations, diagrams, charts, graphs, and pictures, the principles of modern plumbing and pipe-fitting practices. **Price: $26.95**

> Vol. 1—Materials, tools, roughing-in. 320 pages; 5½ x 8¼; hardbound. **Price: $9.95**
>
> Vol. 2—Welding, heating, air-conditioning. 384 pages; 5½ x 8¼; hardbound. **Price: $9.95**
>
> Vol. 3—Water supply, drainage, calculations. 272 pages; 5½ x 8¼; hardbound. **Price: $9.95**

PLUMBERS HANDBOOK

A pocket manual providing reference material for plumbers and/or pipe fitters. General information sections contain data on cast-iron fittings, copper drainage fittings, plastic pipe, and repair of fixtures. 288 pages; 4 x 6; softcover. **Price: $9.95**

QUESTIONS AND ANSWERS FOR PLUMBERS EXAMINATIONS, 2nd Edition

Answers plumbers' questions about types of fixtures to use, size of pipe to install, design of systems, size and location of septic tank systems, and procedures used in installing material. 256 pages; 5½ x 8¼; softcover. **Price: $8.95**

TREE CARE MANUAL

The conscientious gardener's guide to healthy, beautiful trees. Covers planting, grafting, fertilizing, pruning, and spraying. Tells how to cope with insects, plant diseases, and environmental damage. 224 pages; 8½ x 11; softcover. **Price: $8.95**

UPHOLSTERING

Upholstering is explained for the average householder and apprentice upholsterer. From repairing and regluing of the bare frame, to the final sewing or tacking, for antiques and most modern pieces, this book covers it all. 400 pages; 5½ x 8¼; hardbound. **Price: $9.95**

WOOD FURNITURE: Finishing, Refinishing, Repairing

Presents the fundamentals of furniture repair for both veneer and solid wood. Gives complete instructions on refinishing procedures, which includes stripping the old finish, sanding, selecting the finish and using wood fillers. 352 pages; 5½ x 8¼; hardbound. **Price: $9.95**

ELECTRICITY/ELECTRONICS

ELECTRICAL LIBRARY

If you are a student of electricity or a practicing electrician, here is a very important and helpful library you should consider owning. You can learn the basics of electricity, study electric motors and wiring diagrams, learn how to interpret the NEC, and prepare for the electrician's examination by using these books.

Electric Motors, 3rd Edition. 528 pages; 5½ x 8¼; hardbound. **Price: $10.95**

Guide to the 1981 National Electrical Code. 608 pages; 5½ x 8¼; hardbound. **Price: $13.95**

House Wiring, 5th Edition. 256 pages; 5½ x 8¼; hardbound. **Price: $9.95**

Practical Electricity, 3rd Edition. 496 pages; 5½ x 8¼; hardbound. **Price: $10.95**

Questions and Answers for Electricians Examinations, 7th Edition. 288 pages; 5½ x 8¼; hardbound. **Price: $9.95**

ELECTRICAL COURSE FOR APPRENTICES AND JOURNEYMEN

A study course for apprentice or journeymen electricians. Covers electrical theory and its applications. 448 pages; 5½ x 8¼; hardbound. **Price: $10.95**

Use the order coupon on the back page of this book.

All prices are subject to change without notice.

RADIOMANS GUIDE, 4th Edition

Contains the latest information on radio and electronics from the basics through transistors. 480 pages; 5½ x 8¼; hardbound. **Price: $11.95**

TELEVISION SERVICE MANUAL, 4th Edition

Provides the practical information necessary for accurate diagnosis and repair of both black-and-white and color television receivers. 512 pages; 5½ x 8¼; hardbound. **Price: $11.95**

ENGINEERS/MECHANICS/ MACHINISTS

MACHINISTS LIBRARY, 3rd Edition

Covers modern machine-shop practice. Tells how to set up and operate lathes, screw and milling machines, shapers, drill presses, and all other machine tools. A complete reference library. **Price: $29.95**

Vol. 1—Basic Machine Shop. 352 pages; 5½ x 8¼; hardbound. **Price: $10.95**

Vol. 2—Machine Shop. 480 pages; 5½ x 8¼; hardbound. **Price: $10.95**

Vol. 3—Toolmakers Handy Book. 400 pages; 5½ x 8¼; hardbound. **Price: $10.95**

MECHANICAL TRADES POCKET MANUAL

Provides practical reference material for mechanical tradesmen. This handbook covers methods, tools, equipment, procedures, and much more. 256 pages; 4 x 6; softcover. **Price: $8.95**

MILLWRIGHTS AND MECHANICS GUIDE, 2nd Edition

Practical information on plant installation, operation, and maintenance for millwrights, mechanics, maintenance men, erectors, riggers, foremen, inspectors, and superintendents. 960 pages; 5½ x 8¼; hardbound. **Price: $16.95**

POWER PLANT ENGINEERS GUIDE, 2nd Edition

The complete steam or diesel power-plant engineer's library. 816 pages; 5½ x 8¼; hardbound. **Price: $15.95**

QUESTIONS AND ANSWERS FOR ENGINEERS AND FIREMANS EXAMINATIONS, 3RD EDITION

Presents both legitimate and "catch" questions with answers that may appear on examinations for engineers and firemans licenses for stationary, marine, and combustion engines. 496 pages; 5½ x 8¼; hardbound. **Price: $10.95**

WELDERS GUIDE, 2nd Edition

This new edition is a practical and concise manual on the theory, practical operation, and maintenance of all welding machines. Fully covers both electric and oxy-gas welding. 928 pages; 5½ x 8¼; hardbound. **Price: $14.95**

WELDER/FITTERS GUIDE

Provides basic training and instruction for those wishing to become welder/fitters. Step-by-step learning sequences are presented from learning about basic tools and aids used in weldment assembly, through simple work practices, to actual fabrication of weldments. 160 pages· 8½ x 11; softcover. **Price: $7.95**

Use the order coupon on the back page of this book.

All prices are subject to change without notice.

FLUID POWER

PNEUMATICS AND HYDRAULICS, 3rd Edition

Fully discusses installation, operation, and maintenance of both HYDRAULIC AND PNEUMATIC (air) devices. 496 pages; 5½ x 8¼; hardbound. **Price: $10.95**

PUMPS, 3rd Edition

A detailed book on all types of pumps from the old-fashioned kitchen variety to the most modern types. Covers construction, application, installation, and troubleshooting. 480 pages; 5½ x 8¼; hardbound. **Price: $10.95**

HYDRAULICS FOR OFF-THE-ROAD EQUIPMENT

Everything you need to know from basic hydraulics to troubleshooting hydraulic systems on off-the-road equipment. Heavy-equipment operators, farmers, fork-lift owners and operators, mechanics—all need this practical, fully illustrated manual. 272 pages; 5½ x 8¼; hardbound. **Price: $8.95**

HOBBY

COMPLETE COURSE IN STAINED GLASS

Written by an outstanding artist in the field of stained glass, this book is dedicated to all who love the beauty of the art. Ten complete lessons describe the required materials, how to obtain them, and explicit directions for making several stained glass projects. 80 pages; 8½ x 11; softbound. **Price: $6.95**

BUILD YOUR OWN AUDEL DO-IT-YOURSELF LIBRARY AT HOME!

Use the handy order coupon today to gain the valuable information you need in all the areas that once required a repairman. Save money and have fun while you learn to service your own air conditioner, automobile, and plumbing. Do your own professional carpentry, masonry, and wood furniture refinishing and repair. Build your own security systems. Find out how to repair your TV or Hi-Fi. Learn landscaping, upholstery, electronics and much, much more.

All prices are subject to change without notice.

HERE'S HOW TO ORDER

1. Enter the correct title(s) and author(s) of the book(s) you want in the space(s) provided.

2. Print your name, address, city, state and zip code clearly.

3. Detach the order coupon below and mail today to:

Theodore Audel & Company
4300 West 62nd Street
Indianapolis, Indiana 46206
ATTENTION: ORDER DEPT.

All prices are subject to change without notice.

ORDER COUPON

Please rush the following book(s).

Title_____

Author_____

Title_____

Author_____

NAME_____

ADDRESS_____

CITY_____ STATE_____ ZIP_____

☐ Payment enclosed _____
(No shipping and   Total
handling charge)

☐ Bill me (shipping and handling charge will be added)
Add local sales tax where applicable.

Litho in U.S.A.